D0880200

W.D. Wallis

A Beginner's Guide to Discrete Mathematics

Birkhäuser
Boston • Basel • Berlin

W.D. Wallis
Southern Illinois University
Department of Mathematics
Carbondale, IL 62901-4408
U.S.A.

Library of Congress Cataloging-in-Publication Data

Wallis, W. D.
A beginner's guide to discrete mathematics / W.D. Wallis.
p. cm.
Includes bibliographical references and index.
ISBN 0-8176-4269-2 (alk. paper) – ISBN 3-7643-4269-2 (alk. paper)
1. Mathematics. 2. Computer science–Mathematics. I. Title.

QA39.3 .W35 2002
510.6–dc21 2002026043
CIP

AMS Subject Classifications: Primary: 15-01; Secondary: 03-01

Printed on acid-free paper.
 Birkhäuser 

ISBN 0-8176-4269-2 SPIN 10851437
ISBN 3-7643-4269-2

Typeset by the author.
Printed in the United States of America.

9 8 7 6 5 4 3 2 1

Birkhäuser Boston • Basel • Berlin
A member of BertelsmannSpringer Science+Business Media GmbH

Preface

This text is a basic introduction to those areas of discrete mathematics used by students of mathematics and computer science. Introductory courses on this material are now standard at many colleges and universities. Usually these courses are of one semester's duration, and usually they are offered at the sophomore level. Very often this will be the first course where the students see several real proofs. The preparation of the students is very mixed, and one cannot assume a strong background. In particular, the instructor should not assume that the students have seen a linear algebra course, or any introduction to number systems that goes beyond college algebra.

In view of this, I have tried to avoid too much sophistication, while still retaining rigor. I hope I have included enough problems so that the student can reinforce the concepts. Most of the problems are quite easy, with just a few difficult exercises scattered through the text. If the class is weak, a small number of sections will be too hard, while the instructor who has a strong class will need to include some supplementary material. I think this is preferable to a book at a higher mathematical level, which will scare away weaker students.

Outline of topics

The first two chapters include a brief survey of number systems and elementary set theory. Included are discussions of scientific notation and the representation of numbers in computers, topics that were included at the suggestion of computer science instructors. Mathematical induction is treated at this point although the instructor could defer this until later. (There are a few references to induction later in the text, but the student can omit these in a first reading.)

We introduce logic along with set theory. This leads naturally into an introduction to Boolean Algebra, which brings out the commonality of logic and set theory. The latter part of Chapter 3 explains the application of Boolean algebra to circuit theory.

We follow this with a short chapter on relations and functions. The study of relations is an offshoot of set theory, and also lays the foundation for the study of graph theory later. Functions are mentioned only briefly. The student will see them treated extensively in calculus courses, but in discrete mathematics we mostly need basic definitions.

Enumeration, or theoretical counting, is central to discrete mathematics. In Chapter 5 we present the main results on selections and arrangements, and also cover the binomial theorem and derangements. Some of the harder problems here are rather challenging, but we have omitted most of the more sophisticated results.

Counting leads naturally to probability theory. We have included the main ideas of discrete probability, up to Bayes' theorem. There was a conscious decision not to include any real discussion of measures of central tendency (means, medians) or spread (variance, quartiles) because most students will encounter them elsewhere, e.g. in statistics courses.

We study graph theory, including Euler and Hamilton cycles and trees. This is a vehicle for some (easy) proofs, as well as being an important example of a data structure.

Matrices and vectors are defined and discussed briefly. This is not the place for algebraic studies, but matrices are useful for studying other discrete objects, and we illustrate this by a section on adjacency matrices of relations and graphs. A number of students will never study linear algebra, and this chapter will provide some foundation for the use of matrices in programming, mathematical modeling, and statistics. Those who have already seen vectors and matrices can skip most of this chapter, but should read the section on adjacency matrices.

We conclude with an introduction to cryptography, including the RSA cryptosystem, together with the necessary elementary number theory (such as modular arithmetic and the Euclidean algorithm). Cryptography is an important application area and is a good place to show students that discrete mathematics has real world applications. Moreover, most computer science majors will later be presented with electives in this area. The level of mathematical sophistication is higher in parts of this chapter than in most of the book.

Perhaps I should explain the omissions rather than the inclusions. I thought the study of predicates and quantifiers belonged in a course on logic rather than here. I also thought lattice theory was too deep, although it would fit nicely after the section on Boolean forms.

There is no section on recursion and recurrence relations. Again, this is a deep area. I have actually given some problems on recurrences in the induction section, but I thought that a serious study belongs in a combinatorics course. Similarly, the deeper enumeration results, such as counting partitions, belong in higher-level courses.

Another area is linear programming. This was once an important part of discrete mathematics courses. But, in recent years, syllabi have changed. Nowadays, somewhat weaker students are using linear programming, and there are user-friendly computer packages available. I do not think that it will be in the syllabus of many of the courses at which this book is aimed.

Problems and exercises

The book contains a large selection of exercises, collected at the end of sections. There should be enough for students to practice the concepts involved. Most are straightforward; in some sections there are one or two more sophisticated questions at the end.

A number of worked examples, called Sample Problems, are included in the body of each section. Most of these are accompanied by a Practice Exercise, designed primarily to test the reader's comprehension of the ideas being discussed. It is recommended that students work all the Practice Exercises. Complete solutions are provided for all of them, as well as brief answers to the odd numbered problems from the sectional exercise sets.

Gender

In many places a mathematical discussion involves a protagonist — a person who flips a coin or deals a card or traverses a road network. These people used to be exclusively male. In recent years this has rightly been seen to be inappropriate. Unfortunately this has led to frequent repetitions of nouns — "the player's card" rather than "his card" — and the use of the ugly "he or she."

I decided to avoid such problems by a method that was highly appropriate to this text: I flipped a coin to decide whether a character was male or female. If the reader detects an imbalance, please blame the coin.

There were two exceptions to this rule. Cryptographers traditionally write about messages sent from Alice (A) to Bob (B), so I followed this rule in discussing RSA cryptography. And in the discussion of the Monty Hall problem, the game show host is male, in honor of Monty, and the player is female as were most of the contestants in *Let's Make a Deal.*

Acknowledgments

My treatment of Discrete Mathematics owes a great deal to many colleagues and mathematicians in other institutions with whom I have taught or discussed this material. Among my influences are Roger Eggleton, Ralph Grimaldi, Dawit Haile, Fred Hoffman, Bob McGlynn, Nick Phillips, Bill Sticka and Anne Street, although some of them may not remember why their names are here.

I am grateful for the constant support and encouragement of Ann Kostant and the rest of the staff at Birkhäuser.

Contents

A Beginner's Guide
to Discrete Mathematics

1

Properties of Numbers

When we study numbers, many of the problems involve continuous properties. Much of the earliest serious study of mathematics was in geometry, and one essential property of the real world is that between any two points there is a line segment that is continuous and infinitely divisible. All of calculus depends on the continuous nature of the number line. Some of the most famous difficulties of Greek mathematics involved the existence of irrational numbers, and the fact that between any two real numbers one can always find another number.

But the discrete properties of numbers are also very important. The decimal notation in which we usually write numbers depends on properties of the number 10, and we can also study the (essentially discrete) features of representations where other positive whole numbers take the role of 10.

When we write the exact value of a number, or an approximation to its value, we use only the ten integers 0, 1, 2, 3, 4, 5, 6, 7, 8, 9, together with the decimal point. When numbers are represented in a computer, only integers are used. So it is important to understand the discrete properties of numbers in order to talk about their continuous properties.

1.1 Numbers

Sets and number systems

All of discrete mathematics — and, in fact, all of mathematics — rests on the foundations of set theory and numbers. In this first section we remind you of some basic definitions and notations. Further properties of numbers will be explored in the rest of this chapter; sets will be discussed further in Chapter 2.

We use the word *set* in everyday language: a *set of tires*, a *set of saucepans*. In mathematics you have already encountered various sets of numbers. We shall use *set* to mean any collection of objects, provided only that there is a well-defined rule, called the *membership law*, for determining whether a given object belongs to the set. The individual objects in the set are called its *elements* or *members* and are said to belong to the set. If S is a set and s is one of its elements, we denote this fact by writing

$$s \in S$$

which is read as "s belongs to S" or "s is an element of S."

The notation $S \subseteq T$ means that every member of S is a member of T:

$$x \in S \Rightarrow x \in T.$$

Then S is called a *subset* of T.

One way of defining a set is to list all the elements, usually between braces; thus the set of the first three members of the English alphabet is $\{a, b, c\}$. If S is the set consisting of the numbers 0, 1 and 3, we could write $S = \{0, 1, 3\}$. We write $\{1, 2, \ldots, 16\}$ to mean the set of all whole numbers from 1 to 16. This use of a string of dots is not precise, but is usually easy to understand. Another method is the use of the *membership law* of the set: for example, since the numbers 0, 1 and 3 are precisely the numbers that satisfy the equation $x^3 - 4x^2 + 3x = 0$, we could write the set S as

$$S = \{x : x^3 - 4x^2 + 3x = 0\} \text{ or } S = \{x | x^3 - 4x^2 + 3x = 0\}$$

("the set of all x such that $x^3 - 4x^2 + 3x = 0$"). Whole numbers are called *integers*, and *integral* means "being an integer,", so the set of whole numbers from 1 to 16 is

$$\{x : x \text{ integral}, 1 \leq x \leq 16\}.$$

Sample Problem 1.1 *Write three different expressions for the set with elements* 1 *and* -1.

Solution. Three possibilities are $\{1, -1\}$, $\{x : x^2 = 1\}$, and "the set of square roots of 1." There are others.

Practice Exercise. Write three different expressions for the set with elements 1, 2 and 3.

Some sets of numbers are so important that special names are given to them. The set of all integers, or whole numbers, is denoted $\mathbb{Z}$. With this notation, the set $\{1,2,\ldots,16\}$ can be written

$$\{x : x \in \mathbb{Z}, 1 \le x \le 16\}.$$

The *positive integers* or *natural numbers*, usually denoted $\mathbb{Z}^+$ or $\mathbb{N}$, are the integers greater than 0. Another important set of integers is the set $\mathbb{Z}^0$ of *non-negative* integers. We sometimes write $\mathbb{Z}^+ = \{1,2,3,...\}$, $\mathbb{Z}^0 = \{0,1,2,3,...\}$ and $\mathbb{Z} = \{\ldots -3,-2,-1,0,1,2,3,\ldots\}$. The use of a string of dots without a terminating number is understood to mean that the set continues without end. Such sets are called *infinite* (as opposed to *finite* sets like $\{0,1,3\}$).

The *rational numbers* $\mathbb{Q}$ consist of all fractions whose denominator is not 0:

$$\mathbb{Q} = \{\frac{p}{q} : p \in \mathbb{Z}, q \in \mathbb{Z}, q \neq 0\}.$$

Alternatively, $\mathbb{Q}$ is the set of all numbers with a repeating or terminating decimal expansion. Examples are

$$\begin{aligned} \tfrac{1}{2} &= 0.5, \\ \tfrac{-12}{5} &= -2.4, \\ \tfrac{3}{7} &= 0.428571428571\ldots. \end{aligned}$$

In the last example, the digit string 428571 repeats forever, and we usually indicate this by writing

$$\tfrac{3}{7} = 0.\overline{428571}.$$

The denominator q cannot be zero. In fact, division by zero is never possible. This is not an arbitrary rule, but rather it follows from the definition of division. When we write $x = \frac{p}{q}$, we mean "x is the number which, when multiplied by q, gives p." What would $x = 2/0$ mean? There is no number which, when multiplied by 0, gives 2. Similarly, $x = 0/0$ would be meaningless. In this case there *are* suitable numbers x, in fact every number will give 0 when multiplied by 0, but we want a uniquely defined answer.

Different decimal expansions do not always mean different numbers. The exception is an infinite string of 9's. These can be rounded up: $0.\overline{9} = 1$, $0.7\overline{9} = 0.8$, and so on. We prove a small theorem that illustrates this fact.

Theorem 1.1 $0.\overline{9} = 1$.

Proof. Suppose $x = 0.\overline{9}$. Then $10x = 9.\overline{9} = 9 + 0.\overline{9} = 9 + x$. So $9x = 9$ and $x = 1$. □

The integers are all rational numbers, and in fact they are the rational numbers with numerator 1. For example, $5 = 5/1$.

Each rational number has infinitely many representations as a ratio. For example,

$$1/2 = 2/4 = 3/6 = \cdots$$

The final number system we shall use is the set $\mathbb{R}$ of *real numbers*, consisting of all numbers that are decimal expansions. Not all real numbers are rational; one easy example is $\sqrt{2}$. In fact, if n is any natural number other than a perfect square (one of $1, 4, 9, 16, \ldots$), then $\sqrt{n}$ is not rational. Another important number that is not rational is the ratio π of the circumference of a circle to its diameter.

The number systems satisfy $\mathbb{Z}^+ \subseteq \mathbb{Z} \subseteq \mathbb{Q} \subseteq \mathbb{R}$. Rational numbers that are not integers are called *proper fractions*, and real numbers that are not rational are called *irrational numbers*.

It is sometimes useful to discuss number systems with the number 0 omitted from them, especially when division is involved. We denote this with an asterisk: for example, $\mathbb{Z}^*$ is the set of non-zero integers.

Factors and divisors

When x and y are integers, we use the phrase "x divides y" and write $x|y$ to mean "there is an integer z such that $y = x \cdot z$," and we say x is a *divisor* of y. Thus 2 divides 6 (because $6 = 2 \cdot 3$), -2 divides 6 (because $6 = (-2) \cdot (-3)$, 2 divides -6 (because $-6 = 2 \cdot (-3)$). Some students get confused about the case $y = 0$, but according to our definition x divides 0 for any non-zero integer x. We define the *factors* of a positive integer x to be the positive divisors of x (negative divisors are not called factors so -2 is not a factor of 6).

If x divides both y and z, we call x a *common divisor* of y and z. Among the common divisors of y and z there is naturally a greatest one, called (not surprisingly) *the greatest common divisor* of y and z, and denoted (y, z). If $(y, z) = 1$, y and z are called *coprime* or *relatively prime*. For example, $(4, 10) = 2$, so 4 and 10 are not coprime; $(4, 9) = 1$, so 4 and 9 are coprime. In the latter example we also say 4 is *relatively prime to* 9.

A *prime number* is a positive integer x other than 1 whose only factors are 1 and x. The first few primes are 2, 3, 5, 7, 11 and 13. The number 1 is excluded, and is called a *unit*.

Theorem 1.2 *The set of all prime numbers is infinite.*

Proof. Suppose the number of primes is finite. Then there will be some positive integer n such that there exist exactly n primes. Suppose the primes are $p_1, p_2, \ldots, p_n$. Now consider the number

$$x = p_1 \times p_2, \times \ldots \times p_n + 1.$$

Clearly x is not divisible by any of $p_1, p_2, \ldots, p_n$. Either it is prime, or its prime divisors are outside the set of all primes — clearly an impossibility. So the assumption that the number of primes is finite must be false. $\square$

Any positive integer x can be written as a product

$$x = x_1 \times x_2 \times \cdots \times x_k$$

where the x_i are all primes. So every positive integer is the product of prime factors. For example, 36 is the product $2 \times 2 \times 3 \times 3$. We usually collect all the equal factors and use an exponent, so we write $36 = 2^2 \times 3^2$. This prime factor decomposition is unique: for example, if

$$2^a 3^b 5^c = 2^x 3^y 7^z,$$

it must be true that $a = x$, $b = y$, and c and z are both zero. We shall look at this further in Section 9.1.

Sample Problem 1.2 *Use the prime factor decomposition to find the greatest common divisors of the following numbers with* 224*:* 16*,* 53*,* 63*,* 84*,* 97*.*

Solution. $224 = 2^5 \cdot 7$, and $16 = 2^4$, 53 is prime, $63 = 3^2 \cdot 7$, $84 = 2^2 \cdot 3 \cdot 7$, 97 is prime. So $(16, 224) = 16$, $(53, 224) = 1$, $(63, 224) = 7$, $(84, 224) = 28$ and $(97, 224) = 1$.

Practice Exercise. Use the prime factor decomposition to find $(72, 84)$ and $(56, 42)$.

Exponents and logarithms

If x is a positive integer, b^x is the product of x copies of b: $b^x = b \times b \times \cdots \times b$ (x factors). In this expression b is called the *base* and x the *exponent*. It is easy to deduce such properties as

$$b^x b^y = b^{x+y},$$
$$(b^x)^y = b^{xy},$$
$$(ab)^x = a^x b^x.$$

Negative exponents are handled by defining $b^{-x} = \frac{1}{b^x}$, and also $b^0 = 1$ whenever b is non-zero. The multiplication rule leads us to define $b^{\frac{1}{x}}$ to be the x-th root of b. (When x is even, we take the positive root for positive b and say $b^{\frac{1}{x}}$ is not defined for negative b).

Sample Problem 1.3 *Express the following in the simplest form — as decimal numbers if possible:*

$$(612)^0, (x^3)^5, (x^4)^0, (-10)^{-4}, \frac{1}{10^{-2}}.$$

Solution. $(612)^0 = 1$ ($b^0 = 1$ for any b); $(x^3)^5 = x^{3 \cdot 5} = x^{15}$; $(x^4)^0 = 1$ (again, $b^0 = 1$ for any b, or you could also argue that $(x^4)^0 = x^{4 \cdot 0} = x^0 = 1$); $(-10)^{-4} = \frac{1}{(-10)^4} = \frac{1}{(-1)^4 \cdot 10^4} = \frac{1}{10^4} = .0001$; $\frac{1}{10^{-2}} = 10^2 = 100$.

Practice Exercise. Do the same to

$$1^{15}, 0^5, (x^{-1})^0, \frac{x^6}{x^3}, (-2)^{-1}, \frac{1}{5^{-2}}.$$

Sample Problem 1.4 *Express in the simplest possible form, with positive exponents:*

$$\frac{1}{x^{-4}}; \frac{u^{-2}}{v^{-3}}; (3a^2)(5a^{-3}).$$

Solution. $\frac{1}{x^{-4}} = (x^{-4})^{-1} = x^{(-4).(-1)} = x^4$;
$\frac{u^{-2}}{v^{-3}} = (u^{-2})(v^{-3})^{-1} = u^{-2}v^3 = \frac{v^3}{u^{-2}}$;
$(3a^2)(5a^{-3}) = 3 \cdot 5 \cdot a^2 \cdot a^{-3} = 15a^{-1} = \frac{15}{a}$.

Practice Exercise. Express in the simplest possible form, with positive exponents:

$$\frac{t^{-2}}{t^{-3}}; y^{5-2}, (4x^{-2})(3x^4).$$

The process of taking powers can be inverted. The *logarithm* of x to base b, or $\log_b x$, is defined to be the number y such that $b^y = x$. Clearly $\log_b x$ is not defined if b is 1 (if $x = 1$, any y would be suitable, and if $x \neq 1$, no y would work).

Sample Problem 1.5 *What are* $\log_2 8$ *and* $\log_{16} 4$?

Solution. $2^3 = 8$, so $\log_2 8 = 3$. $\sqrt{16} = 16^{\frac{1}{2}} = 4$, so $\log_{16} 4 = \frac{1}{2}$.

Practice Exercise. What are $\log_3 9$, $\log_5 125$ and $\log_4 2$?

Several properties of logarithms follow from the elementary properties of exponents. In particular

$$\begin{aligned} \log_b uv &= \log_b u + \log_b v, \\ \log_b(u^{-1}) &= -\log_b u, \\ \log_b(x^y) &= y \log_b x, \\ \log_b 1 &= 0 \text{ for any } b. \end{aligned}$$

Theorem 1.3 *For any x and any base b, $x = \log_b b^x$.*

Proof. By definition, $u = b\log_b u$. Putting $u = b^x$ we have $bx = b(\log_b b^x)$. So, comparing the exponents, $x = \log_b b^x$. □

Sample Problem 1.6 *Evaluate* $125^{\log_5 2}$.

Solution.

$$\begin{aligned} 125^{\log_5 2} &= (5^3)^{\log_5 2} \\ &= 5^{3\log_5 2} \\ &= 5^{\log_5 2^3} \\ &= 5^{\log_5 8} \\ &= 8. \end{aligned}$$

Practice Exercise. Evaluate $\log_3(9\sqrt{3})^5$.

Absolute value, floor and ceiling

The *absolute value* or *modulus* of the number x, which is written $|x|$, is the non-negative number equal to either x or $-x$. For example, $|5.3| = 5.3, |-7.2| = 7.2$.

The *floor* $\lfloor x \rfloor$ of x is the largest integer not greater than x. If x is an integer, $\lfloor x \rfloor = x$. Some other examples are $\lfloor 6.1 \rfloor = 6$, $\lfloor -6.1 \rfloor = -7$. It is easy to deduce the following properties:

(1) $\lfloor x \rfloor = n$ if and only if n is an integer and $n \leq x < n+1$.

(2) If x is non-negative, then $\lfloor x \rfloor$ equals the integer part of x. If x is negative and non-integral, then $\lfloor x \rfloor$ is 1 less than the integer part of x.

(3) If n and k are integers, then k divides n if and only if $\frac{n}{k} = \lfloor \frac{n}{k} \rfloor$.

The *ceiling* $\lceil x \rceil$ is defined analogously as the smallest integer not less than x.

Sample Problem 1.7 *What are* $\lfloor 6.3 \rfloor$, $\lceil 7.2 \rceil$, $\lceil -2.4 \rceil$, $|3.4|$, $|-3.4|$?

Solution. $6, 8, -2, 3.4, 3.4$.

Practice Exercise. What are $\lfloor 2.6 \rfloor$, $\lfloor -1.7 \rfloor$, $\lceil 4.8 \rceil$, $\lceil -4 \rceil$, $|3.1|$, $|-4.4|$?

Exercises 1.1

Are the given statements true or false in Exercises 1 *to* 10?

1. $3 \in \{2,3,4,6\}$.

2. $4 \notin \{2,3,4,6\}$.

3. $5 \in \{2,3,4,6\}$.

4. $\{3,2\} = \{2,3\}$.

5. $\{1,2\} \in \{1,2,3\}$.

6. $\{1,2\} = \{1,2,3\}$.

7. $5 \in \{1,3,4,7\}$.

8. $6 \notin \{1,3,4,7\}$.

9. $4 \in \{1,3,4,7\}$.

10. $\{1,3\} = \{1,2,3\}$.

In Exercises 11 *to* 15, *write the list of all members of the set.*

11. $\{x : x$ is a month whose name starts with J$\}$.

12. $\{x : x$ is an odd integer between -6 and $6\}$.

13. $\{x : x$ is a letter in the word "Mississippi"$\}$.

14. $\{x : x$ is an even positive integer less than $12\}$.

15. $\{x : x$ is a color on the American flag $\}$.

16. Which of the following are true?
(i) All natural numbers are integers.
(ii) All integers are natural numbers.

17. For each of the following numbers, to which of the sets $\mathbb{Z}^+, \mathbb{Z}, \mathbb{Q}, \mathbb{R}$ does it belong?

(i) -2.13
(ii) $\sqrt{5}$
(iii) $\sqrt{4}$
(iv) 2π
(v) $1.8\overline{34}$
(vi) $2+\sqrt{2}$
(vii) 1.308
(viii) $1.\overline{3}$
(ix) -7
(x) $\sqrt{3}$
(xi) 10508
(xii) $1-\sqrt{7}$

18. In each case, list all the factors of the number.

(i) 36
(ii) 50
(iii) 72
(iv) 61
(v) 63
(vi) 24
(vii) 29
(viii) 70
(ix) 48

In Exercises 19 *to* 26, *decompose the two numbers into primes, and then compute their greatest common divisor.*

19. 231 and 275

20. 444 and 629

21. 95 and 125

22. 462 and 252

23. 88 and 132

24. 256 and 224

25. 1080 and 855

26. 168 and 231

In Exercises 27 *to* 34, *simplify the expression, writing the answer using positive exponents only.*

27. t^3t^{-3}

28. $(2x^2y^{-3})^2$

29. $\dfrac{(xy)^3}{xy^2}$

30. $\dfrac{5^6 2^4}{10^3}$

31. $(2xy)^{-2}$

32. x^4x^{-4}

33. $5y^2z^{-3}$

34. $(3x^2)^3(2x)^{-4}$

In Exercises 35 *to* 46, *evaluate the expression.*

35. $\log_3 9$

36. $\log_2 \frac{1}{4}$

37. $\log_{25} 5$

38. $\log_{10}(.1)$

39. $\lfloor 28.4 \rfloor$

40. $\lceil -107.7 \rceil$

41. $|\lfloor -11.9 \rfloor|$

42. $|11.4|$

43. $\lceil 77.7 \rceil$

44. $\lfloor |-11.9| \rfloor$

45. $|1.73|$

46. $\lfloor \sqrt{2} \rfloor$

1.2 Sums

Sum notation

The sum of the first sixteen positive integers can be written

$$1+2+3+4+5+6+7+8+9+10+11+12+13+14+15+16,$$

or more briefly $1+2+\cdots+16$. It should be clear that each number in the sum is obtained by adding 1 to the preceding number. A more precise notation is

$$\sum_{i=1}^{16} i.$$

In the same way,

$$\sum_{i=1}^{6} f(i) = f(1)+f(2)+f(3)+f(4)+f(5)+f(6);$$

the notation means "first evaluate the expression after the $\sum$ (that is, $f(i)$) when $i=1$, then when $i=2,\ldots$, then when $i=6$, and add the results."

Definition. A *sequence* (a_i) of length n is a set of n numbers $\{a_1, a_2, \ldots, a_n\}$, or $\{a_i : 1 \le i \le n\}$. The set of numbers is *ordered* — a_1 is first, a_2 is second, and so on — and a_i, where i is any one of the positive integers $1, 2, \ldots, n$, is called the i-th *member of the sequence*.

Definition. If (a_i) is a sequence of length n or longer, then $\sum_{i=1}^{n} a_i$ is defined by the rules

$$\sum_{i=1}^{1} a_i = a_1,$$

$$\sum_{i=1}^{n} a_i = \left(\sum_{i=1}^{n-1} a_i\right) + a_n.$$

Usually the sigma notation is used with a formula involving i for the term following $\sum$, as in the following examples. Notice that the range need not start at 1; we can write $\sum_{i=j}^{n}$ when j and n are any integers, provided $j \leq n$.

Sample Problem 1.8 *Write the following as sums and evaluate them:*

$$\sum_{i=1}^{4} i^2; \quad \sum_{i=3}^{6} i(i+1).$$

Solution.

$$\begin{aligned}\sum_{i=1}^{4} i^2 &= 1^2+2^2+3^2+4^2 \\ &= 1+4+9+16 \\ &= 30; \\ \sum_{i=3}^{6} i(i+1) &= 3\cdot 4+4\cdot 5+5\cdot 6+6\cdot 7 \\ &= 12+20+30+42 \\ &= 104.\end{aligned}$$

Practice Exercise. Write the following as sums and evaluate them:

$$\sum_{i=3}^{5} i(i-1); \quad \sum_{i=2}^{6} i.$$

Sample Problem 1.9 *Write the following in sigma notation:*

$$2+6+10+14; \quad 1+16+81.$$

Solution. $\sum_{i=1}^{4} (4i-2); \quad \sum_{i=1}^{3} i^4.$

Practice Exercise. Write the following in sigma notation:

$$1+3+5+7+9; \quad 8+27+64+125.$$

Some properties of sums

It is easy to see that the following properties of sums are true.

(1) If c is any given number, then $\sum_{i=1}^{n} c = nc$.

(2) If c is any given number and (a_i) is any sequence of length n, then

$$\sum_{i=1}^{n}(ca_i) = c\left(\sum_{i=1}^{n} a_i\right).$$

(3) If (a_i) and (b_i) are any two sequences, both of length n, then

$$\sum_{i=1}^{n}(a_i + b_i) = \left(\sum_{i=1}^{n} a_i\right) + \left(\sum_{i=1}^{n} b_i\right).$$

(4) If (a_i) is any sequence of length n, $1 \le j < n$, then

$$\sum_{i=1}^{n} a_i = \sum_{i=1}^{j} a_i \sum_{i=j+1}^{n} a_i.$$

The following is a standard result on sums.

Theorem 1.4 *The sum of the first n positive integers is* $\frac{1}{2}n(n+1)$, *or*

$$\sum_{i=1}^{n} i = \frac{n(n+1)}{2}.$$

Proof. Let us write s for the answer. Then $s = \sum_{i=1}^{n} i$. We shall define two very simple sequences of length n. Write $a_i = i$ and $b_i = n+1-i$. Then (a_i) is the sequence $(1,2,\ldots,n)$ and (b_i) is $(n,n-1,\ldots,1)$. They both have the same elements, although they are written in different order, so they have the same sum,

$$s = \sum_{i=1}^{n} a_i = \sum_{i=1}^{n} b_i.$$

So

$$\begin{aligned}
2s &= \sum_{i=1}^{n} a_i + \sum_{i=1}^{n} b_i \\
&= \sum_{i=1}^{n}(a_i + b_i) \\
&= \sum_{i=1}^{n}(i + (n+1-i) \\
&= \sum_{i=1}^{n}(n+1) \\
&= n(n+1).
\end{aligned}$$

Therefore, dividing by 2, we get $s = n(n+1)/2$. □

Since adding 0 does not change any sum, this result could also be stated as

$$\sum_{i=0}^{n} i = \frac{n(n+1)}{2}.$$

That form is sometimes more useful.

Sample Problem 1.10 *Find the sum of the numbers from* 11 *to* 30 *inclusive.*

Solution. $\sum_{i=11}^{30} i$ is required. Now

$$\sum_{i=1}^{30} i = \sum_{i=1}^{10} i + \sum_{i=11}^{30} i,$$

and, using the theorem and substituting, we have

$$\begin{aligned} \frac{30 \cdot 31}{2} &= \frac{10 \cdot 11}{2} + \sum_{i=11}^{30} i, \\ 465 &= 55 + \sum_{i=11}^{30} i, \end{aligned}$$

so $\sum_{i=11}^{30} i = 465 - 55 = 410$.

Sample Problem 1.11 *Find* $2+6+10+14+18+22+\cdots+122$.

Solution. This is the sum of term $2+4i$, where i goes from 0 to 30.

$$\begin{aligned} \sum_{i=0}^{30} (2+4i) &= \left(\sum_{i=0}^{30} 2\right) + \left(\sum_{i=0}^{30} 4i\right) \\ &= \left(\sum_{i=0}^{30} 2\right) + 4 \cdot \left(\sum_{i=0}^{30} (i)\right) \\ &= 31 \cdot 2 + 4 \cdot \frac{30 \cdot 31}{2} \\ &= 62 + 1860 \\ &= 1922. \end{aligned}$$

Practice Exercise. Find $3+7+\cdots+43$.

Two other standard results that will be proved in the exercises are:

Theorem 1.5 $\sum_{i=1}^{n} i^2 = \frac{n(n+1)(2n+1)}{6}$.

Theorem 1.6 $\sum_{i=1}^{n} x^2 = \frac{x^{n+1} - 1}{x - 1}$ *unless* $x = 1$.

Exercises 1.2

In Exercises 1 *to* 12, *write the expression as a sum and evaluate it.*

1. $\sum_{i=1}^{6}(i^2+1)$

2. $\sum_{i=3}^{9} i(i-3)$

3. $\sum_{i=2}^{4}\sqrt{i}$

4. $\sum_{i=2}^{5}(2i-1)$

5. $\sum_{i=1}^{3} i^3$

6. $\sum_{i=1}^{4}\frac{1}{\sqrt{i}}$

7. $\sum_{i=3}^{6} i(i-2)$

8. $\sum_{i=2}^{4} i^2(i-1)$

9. $\sum_{i=10}^{12} i+\frac{1}{i}$

10. $\sum_{i=1}^{3} i^2-\frac{1}{i^2}$

11. $\sum_{i=2}^{5} i+(-1)^i$

12. $\sum_{i=2}^{6}(1+i^2)$

13. (i) Say $b_i = i^3-(i-1)^3$. Prove that $\sum_{i=1}^{n} b_i = n^3$.

(ii) Prove that $b_i = 3i^2-3i+1$, and therefore

$$\sum_{i=1}^{n} b_i = 3\sum_{i=1}^{n} i^2 - 3\sum_{i=1}^{n} i + n.$$

(iii) Use this to prove that

$$\sum_{i=1}^{n} i^2 = \frac{n(n+1)(2n+1)}{6}.$$

14. Suppose $a_1 = 1, a_2 = x, a_3 = x^2$, and, in general, $a_i = x^{i-1}$.

(i) Prove that $\sum_{i=2}^{n+1} a_i = x\sum_{i=1}^{n} a_i$.

(ii) Use this to show that $\sum_{i=1}^{n} a_i = 1 + x\sum_{i=1}^{n} a_i - x^n$.

(iii) Use part (b) to find the value of $\sum_{i=1}^{n} a_i$ when $x \neq 1$.

(iv) Why did we have to require "$x \neq 1$" in the preceding part?

In Exercises 15 *to* 18, $\sum_{i=1}^{n} a_i = A$ *and* $\sum_{i=1}^{n} b_i = B$. *Evaluate* $\sum_{i=1}^{n} c_i$ *in each case.*

15. $c_i = 2a_i + 1.$

16. $c_i = 5a_i.$

17. $c_i = 3a_i - b_i.$

18. $c_i = a_i + 2b_i.$

Use the standard results of Theorems 1.4, 1.5, *and* 1.6 *and the four properties of sums to evaluate the expressions in Exercises* 19 *to* 30.

19. $\sum_{i=7}^{12} (i-3)$

20. $\sum_{i=3}^{8} (5i^2 - 4)$

21. $\sum_{i=2}^{6} 2^i$

22. $\sum_{i=1}^{n} 3^i$

23. $\sum_{i=5}^{20} (2i+1)$

24. $\sum_{i=4}^{10} (i^2 + 1)$

25. $\sum_{i=1}^{n} (i^2 + 1)$

26. $\sum_{i=1}^{n} (2i^2 - 1)$

27. $\sum_{i=1}^{9} 2^{-i}$

28. $\sum_{i=1}^{n} (i^2 - i)$

29. $\sum_{i=1}^{n} (1+i)^2$

30. $\sum_{i=0}^{n} 2i^{-i}$

1.3 Bases

Arithmetic in various bases

In ordinary arithmetic we use ten digits or one-symbol numbers. $\{0,1,2,3,4,5,6,7,8,9\}$ to write all the possible numbers. The symbol for "ten" is 10, meaning "once ten plus zero times one." For example, 243 means "twice ten-squared plus four times ten plus three." In general, suppose $a_0, a_1, a_2, \ldots$ and $b_1, b_2, b_3, \ldots$ are any digits. When we write the number $\ldots a_2a_1a_0.b_1b_2b_3\ldots$ it means $\cdots + a_2 \cdot 10^2 + a_1 \cdot 10^1 + a_0 \cdot 10^0 + b_1 \cdot 10^{-1} + b_2 \cdot 10^{-2} + b_3 \cdot 10^{-3} + \cdots$ or in sigma notation,

$$\sum_{i \geq 0} a_i \cdot 10^i + \sum_{j > 0} b_j \cdot 10^{-j}.$$

There is no special reason for choosing ten. If the base b were used, we might use 243 to mean "twice b-squared plus four times b plus three." We shall refer to this as an *expression in base b*. To avoid confusion, it will be necessary to write down the base. We shall write $(243)_b$ to mean the number in base b, so that

$$(243)_b = 2b^2 + 4b + 3.$$

Similarly,

$$(0.73)_b = 7b^{-1} + 3b^{-2}.$$

When no subscript is used, the usual base (base 10) is intended. Another common notation when the base is 2 is to write B after the number, because numbers written in base 2 are called binary numbers. So $101.11B$ means the same as $(101.11)_2$.

To write regular (base 10) numbers, ten digits are used, but when we write binary numbers, only the two digits 0 and 1 are necessary. Similarly, in base b, we need b digits. If b is greater than 10, some new symbols must be invented—for example, in the base 16, which is called *hexadecimal* and is often used in computer applications, A, B, C, D, E and F are used for 10, 11, 12, 13, 14 and 15.

Change of base

Sample Problem 1.12 *Convert* $(243)_7$ *and* $(104)_5$ *to base ten.*

Solution.

$$\begin{aligned}
(243)_7 &= 2\cdot 7^2 + 4\cdot 7 + 3 \\
&= 98 + 28 + 3 \\
&= 129; \\
(104)_5 &= 1\cdot 5^2 + 0\cdot 5 + 4 \\
&= 25 + 4 \\
&= 29.
\end{aligned}$$

Practice Exercise. What are $(144)_5, (203)_7, (112)_3$?

The process is just the same for non-integers, as the following sample problem shows.

Sample Problem 1.13 *Convert* $(0.231)_5$ *and* $(104.231)_5$ *to base ten.*

Solution.

$$\begin{aligned}
(.231)_5 &= 2\cdot 5^{-1} + 3\cdot 5^{-2} + 1\cdot 5^{-3} \\
&= 2\cdot .2 + 3\cdot 0.04 + 1\cdot 0.008 \\
&= 0.4 + 0.12 + 0.008 \\
&= 0.528.
\end{aligned}$$

Using this and the preceding sample problem,

$$\begin{aligned}
(104.231)_5 &= (104)5 + (0.231)5 \\
&= 29 + 0.528 \\
&= 29.528.
\end{aligned}$$

Practice Exercise. What are $(0.242)_5, (144.242)_5$?

In order to convert from base ten to another base, use continued division.

Sample Problem 1.14 *What is* 108 *in base* 7*?*

Solution.
$$\begin{array}{r|ccc} 7 & 108 & & \\ \hline 7 & 15 & + & 3 \\ \hline & 2 & + & 1 \end{array}$$
So $108 = (213)_7$.

Practice Exercise. What is 54 in base 5? What is 103 in base 6?

To discuss conversion of non-integers from another base to base 10, look again at the second sample problem. Say $x = (0.231)_5$. Then $5x = 2 + (0.31)_5$. So when we multiply by the base (in this case 5), the integer part of the result is the first digit of the expansion. Then

$$\begin{aligned} 5x - 2 &= (0.31)_5, \\ 5(5x-2) &= 3 + (0.1)_5. \end{aligned}$$

So multiplying the remainder by the base gives the second digit. And so on.

Sample Problem 1.15 *Express* 0.71875 *in base* 2.

Solution.

$$\begin{aligned} 2 \cdot (0.71875) &= 1.4375, \text{ first digit } 1 \\ 2 \cdot (0.4375) &= 0.875, \text{ second digit } 0 \\ 2 \cdot (0.875) &= 1.75, \text{ third digit } 1 \\ 2 \cdot (0.75) &= 1.5, \text{ fourth digit } 1 \\ 2 \cdot (0.5) &= 1.0, \text{ fifth digit } 1. \end{aligned}$$

So $0.71875 = (0.10111)_2 = 0.10111B$.

Practice Exercise. Express 0.40625 in base 2.

Just as in base ten, some numbers have terminating expressions and others recur. See the next sample problem.

Sample Problem 1.16 *Express* 33.7125 *in base* 4.

Solution.
$$\begin{array}{r|ccc} 4 & 33 & & \\ \hline 4 & 8 & + & 1 \\ \hline & 2 & + & 0 \end{array}$$
so $33 = (201)_4$.

$$\begin{aligned} 4 \cdot 0.7125 &= 2.85, \text{ first digit } 2 \\ 4 \cdot 0.85 &= 3.4, \text{ second digit } 3 \\ 4 \cdot 0.4 &= 1.6, \text{ third digit } 1 \\ 4 \cdot 0.6 &= 2.4, \text{ fourth digit } 2 \\ 4 \cdot 0.4 &= \ldots \end{aligned}$$

At this point we have been asked to repeat an earlier calculation, so the whole process will recur:

$$0.7125 = (0.231212121212\ldots)_4 = (0.23\overline{12})_4.$$

Adding, $33.7125 = (201.23\overline{12})_4$.

Practice Exercise. Express 53.12 in base 6.

Conversion between binary and hexdecimal numbers is particularly easy. Each hexadecimal number can be expressed as a 4-digit binary number, provided leading zeros are included. The conversion table is as follows:

$0_{16} = 0000_2$	$4_{16} = 0100_2$	$8_{16} = 1000_2$	$C_{16} = 1100_2$
$1_{16} = 0001_2$	$5_{16} = 0101_2$	$9_{16} = 1001_2$	$D_{16} = 1101_2$
$2_{16} = 0010_2$	$6_{16} = 0110_2$	$A_{16} = 1010_2$	$E_{16} = 1110_2$
$3_{16} = 0011_2$	$7_{16} = 0111_2$	$B_{16} = 1011_2$	$F_{16} = 1111_2$

To change hexadecimal to binary, simply use the table to make replacements, and then delete excess zeros. To convert binary to hexadecimal, first add zeros at both ends to make the number of symbols before the decimal point and the number after both multiples of 4, and use the table again.

Sample Problem 1.17 *Convert* 3A04.A4 *from hexadecimal to binary.*

Solution. Replacing term by term we get

$$0011\ 1010\ 0000\ 0100.1010\ 0100,$$

so the answer is

$$11\ 1010\ 0000\ 0100.1010\ 01.$$

(Notice the use of spaces to make the numbers more readable.)

Practice Exercise. Convert $5B3.76$ from hexadecimal to binary.

Sample Problem 1.18 *Convert* $10\ 1101.111_2$ *to hexadecimal.*

Solution. $10\ 1101.111_2 = 0010\ 1101.1110_2 = 2D.E$.

Practice Exercise. Convert $10\ 1101\ 1011.01_2$ to hexadecimal.

Exercises 1.3

In Exercises 1 *to* 39, *express the number in base* 10.

1. 110111_2 **2.** 1314_7 **3.** 2081_9

4. 101101_3 **5.** 6443_8 **6.** 110011_2

7. 11111111_2 **8.** 102_7 **9.** 115_9

10. 10101_2 **11.** 12121_3 **12.** 4041_8

13. 210_8 **14.** 231_5 **15.** $11.\overline{3}_5$

16. 17.46_9 **17.** $1.\overline{1}_3$ **18.** 110111.01_2

19. 22.6_7 **20.** 0.04_8 **21.** 210.04_8

22. $101.\overline{14}_6$ **23.** 32.03_6 **24.** 10001000.0001_2

25. 141.22_5 **26.** 50.2181_6 **27.** 0.01_2

28. 0.101_2 **29.** 10101.101_2 **30.** $11.2\overline{1}_3$

31. 112.2_3 **32.** 2.031_4 **33.** 101.31_6

34. $1.\overline{1}_6$ **35.** 1011.1_5 **36.** 44041.22_8

37. $11.\overline{2}_4$ **38.** 2.005_8 **39.** 7.41_9

In Exercises 40 *to* 54, *express the number in base* 2.

40. 253 **41.** 43 **42.** 0.2

43. 55.75 **44.** 8 **45.** $14.8\overline{3}$

46. 228 **47.** 126 **48.** 13.13

49. 111 **50.** 0.25 **51.** 126.25

52. 12.45 **53.** 1819 **54.** 517.8

In Exercises 55 *to* 69, *express the number in base* 6.

55. $4.\overline{1}$ **56.** 1011 **57.** 11.735

58. 2.2 **59.** $0.\overline{3}$ **60.** 75.8

61. 119 **62.** 51.4 **63.** 1104

64. 3.205 **65.** $14.\overline{4}$ **66.** $112.\overline{2}$

67. 26.3 **68.** 77 **69.** 28.4

In Exercises 70 *to* 84, *express the number in hexadecimal (base* 16*) using the symbols* $0, 1, \ldots, 9, A, B, C, D, E, F$.

70. 723 **71.** 108 **72.** 88.91

73. $11.\overline{5}$ **74.** 7.16 **75.** 14.03

76. 2.12 **77.** 255 **78.** 104

79. 27 **80.** 10.5 **81.** $1.\overline{12}$

82. 7.8 **83.** 64.6 **84.** 257

In Exercises 85 *to* 92, *convert the binary number to hexadecimal.*

85. 1001 0010 0100 **86.** 1001.1101 101

87. 101 1010 1101.11 **88.** 110 1101 0110.111

89. $101.0010\ 1110\ \overline{011}$ **90.** 1.0110 111

91. 111 0110 1011.01 **92.** $1\ 0101.1100\ 1\overline{101}$

In Exercises 93 *to* 100, *convert the hexadecimal number to binary.*

93. $1A01$ **94.** $A.0B$ **95.** 1101

96. $912.B\overline{5}$ **97.** $AE.FE$ **98.** $213F.EE$

99. $1E4.\overline{A}$ **100.** $5EE.C$

1.4 Scientific Notation

Floating point numbers

It is common to write very large or very small numbers in *scientific* (or *exponential*) notation — as an example, two million million million million is written as 2×10^{24}, rather than 2 followed by 24 zeroes. There can be various numbers of digits before the decimal point — for example, 24.53 could be written as 0.2453×10^2 or 2453×10^{-2}, or even 24.53×10^0. In any of these expressions we refer to the first number as the *mantissa* and the power as the *exponent* — so 2.453×10^3 has mantissa 2.453 and exponent 3. The part of the mantissa to the right of the decimal point is called the *fraction*. All computers use some form of exponential notation to represent non-integers.

Floating point notation is a special form of scientific notation. The mantissa has a fixed number of digits, which we shall call the *length*, and the exponent is chosen so that there is exactly one digit to the left of the decimal point. The absolute value $|f|$ of the mantissa satisfies $1 \leq |f| < 10$. A number whose mantissa satisfies this equation is called *normalized* — for example, the normalized form of 211.7 is 2.117×10^2. The forms 211.7×10^0 and 0.2117×10^3 would not be used — in the first one the mantissa is too large, and in the second it is too small. An exception to the rule is the number zero, which has 0 as its mantissa.

An alternative convention that is widely used is to require the mantissa f to satisfy $0.1 \leq |f| < 1$, so that the part before the decimal point is 0 and the first digit after the point is non-zero. We shall not use that method here, but you may encounter it in other books.

Sample Problem 1.19 *In a floating point system of length* 4, *what are the representations of* $12.34, 101.2, -0.0012$*?*

Solution. $1.234 \times 10^1, 1.012 \times 10^2, -1.200 \times 10^{-3}$.

Practice Exercise. In a floating point system of length 4, what are the representations of $17.5, -11.12, 1.401$?

Rounding and dropping digits

Sometimes the number of digits in the mantissa of a number is greater than the number of digits allowed by the floating point system. For example, in a floating point system of length 4, how are we to write the numbers 101.73 and 21.468? We would represent the first as 1.017×10^2 and the second as 2.147×10^1. In the first case, where the last digit was less than 5, we rounded down, and ignored it; in the second case, where it was greater, we rounded up and added 1 to the second to last digit. In the extreme case, 123.99 would become 1.240×10^2 — rounding up 9 yields 10 ("carry the 1"). In the middle, with a 5, one rounds up.

If you check by using several calculators, you will find that some of them round up and down according to the above rules, whereas others simply ignore the last digit — this is called *dropping*. For example, if the exact answer to a calculation is 27.73358, and your calculator only shows six digits, it might give you 27.7335, or 2.77335×10^1 in floating point form. (Usually the more expensive calculators round, the cheaper ones drop!)

In the case of negative numbers, one rounds or drops on the absolute value. To round -2.3149 to length four, first look at the absolute value 2.3149, which rounds to 2.315. So -2.3149 rounds to -2.315.

Sample Problem 1.20 *Write the following numbers in floating point form, of length* 3.

(1) 11.48
(2) 5.302
(3) -77.335
(4) $3/11$.

Solution. (1) 1.15×10^1, (2) 5.30×10^0, (3) -7.73×10^1, (4) 2.73×10^{-1} (since $3/11 = .2727\ldots$).

Practice Exercise. Write the following numbers in floating point form, of length 4.

(1) -804.955
(2) 108.798
(3) 1144.114
(4) $4/13$

Simplified floating point arithmetic

To illustrate how floating point numbers are used in calculations, we shall use a system in which the mantissa is limited to four places. Typical numbers are 2.123×10^3, 1.584×10^{-1}, -3.113×10^2.

If two numbers have the same exponent, they are added by adding the mantissas. If the two exponents are different they must first be adjusted by increasing the smaller one.

Sample Problem 1.21 *Find* $8.348 \times 10^3 + 2.212 \times 10^2$.

Solution. First adjust the exponents: instead of 2.212×10^2, use 0.221×10^3. (Notice that one digit is rounded; this is called *truncation*.) Then add the mantissas:

$$8.438 + .221 = 8.569.$$

So the answer is 8.569×10^3.

Practice Exercise. Find $1.043 \times 10^2 + 3.223 \times 10^3$.

Sample Problem 1.22 *Find* $7.124 \times 10^{-2} + 6.004 \times 10^{-2}$.

Solution. $7.124 + 6.004 = 13.128$. Since 13.128×10^{-2} is not normalized, divide by 10 (and add 1 to the exponent).

$$13.128 \times 10^{-2} = 1.3128 \times 10^{-1},$$

and the answer is 1.313×10^{-1}. Notice that again a place is lost by truncation, and rounding to the nearer four-digit mantissa occurs.

Practice Exercise. Calculate in floating point arithmetic:

$$\begin{aligned} 6.041 \times 10^2 &\quad + \quad 3.303 \times 10^3; \\ 7.007 \times 10^{-4} &\quad + \quad 4.644 \times 10^{-4}. \end{aligned}$$

Sometimes adjustment of the exponent can have extreme results. For example, in the addition

$$1.753 \times 10^3 + 4.004 \times 10^{-1}$$

the exponent required is 3. When 4.004×10^{-1} is adjusted to have exponent 3, the result is 0.000×10^3.

Multiplication is carried out by multiplying mantissas and adding exponents — normalization may then be necessary.

Sample Problem 1.23 *Multiply* $(5.045 \times 10^2) \times (2.123 \times 10^3)$.

Solution. $5.045 \times 2.123 = 10.710535$. So the required product equals 10.710535×10^5. In normalized form the answer is 1.071×10^4.

Practice Exercise. Multiply 4.640×10^2 by 3.020×10^4.

Exercises 1.4

1. In each case identifythe mantissa and the exponent. Then write the number in floating point form, of length 3.

(i) 104.53×10^4 (iii) 11.11×10^{-5}

(ii) -11×10^{-3} (iv) 104.3×10^7

2. In each case write down the mantissa and the exponent. Then write the number in floating point form, of length 4.

(i) 72.11×10^4 (iii) 1104.7×10^3

(ii) -423×10^{-2} (iv) 104.33×10^{-3}

In Exercises 3 *to* 20, *express the number in floating point form, of length* 5.

3. 117.4 **4.** 48.1179 **5.** $\frac{3}{7}$

6. -23.08 **7.** 3834.992 **8.** -56.44

9. 224.113 **10.** -14.7909 **11.** -21.2

12. 0.011212 **13.** 208.3 **14.** $1/3$

15. 107.986 **16.** -15.55 **17.** -73.48

18. 108.10888 **19.** 1122.99 **20.** 52.1713

In Exercises 21 *to* 36, *carry out the additions and subtractions in a floating point system of length* 5.

21. $3.2084 \times 10^3 + 2.7941 \times 10^3$

22. $1.1081 \times 10^2 + 4.9392 \times 10^3$

23. $7.5493 \times 10^3 + 8.7143 \times 10^3$

24. $7.4083 \times 10^2 - 3.9214 \times 10^2$

25. $6.4132 \times 10^3 - 3.1102 \times 10^2$

26. $7.8101 \times 10^3 - 6.4103 \times 10^4$

27. $7.2163 \times 10^{-3} - 8.5123 \times 10^{-3}$

28. $8.6141 \times 10^{-1} - 2.7814 \times 10^{-2}$

29. $1.4083 \times 10^2 + 7.7411 \times 10^2$

30. $2.7174 \times 10^3 + 1.3908 \times 10^2$

31. $7.4192 \times 10^2 + 6.3142 \times 10^2$

32. $3.6131 \times 10^4 - 2.1142 \times 10^4$

33. $4.2401 \times 10^4 - 7.1432 \times 10^3$

34. $5.8974 \times 10^3 - 2.8904 \times 10^4$

35. $7.1384 \times 10^{-3} + 7.2241 \times 10^{-3}$

36. $7.8418 \times 10^{-3} - 6.6616 \times 10^{-4}$

In Exercises 37 *to* 44, *carry out the multiplication in a floating point system of length* 3.

37. Multiply 1.14×10^2 by 1.48×10^2.

38. Multiply -1.73×10^2 by 2.18×10^3.

39. Multiply -2.85×10^5 by 1.17×10^{-4}.

40. Multiply 5.13×10^4 by 1.16×10^{-5}.

41. Multiply 7.20×10^{-2} by 1.08×10^2.

42. Multiply 4.44×10^2 by 1.13×10^3.

43. Multiply -3.14×10^3 by 1.13×10^4.

44. Multiply -5.92×10^2 by 8.14×10^{-3}.

1.5 Arithmetic In Computers

Storing numbers in computers

There are two facts about computers that you should bear in mind when thinking about how computers store and use numbers. First of all, a computer uses binary arithmetic because a computer recognizes two states — on or off, electricity flowing or electricity not flowing. (There are exceptions to this rule, but the computers you will see about you are binary.) The computer converts your input (in ordinary decimal notation) to binary before doing any arithmetic.

Second, the computer is limited in size. It cannot arbitrarily decide to give more digits in an answer the way a human being can. A number might for example be restricted to 8 binary digits, or 16, or 64. The degree of restriction is usually built into the computer hardware. Binary digits are called *bits*, so we refer to 8-bit, or 16-bit, or 64 bit numbers.

Integers

We shall illustrate integer arithmetic in a computer using 8-bit numbers although in practice the range of numbers available in a computer is much larger. So a number like 53 is stored as 00110101. (This is because $53 = 110101_2$; the computer has enough room to put in 8 bits for each number it processes, so it includes two extra zeros at the beginning (called "leading zeros"). To store the negative -53, the computer first calculates what is called the *complement* (or *one's complement*) of 00110101 by changing every 0 to a 1 and every 1 to a 0, so it calculates 11001010. Then it calculates the *two's complement* by adding 1 to the one's complement. So the two's complement is 11001011. This is used to represent -53. In the remainder of this section we shall show you how this "two's complement" representation is used.

Exercises 1.5

In Exercises 1 *to* 18, *assume you have an* 8*-bit computer using two's complement arithmetic. In each question, how is the given number represented? How would the representation be written in hexadecimal?*

1. 46 **2.** −46 **3.** 127

4. 0 **5.** −1 **6.** −3

7. −126 **8.** 21 **9.** −14

10. 57 **11.** −57 **12.** −127

13. −128 **14.** −2 **15.** −4

16. −125 **17.** 17 **18.** −23

In Exercises 19 *to* 30, *what numbers are represented by the given number in an* 8*-bit computer using two's complement arithmetic?*

19. 00000000 **20.** 11111111 **21.** 10001010

22. 10000110 **23.** 01110001 **24.** 01011111

25. 00001111 **26.** 11110000 **27.** 11000110

28. 10000010 **29.** 10000011 **30.** 01101111

Use two's complement arithmetic to carry out the calculations in Exercises 31 *to* 38.

31. $17 - 4$ **32.** $26 - 38$

33. $-4 - 12$ **34.** $38 - 14$

35. $15 - 7$ **36.** $28 - 32$

37. $44 - 4$ **38.** $-16 - (-29)$

*What is the IEEE*754 *representation of the numbers in Exercises* 39 *to* 46*? Give both binary and hexadecimal forms.*

39. 19.625 **40.** −19.625

41. 48 **42.** 0.1

43. 27

44. 1033.75

45. -1033.75

46. -0.3

The hexadecimal numbers in Exercises 47 *to* 52 *are in* $IEEE754$ *format. What real numbers do they represent (in base* 10*)? (Give your answer as an integer multiplied by a power of* 2*.)*

47. 02*A*80000

48. 9350*C*000

49. 86*BC*0000

50. 03400000

51. 43*D*00000

52. 6*AB*80000

53. Suppose a computer uses two's complement arithmetic and 16 bits are available. What are the smallest and largest integers that can be represented? What are the approximate answers if 32 bits and 64 bits are available? (Use scientific notation.)

54. In $IEEE754$ format (32 bit words), what are the largest real number and the smallest positive real number that can be represented? What are these numbers in hexadecimal form?

55. Can the string 10000000 ever arise in two's complement form? If "yes," what does it represent? If "no," why not?

2

Sets and Data Structures

Sets are of fundamental importance in mathematics. The very number systems that we studied in the preceding chapter were all discussed in terms of a *set of numbers*. Many problems of discrete mathematics can conveniently be expressed in terms of sets, especially finite sets. For this reason, we need to discuss the properties of sets and develop language to talk about them.

Along with sets, it is appropriate to discuss elementary logic. We shall observe a parallel between the logic of propositions and set theory.

Both mathematical logic and set theory are broad areas of mathematics, and we only discuss them briefly here; moreover our interest is in the discrete case, and we emphasize finite cases. The interested reader will find that there is a wide literature on both these topics, and they contain very deep problems.

2.1 Propositions and Logic

Propositions and truth tables

We shall define a *proposition* to be a statement that has a well-defined *truth value*, that is, it is either true (T) or false (F). Some statements in English are not propositions — one example is matters of opinion, such as "I like apples"; these are not propositions. On the other hand, "it will rain on this day next year" is a proposition: it is either true or false (we are not worried about whether or not we know the truth value, or even if it is possible to know it).

Simple propositions, like "today is Tuesday" and "it is raining," can be combined to form *compound* propositions, like "today is Tuesday and it is raining," by using a *connective* ("and" in the example). The truth value of a compound proposition can be calculated once we know the truth values of the simple propositions from which it is formed, and the connective used to combine these simple propositions together.

The simplest connectives are "not," "and," "or," denoted by $\sim$, $\wedge$, $\vee$ respectively. If p and q denote propositions, then the proposition "not p" (denoted $\sim p$) is true precisely when p is false, the proposition "p and q" (denoted $p \wedge q$) is true precisely when p is true and q is true, and the proposition "p or q" (denoted $p \vee q$) is true precisely when p is true or when q is true or when both p and q are true. (This is the meaning called "inclusive or," the usual usage in mathematics.) The truth values of these compound propositions are shown in Table 2.1. Such tables are called *truth tables*.

Formally, $\sim p$ is called the *negation* of p, $p \vee q$ is the *disjunction* of propositions p and q, and $p \wedge q$ is the *conjunction* of p and q.

p	$\sim p$
T	F
F	T

p	q	$p \vee q$	$p \wedge q$
T	T	T	T
T	F	T	F
F	T	T	F
F	F	F	F

Table 2.1: Truth tables of $\sim, \wedge, \vee$

Often alternate phrases are used. For example, we sometimes use "as well (as)" instead of "and." In English, we often use "but" instead of "and" when one of the two propositions is negative. Both these connectives are represented by $\wedge$: if p means "today is cold" and q means "today is sunny," then $p \wedge q$ could be translated as "today is cold and sunny," "today is cold but sunny," or "today is cold as well as sunny."

Sample Problem 2.1 *Let p denote the proposition "the sun is shining" and q the proposition "the wind is blowing." Write expressions for "the sun is not shining," "the sun is shining and the wind is blowing," and "the sun is shining but the wind is not blowing."*

Solution. $\sim p$ denotes "the sun is not shining," $p \vee q$ denotes "the sun is shining or the wind is blowing," and $p \wedge q$ denotes "the sun is shining and the wind is blowing."

Practice Exercise. Write expressions for "the wind is not blowing," "the sun is shining or the wind is blowing (maybe both)," and "the sun is not shining but the wind is blowing."

Sample Problem 2.2 *Suppose p, q and r mean "Joseph is here," "Nancy is here" and "Donna is here." Interpret* $p\vee\sim q$ *and* $(p\wedge q)\wedge r$.

Solution. $p\vee\sim q$ means "Joseph is here but Nancy is not"; $(p\wedge q)\wedge r$ means "Joseph, Nancy and Donna are here."

Practice Exercise. In this situation, interpret $(p\wedge r)\wedge\sim q$ and $q\vee r$.

We sometimes think of a truth table as showing the truth or falsity of different outcomes of an experiment or set of events, as the following sample problem shows.

Sample Problem 2.3 *Suppose one card is drawn from a standard deck. Let p represent the statement "the card is a heart" and q represent "the card is an honor" (ace, king, queen, jack or ten). For which draws are* $p\wedge q$ *and* $p\vee q$ *true?*

Solution. For $p\wedge q$ to be true, the card must be a heart and it must be an honor. So it is true when the draw is the ace, king, queen, jack or ten of hearts (5 cases), and is false for all other cards (the other 47 cases). $p\vee q$ will be true for all thirteen hearts and all five honors in clubs, diamonds or spades (28 cases in all) and false in the other 24 cases.

Practice Exercise. In the same situation, which draws make $p\wedge\sim q$ true? Which make $p\vee\sim q$ true?

To find the truth table of a statement with several connectives, we work one step at a time. For instance, to find the truth table of

$$(p\vee\sim q)\wedge(q\vee\sim p),$$

we consider $\sim p$, $\sim q$, $p\vee\sim q$, $q\vee\sim p$, and finally the whole expression, as shown in Table 2.2.

p	q	$\sim p$	$\sim q$	$p\vee\sim q$	$q\vee\sim p$	$(p\vee\sim q)\wedge(\vee\sim p)$
T	T	F	F	T	T	T
T	F	F	T	T	F	F
F	T	T	F	F	T	F
F	F	T	T	T	T	T

Table 2.2: Truth table of $(p\vee\sim q)\wedge(q\vee\sim p)$

Sample Problem 2.4 *Find the truth table of* $(p \wedge q) \vee (q \wedge \sim r)$.

Solution.

p	q	r	$\sim r$	$(p \wedge q)$	$(q \wedge \sim r)$	$(p \wedge q) \vee (q \wedge \sim r)$
T	T	T	F	T	F	T
T	T	F	T	T	T	T
T	F	T	F	F	F	F
T	F	F	T	F	F	F
F	T	T	F	F	F	F
F	T	F	T	F	T	T
F	F	T	F	F	F	F
F	F	F	T	F	F	F

Practice Exercise. Find the truth table of $(p \vee (q \vee (\sim p \wedge \sim r)))$.

Two other connectives we use frequently are $p \rightarrow q$, meaning "if p then q," or "p implies q," and $p \leftrightarrow q$, meaning "p if and only if q," or in other words "if p then q and if q then p." These are called the *conditional* ($\rightarrow$) and the *biconditional* ($\leftrightarrow$). Their truth tables are shown in Table 2.3.

p	q	$p \rightarrow q$	$p \leftrightarrow q$
T	T	T	T
T	F	F	F
F	T	T	F
F	F	T	T

Table 2.3: Truth tables of $p \rightarrow q$ and $p \leftrightarrow q$

Sample Problem 2.5 *Find the truth table for*

$$(p \wedge \sim q) \rightarrow (q \vee p).$$

Solution. We proceed in steps as before. The result is

p	q	$\sim q$	$p \wedge \sim q$	$q \vee p$	$(p \wedge \sim q) \rightarrow (q \vee p)$
T	T	F	F	T	T
T	F	T	T	T	T
F	T	F	F	T	T
F	F	T	F	F	T

Practice Exercise. Find the truth table for

$$(p \vee \sim q) \equiv (q \rightarrow p).$$

Tautologies, theorems and logical equivalence

The statement Sample Problem 2.5 is in fact a *tautology*. A compound statement is a tautology if it is always true, regardless of the truth values of the simple statements from which it is constructed. A statement that is always false is called a *contradiction*; a very simple example is $p \wedge \sim p$. Other statements that do not fall into either category are called *contingent*.

One of the main aims of logical deduction is to establish tautologies. For example, what we call theorems in mathematics are actually tautologies. The word "theorem" usually denotes a tautology whose essential truth is not immediately obvious, so that some proof is required to establish it.

One very easy example is the fact that every integer is a rational number. This requires a proof with only one step: we need to observe that any integer x can be written as the ratio of two integers, namely $x/1$. The truth of the theorem does not depend on the value of the rational number x.

Sample Problem 2.6 *Show that* $p \vee \sim(p \wedge q)$ *is a tautology.*

Solution. We use the following truth table.

p	q	$(p \wedge q)$	$\sim(p \wedge q)$	$p \vee \sim(p \wedge q)$
T	T	T	F	T
T	F	F	T	T
F	T	F	T	T
F	F	F	T	T

Practice Exercise. Show that $(p \wedge q) \wedge \sim(p \vee q)$ is a contradiction.

If $p \rightarrow q$ is a tautology, we say "p implies q," and write "$p \Rightarrow q$." If $p \leftrightarrow q$ is a tautology, we say that "p is equivalent to q" and write "$p \Leftrightarrow q$" or "$p \equiv q$." In order to prove that $p \equiv q$, it is sufficient to prove that p and q have the same truth table.

The laws of logic

A number of theorems (tautologies) about propositions may be deduced from truth tables, and together they form an algebraic system that is called *mathematical* (or *symbolic*) *logic*. (An alternative view is to take some of the tautologies as axioms, and deduce the truth tables for the standard connectives.) Some of them are very reminiscent of the usual arithmetical laws, with $\equiv$ taking the place of equality. Among these we have:

Commutative laws:

$$p \vee q \equiv q \vee p,$$
$$p \wedge q \equiv q \wedge p,$$

Associative laws:

$$p \vee (q \vee r) \equiv (p \vee q) \vee r,$$
$$p \wedge (q \wedge r) \equiv (p \wedge q) \wedge r,$$

Distributive laws:

$$p \vee (q \wedge r) \equiv (p \vee q) \wedge (p \vee r),$$
$$(p \wedge q) \vee r \equiv (p \vee r) \wedge (q \vee r),$$

$$p \wedge (q \vee r) \equiv (p \wedge q) \vee (p \wedge r),$$
$$(p \vee q) \wedge r \equiv (p \wedge r) \vee (q \wedge r),$$

where each statement is true for all propositions p, q, r. This reminds us of the behavior of addition and multiplication, except that only one pair of distributive laws holds for ordinary arithmetic.

In view of the associative laws, stated above, we can simply write $p \vee q \vee r$ whenever either $p \vee (q \vee r)$ or $(p \vee q) \vee r$, is intended, and similarly $p \wedge q \wedge r$ means either $p \wedge (q \wedge r)$ or $(p \wedge q) \wedge r$.

If t is a proposition that is always true, and if q is always false, then p acts like an identity element for the operation $\wedge$ and q acts like an identity element for $\vee$:

$$p \wedge t \equiv p, \quad p \vee f \equiv p$$

for all p. This is like the behavior of 1 under multiplication or 0 for addition. There are also *zero laws*:

$$p \vee t \equiv t, \quad p \wedge f \equiv f$$

for all p. This reminds us of 0 under multiplication, but there is no corresponding element for addition. Finally, there are two laws called *de Morgan's laws*:

$$\sim(p \vee q) \quad \equiv \quad (\sim p) \wedge (\sim q);$$
$$\sim(p \wedge q) \quad \equiv \quad (\sim p) \vee (\sim q)$$

for all p and q.

Exercises 2.1

In Exercises 1 *to* 12, *find the truth table for the given compound statement.*

1. $\sim p \wedge q$

2. $\sim(p \rightarrow q)$

3. $p \rightarrow (p \rightarrow q)$

4. $\sim(\sim p \vee \sim q)$

5. $p \vee (\sim p \rightarrow q)$

6. $p \vee (\sim p \wedge q)$

7. $\sim p \vee \sim q$

8. $\sim(p \wedge q)$

9. $(p \wedge q) \vee r$

10. $(p \vee r) \wedge (q \vee r)$

11. $p \vee (q \vee r)$

12. $(p \vee q) \vee r$

13. Use the results of Exercises 7 to 12 to prove the following equivalences.

(i) $\sim p \vee \sim q \Leftrightarrow \sim(p \wedge q)$.

(ii) $(p \wedge q) \vee r \Leftrightarrow (p \vee r) \wedge (q \vee r)$.

(iii) $p \vee (q \vee r) \Leftrightarrow (p \vee q) \vee r$.

Prove the equivalences in Exercises 14 *to* 16.

14. $(p \rightarrow q) \Leftrightarrow (\sim p \vee q)$.

15. $(p \leftrightarrow q) \Leftrightarrow (\sim p \wedge \sim q) \vee (p \wedge q)$.

16. $(p \rightarrow q) \Leftrightarrow (\sim q \rightarrow \sim p)$.

17. Prove that $(q \rightarrow p)$ is not equivalent to $(p \rightarrow q)$ in general.

Find the truth tables for the compound statements in Exercises 18 *to* 21.

18. $(p \rightarrow q) \rightarrow r$

19. $p \rightarrow (q \rightarrow r)$

20. $(p \rightarrow r) \rightarrow (q \rightarrow r)$

21. $(p \vee q) \wedge \sim(p \wedge q)$

22. Consider four possible definitions of a connective $p \uparrow q$, given by the following table.

		Definition of $\uparrow$			
p	q	A	B	C	D
T	T	T	T	T	T
T	F	F	F	F	F
F	T	F	F	T	T
F	F	F	T	F	T

Show that definition D, and only definition D, makes the statement "$p \uparrow p \vee q$" a tautology. Also show that definition A is that of $p \wedge q$, B is that of $p \leftrightarrow q$, C is that of q itself, and D is $p \rightarrow q$.

23. Prove that the following statements are equivalent.

(i) $(p \rightarrow q)$

(ii) $(p \wedge \sim q) \rightarrow \sim p$

(iii) $(p \wedge \sim q) \rightarrow q$

24. Prove that $(p \rightarrow q) \rightarrow r$ and $p \rightarrow (q \rightarrow r)$ are not equivalent.

25. Prove the two commutative laws.

26. Prove the two associative laws.

27. Prove the two distributive laws.

28. Prove de Morgan's laws.

29. Let $p \underline{\vee} q$ denote the compound statement "p or q but not both," which is often called *exclusive or*. Find the truth table for $p \underline{\vee} q$. Compare it with Table 2.2 and Exercise 21.

In Exercises 30 *to* 39, *find the truth table for the given statement.*

30. $(p \to p)$

31. $(p \to \sim p)$

32. $p \wedge \sim p$

33. $(p \wedge q) \to (p \vee q)$

34. $(p \to q) \to (p \wedge q)$

35. $((p \to q) \to (p \wedge q)) \vee (\sim p)$

36. $(\sim p) \wedge (p \vee q) \to (\sim q)$

37. $q \to (p \to q)$

38. $(p \vee q) \to (p \wedge q)$

39. $\sim p \to (q \to p)$

40. Find truth tables for the following propositions. Are any of them equivalent?

(i) $(p \to q) \wedge (\sim r \to \sim q)$

(ii) $r \to \sim p$

(iii) $p \to \sim r$

(iv) $\sim((\sim q \to \sim p) \wedge (q \to \sim r))$

2.2 Elements of Set Theory

Sets

We saw *sets* in Section 1.1. The notations $\mathbb{R}$, $\mathbb{Q}$, $\mathbb{Z}$, $\mathbb{Z}^+$ and $\mathbb{Z}^*$ were introduced for the sets of real numbers, rational numbers, integers, positive integers and non-negative integers respectively.

As we noted, a set can be finite or infinite. We saw that it is often convenient to specify a finite set by listing its elements between braces, but most infinite sets must be defined by explicitly stating the membership law. The set of all objects x for which the statement $S(x)$ is true was written $\{x \mid S(x)\}$, or sometimes as $\{x : S(x)\}$. For example, the set Π of prime numbers could be denoted (trivially) as

$$\Pi = \{x \mid x \text{ is a prime }\}.$$

The definition of a set does not allow for ordering of its elements, or for repetition of its elements. Thus $\{1,2,3\}, \{1,3,2\}$ and $\{1,2,3,1\}$ all represent the same set (which could be written $\{x \mid x \in \mathbb{Z}^* \text{ and } x \leq 3\}$, or $\{x \in \mathbb{Z}^* \mid x \leq 3\}$). To handle problems that involve ordering, we define a *sequence* to be an ordered set. Sequences can be denoted by parentheses; $(1,3,2)$ is the sequence with first element 1, second element 3 and third element 2, and is different from $(1,2,3)$. Sequences may contain repetitions, and $(1,2,1,3)$ is quite different from $(1,2,3)$; the two occurrences of object 1 are distinguished by the fact that they lie in different positions in the ordering.

We defined the notation

$$s \in S$$

to mean "s belongs to S" or "s is an element of S," and $S \subseteq T$ to mean S is a subset of T. If $S \subseteq T$ we also say that T contains S or T is a *superset* of S, and write $T \supseteq S$. Sets S and T are equal, $S = T$, if and only if $S \subseteq T$ and $T \subseteq S$ are both true. We can represent the situation where S is a subset of T but S is not equal to T — there is at least one member of T that is not a member of S — by writing $S \subset T$.

An important concept is the *empty* set, or *null* set, which has no elements. This set, denoted by $\emptyset$, is unique and is a subset of every other set.

In all the discussions of sets in this book, we shall assume (usually without bothering to mention the fact) that all the sets we are dealing with are subsets of some given universal set U. U may be chosen to be as large as necessary in any problem we deal with; in most of our discussion so far we could have chosen $U = \mathbb{Z}$ or $U = \mathbb{R}$. U can often be chosen to be a finite set.

The *power set* of any set S consists of all the subsets of S (including S itself and $\emptyset$), and is denoted by $\mathcal{P}(S)$:

$$\mathcal{P}(S) = \{T : T \subseteq S\}. \tag{2.1}$$

The power set is a set whose elements are *themselves* sets.

Sample Problem 2.7 *Write down all elements of the power set of* $\{1,2,3\}$. *How many elements are there?*

Solution. There are eight elements: $\{1,2,3\}$, $\{1,2\}$, $\{1,3\}$, $\{2,3\}$, $\{1\}$, $\{2\}$, $\{3\}$, and $\emptyset$.

Practice Exercise. Write down all elements of the power set of $\{x,y,z\}$.

Operations on sets

Given sets S and T, we define three operations: the *union* of S and T is the set

$$S \cup T = \{x : x \in S \text{ or } x \in T \text{ (or both) }\};$$

the *intersection* of S and T is the set

$$S \cap T = \{x : x \in S \text{ and } x \in T\};$$

the *relative complement* of T with respect to S (or alternatively the *set-theoretic difference* or *relative difference* between S and T) is the set

$$S\backslash T = \{x : x \in S \text{ and } x \notin T\}.$$

In particular, the relative complement $U\backslash T$ with respect to the universal set U is denoted by $\overline{T}$ and called the *complement* of T. We could also write $R\backslash S = R \cap \overline{S}$, since each of these sets consists of the elements belonging to R but not to S. Hence we see that $R \subseteq S$ if and only if $R\backslash S = \emptyset$.

Sample Problem 2.8 *If* $\mathbb{E}$ *is the set of all even integers, what are* $\mathbb{E} \cup \Pi$, $\mathbb{E} \cap \Pi$, $\mathbb{E}\backslash\Pi$, $\mathbb{Z} \cup \mathbb{Z}^+$, $\mathbb{Z}^*\backslash\Pi$?

Solution.

$$\begin{aligned}
\mathbb{E} \cup \Pi &= \{\ldots, -8, -6, -4, 2, 0, 2, 3, 4, 5, 6, 7, 8, 10, 11, \ldots\}, \\
\mathbb{E} \cap \Pi &= \{2\}, \\
\mathbb{E}\backslash\Pi &= \{\ldots, -8, -6, -4, -2, 0, 4, 6, 8, \ldots\}, \\
\mathbb{Z} \cup \mathbb{Z}^+ &= \mathbb{Z}, \\
\mathbb{Z}^*\backslash\Pi &= \{0, 1, 4, 6, 8, 9, 10, 13, 14, 15, 16, 18, \ldots\}.
\end{aligned}$$

Practice Exercise. What are $\mathbb{Z}\backslash\mathbb{Z}^+$, $\mathbb{Z} \cap \mathbb{Z}^+$, $(\mathbb{Z}^+\backslash\mathbb{E}) \cup \Pi$?

If two sets, S and T, have no common element, so that $S \cap T = \emptyset$, then we say that S and T are *disjoint*. Observe that $S\backslash T$ and T must be disjoint sets; in particular, T and $\overline{T}$ are disjoint. If n sets $S_1, S_2, \ldots, S_n$ are such that each pair of them are disjoint, so that

$$S_i \cap S_j \text{ for } 1 \leq i, j \leq n \text{ and } i \neq j,$$

then we say that these n sets are *pairwise disjoint* or *mutually disjoint*. By a *partition* of a set S we mean a collection of pairwise disjoint non-empty sets $S_1, S_2, \ldots, S_n$ whose union is S.

In general, to prove that the set S is a subset of the set T, we start with the statement, "suppose x is any element of S," and finish with "therefore x is an element of T." To show that S and T are equal, prove both $S \subseteq T$ and $S \supseteq T$. Another method of proving $S = T$ is to work as follows. Find an exact description of the elements of S — something of the form "S is precisely the set of all elements x with the following properties ...," and prove that this description is also precisely the description of elements of T.

Sometimes proofs of the form "suppose x is any element of S" are simpler if the argument is broken into two parts: first consider all elements x with a certain property, then all those without that property. For example, to prove that $(R\backslash S) \cup S = R \cup S$, for any two sets R and S, first observe that if x is a member of S, then it belongs to both $(R\backslash S) \cup S$ and $R \cup S$. So we need only discuss x not in S. The elements of $(R\backslash S) \cup S$ not in S are precisely the elements of $R\backslash S$, while the elements of $R \cup S$ not in S are the members of R not in S — precisely the same elements. So $(R\backslash S) \cup S = R \cup S$.

Properties of the operations

We now investigate some of the easier properties of the operations $\cup$, $\cap$ and $\backslash$; for the more difficult problems, we shall introduce some techniques in the next section.

Union and intersection both satisfy *idempotence laws*: for any set S,

$$S \cup S = S \cap S = S.$$

Both operations satisfy commutative laws; in other words

$$S \cup T = T \cup S$$

and

$$S \cap T = T \cap S,$$

for any sets S and T. Similarly, the associative laws

$$R \cup (S \cup T) = (R \cup S) \cup T$$

and

$$R \cap (S \cap T) = (R \cap S) \cap T$$

are always satisfied. The associative law means that we can omit brackets in a string of unions (or a string of intersections); expressions like $(A \cup B) \cup (C \cup D)$, $((A \cup B) \cup C) \cup D$ and $(A \cup (B \cup C)) \cup D$, are all equal, and we usually omit all the parentheses and simply write $A \cup B \cup C \cup D$. But be careful not to mix operations. $(A \cup B) \cap C$ and $A \cup (B \cap C)$ are quite different. Combining the commutative and associative laws, we see that any string of unions can be rewritten in any order: for example,

$$(D \cup B) \cup (C \cup A) = C \cup (B \cup (A \cup D)) = (A \cup B \cup C \cup D).$$

Sample Problem 2.9 *Prove that* $(A \cup B) \cap C = A \cup (B \cap C)$ *is not always true.*

Solution. To prove that a general rule is not true, it suffices to find just one case in which it is false. As an example we take the case $A = \mathbb{R}$, $B = \mathbb{Z}$, $C = \{0\}$. Then $(A \cup B) \cap C = \{0\}$, while $A \cup (B \cap C) = \mathbb{R}$.

The following distributive laws hold:

$$R \cup (S \cap T) = (R \cup S) \cap (R \cup T); \quad (2.2)$$

$$R \cap (S \cup T) = (R \cap S) \cup (R \cap T); \quad (2.3)$$

$$R \cup (S \backslash T) = (R \cup S) \backslash (R \cup T); \quad (2.4)$$

$$(R \cup S) \backslash T = (R \backslash S) \cap (R \backslash T). \quad (2.5)$$

Sample Problem 2.10 *Prove the distributive law* (2.2).

Solution. Suppose $x \in R \cup (S \cap T)$. It may be that $x \in R$; in that case, both $x \in (R \cup S)$ and $x \in (R \cup T)$ are true (in fact, $x \in (R \cup A)$ is true for *any* set

A), so $x \in (R \cup S) \cap (R \cup T)$. On the other hand, if $x \notin R$, then $x \in (S \cap T)$, and x belongs both to S and to T. Now $x \in S \Rightarrow x \in (R \cup S)$, and $x \in T \Rightarrow x \in (R \cup T)$, so $x \in (S \cap T) \Rightarrow x \in (R \cup S) \cap (R \cup T)$. So in either case,

$$x \in R \cup (S \cap T) \Rightarrow x \in (R \cup S) \cap (R \cup T),$$

and $R \cup (S \cap T) \subseteq (R \cup S) \cap (R \cup T)$.

Conversely, suppose $x \in (R \cup S) \cap (R \cup T)$. If $x \in R$, then certainly $x \in R \cup (S \cap T)$. If $x \notin R$, then $x \in (R \cup S) \Rightarrow x \in S$, and $x \in (R \cup T) \Rightarrow x \in T$. So

$$x \in (R \cup S) \cap (R \cup T) \Rightarrow x \in (S \cap T) \Rightarrow x \in R \cup (S \cap T)$$

and $(R \cup S) \cap (R \cup T) \subseteq R \cup (S \cap T)$. So the two sets are equal.

Practice Exercise. Prove the distributive law (2.3).

We have also the equation

$$R \backslash (S \cup T) = (R \backslash S) \cap (R \backslash T), \tag{2.6}$$

and the analogous

$$R \backslash (S \cap T) = (R \backslash S) \cup (R \backslash T). \tag{2.7}$$

Sample Problem 2.11 *Prove equation* (2.6) *from the definition.*

Solution. $R \backslash (S \cup T)$ consists of precisely those members of R that are not members of $S \cup T$, in other words those elements of R that do not belong to S or to T. That is,

$$R \backslash (S \cup T) = \{x \mid x \in R \text{ and } x \notin S \text{ and } x \notin T\}.$$

On the other hand, $(R \backslash S)$ consists of all the things in R that are not in S, and $(R \backslash S) \cap (R \backslash T)$ consists of all the things in R that are not in T; the common elements of these sets are all the things in R but not in S and not in T, which is the same as the description of $R \backslash (S \cup T)$.

Equation (2.6) can also be verified using the idempotence, associative and commutative laws. From

$$R \backslash (S \cup T) = \{x \mid x \in R \text{ and } x \notin S \text{ and } x \notin T\}$$

we have

$$\begin{aligned} R \backslash (S \cup T) &= R \cap (\overline{S} \cap \overline{T}) \\ &= (R \cap R) \cap (\overline{S} \cap \overline{T}) \quad \ldots \text{idempotence} \\ &= (R \cap \overline{S}) \cap (R \cap \overline{T}) \quad \ldots \text{associativity, commutativity} \\ &= (R \backslash S) \cap (R \backslash \overline{T}). \end{aligned}$$

When we take the particular case where R is the universal set in (2.6) and (2.7), those two equations become *de Morgan's laws*:

$$\overline{S \cup T} = \overline{S} \cap \overline{T}, \tag{2.8}$$

$$\overline{S \cap T} = \overline{S} \cup \overline{T}. \tag{2.9}$$

Exercises 2.2

1. Consider the sets

$$\begin{aligned} S_1 &= \{2,5\}, \\ S_2 &= \{1,2,4\}, \\ S_3 &= \{1,2,4,5,10\}, \\ S_4 &= \{x \in \mathbb{Z}^+ : x \text{ is a divisor of } 20\}, \\ S_5 &= \{x \in \mathbb{Z}^+ : x \text{ is a power of 2 and a divisor of } 20\}. \end{aligned}$$

For which i and j, if any, is $S_i \subseteq S_j$? For which i and j, if any, is $S_i = S_j$?

2. Prove the commutative and associative laws for $\cup$.

3. Prove the commutative and associative laws for $\cap$.

In Exercises 4 *to* 14, *U is a universal set and S and T are any sets. Prove the given result.*

4. $S \cup \emptyset = S$. **5.** $S \cup U = U$. **6.** $S \cup S = S$.

7. $S \cup \overline{S} = U$. **8.** $S \cap U = S$. **9.** $S \cap \emptyset = \emptyset$.

10. $S \cap S = S$. **11.** $S \cap \overline{S} = \emptyset$. **12.** $(S \cap T) \subseteq S$.

13. $S \subseteq (S \cup T)$.

14. If $S \cup T = U$ and $S \cap T = \emptyset$, then $T = \overline{S}$.

15. Prove the commutative and associative laws for $\cup$.

16. Prove the commutative and associative laws for $\cap$.

17. Prove the distributive laws (2.4) and (2.5).

18. Prove de Morgan's laws.

19. Suppose the sets A, B, C, D, S are defined in terms of $\emptyset$ as follows.

$A = \{\emptyset\}, B = \{A\}, C = \{\emptyset, A\}, D = \{\emptyset, A, C\}$, $S = \{\emptyset, A, B, C, D\}$.

Show that:

(i) $\{x \mid x \in S \text{ and } x \subseteq D\} = S$;

(ii) $\{x \mid x \in S \text{ and } x \in D\} = D$.

In Exercises 20 *to* 26, *R, S and T are any sets, and U is the universal set.*

20. Prove: if $R \subseteq S$ and $R \subseteq T$, then $R \subseteq (S \cap T)$.

21. Prove: if $R \subseteq T$ and $S \subseteq T$, then $(R \cup S) \subseteq T$.

22. Prove: if $R \subseteq S$, then $R \cap T \subseteq S \cap T$ and $R \cup T \subseteq S \cup T$.

23. Show that $R \subseteq S$ if and only if $R \cap \overline{S} = \emptyset$.

24. Show that if $S \cup T = \emptyset$, then $S = T = \emptyset$ and that if $S \cap T = U$, then $S = T = U$.

25. Prove that $R \backslash (S \backslash T)$ contains all members of $R \cap T$, and hence prove that

$$(R \backslash S) \backslash T = R \backslash (S \backslash T)$$

is *not* a general law (in other words, relative difference is *not* associative).

26. Show that the following three statements are equivalent: $S \subseteq T, S \cup T = T, S \cap T = S$.

In Exercises 27 *to* 38, *the sets* A, B, C, D *are defined as follows:* $A = \{\emptyset\}$, $B = \{A\}$, $C = \{\emptyset, A\}$, $D = \{B\}$. *Determine whether the given statement is true or false.*

27. $\emptyset \subseteq A$.

28. $\emptyset \in A$.

29. $\emptyset \in B$.

30. $\emptyset \subseteq B$.

31. $A \subseteq B$.

32. $A \in B$.

33. $A \subseteq C$.

34. $A \in C$.

35. $B \subseteq C$.

36. $B \in C$.

37. $B \subseteq D$.

38. $B \in D$.

2.3 Proofs in Set Theory

Method of truth tables

Set-theoretic propositions can often be proved using truth tables. The reasoning is as follows. To prove that sets S and T are equal, it suffices to show that, for any element x of the universal set, the statement "$x \in S \leftrightarrow s \in T$" is a tautology. This can be proved (or disproved) using a truth table.

To illustrate this, consider the distributive law

$$R \cup (S \cap T) = (R \cup S) \cap (R \cup T). \tag{2.2}$$

The truth table might look like

R	S	T	$S \cap T$	$R \cup S$	$R \cup T$	$R \cup (S \cap T)$	$(R \cup S) \cap (R \cup T)$
T	T	T	T	T	T	T	T
T	T	F	F	T	T	T	T
T	F	T	F	T	T	T	T
T	F	F	F	T	T	T	T
F	T	T	T	T	T	T	T
F	T	F	F	T	F	F	F
F	F	T	F	F	T	F	F
F	F	F	F	F	F	F	F

In writing the table, it is implicitly assumed that a general element x is being considered; the notation T (or F) under the name of a set means that the statement that x is a member of that set is true (or false). For example, the third line of the table could be interpreted as saying, "in all cases where $x \in R$ is true, $x \in S$ is false, and $x \in T$ is true, it is always false that $x \in S \cap T$, true that $x \in R \cup S$, true that $x \in R \cup T$, true that $x \in R \cup (S \cap T)$ and true that $x \in (R \cup S) \cap (R \cup T)$."

The statement (2.2) is a tautology provided the last two columns are equal, which is seen to be true.

Compare this proof with the proof given in Sample Problem 2.2.2.10.

In order to apply the truth table method, it is necessary to know the truth tables of the main binary operations. The truth tables for complement, union, intersection and relative complement are shown in Table 2.4.

R	S	$\overline{R}$	$R \cup S$	$R \cap S$	$(R \backslash S)$
T	T	F	T	T	F
T	F	F	T	F	T
F	T	T	T	F	F
F	F	T	F	F	F

Table 2.4: Truth tables for $R \cup S$, $R \cap S$ and $(R \backslash S)$

The method of truth tables can also be used to prove that one set is a subset of another. If you need to prove that $R \subseteq S$, then the last two columns will be labeled R and S, and it is required that no row has T in the R column and F in the S column.

Sample Problem 2.12 *Prove that* $A \cap C \subseteq (\overline{A} \cap B) \cup C$.

Solution. The truth value of $x \in \overline{A}$ is opposite that of $x \in A$, so the table is

A	B	C	$\overline{A}$	$\overline{A} \cap B$	$A \cap C$	$(\overline{A} \cap B) \cup C$
T	T	T	F	F	T	T
T	T	F	F	F	F	T
T	F	T	F	F	T	T
T	F	F	F	F	F	T
F	T	T	T	T	F	T
F	T	F	T	T	F	T
F	F	T	T	F	F	T
F	F	F	T	F	F	F

There is no line with T in the second-last column and F in the last, so inclusion is proved.

Practice Exercise. Prove that $A \cap B \cap C \subseteq B \cap (A \cup C)$.

Venn diagrams

It is common to illustrate sets and operations on sets by diagrams. A set A is represented by a circle, and it is assumed that the elements of A correspond to the points (or some of the points) inside the circle. The universal set is usually shown as a rectangle enclosing all the other sets; if it is not needed, the universal set is often omitted. Such an illustration is called a *Venn diagram*.

Here are Venn diagrams representing $A \cup B$, $A \cap B$, $\overline{A}$ and $A \backslash B$; in each case, the set represented is shown by the shaded area. The universal set is shown in each case.

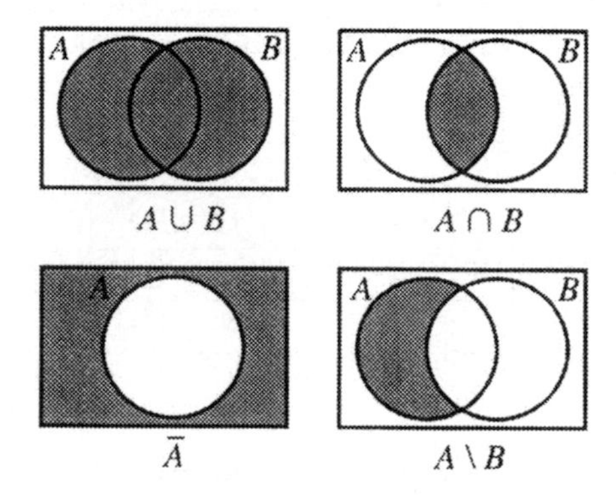

Two sets are equal if and only if they have the same Venn diagram. In order to illustrate this, we again consider the distributive law

$$R \cup (S \cap T) = (R \cup S) \cap (R \cup T). \tag{2.2}$$

The Venn diagram for $R \cup (S \cap T)$ is constructed in the upper half of Figure 2.1, and that for $(R \cup S) \cap (R \cup T)$ is constructed in the lower half. The two are obviously identical.

In the first part of this section, we applied the method of truth tables (developed for use with propositions) to set identities. We can also apply the methods of set theory to the analysis of propositions. If s is any proposition, we define S to be the set of all sets of circumstances in which proposition s is true; similarly we make proposition t correspond to set T. Then $s \Leftrightarrow t$ is equivalent to $S = T$, and Venn diagram proofs can be used.

For example, the preceding Venn diagram proof that

$$R \cup (S \cap T) = (R \cup S) \cap (R \cup T)$$

for all sets R, S and T is also a proof of the distributive law for propositions,

$$r \vee (s \wedge t) \leftrightarrow (r \vee s) \wedge (r \vee t).$$

Sample Problem 2.13 *Write down a statement involving propositions that can be proven by establishing the set-theoretic identity*

$$(R \backslash S) \backslash T = R \backslash (S \backslash T).$$

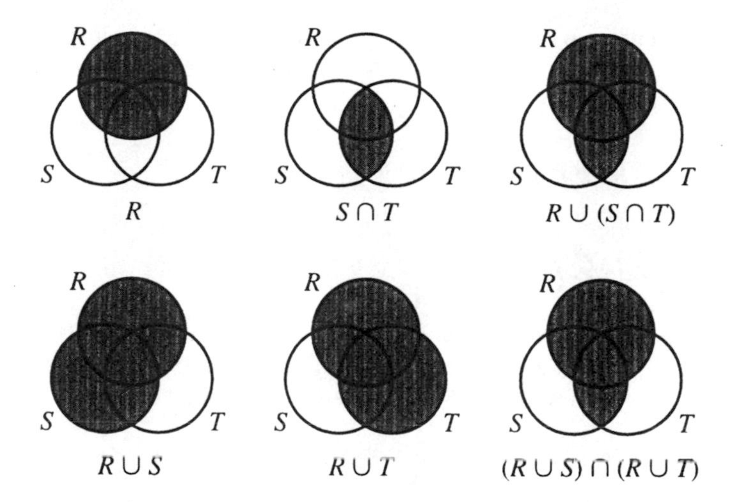

Figure 2.1: $R\cup(S\cap T)=(R\cup S)\cap(R\cup T)$

Solution. We let r correspond to set R, and so on. As $R\backslash S$ corresponds to the proposition $r\wedge\sim s$, the answer is

$$((r\wedge\sim s)\wedge\sim t)\leftrightarrow(r\wedge\sim(s\wedge\sim t)).$$

To prove $A\subseteq B$, it is sufficient to show that the diagram for A contains no shaded area that is not shaded in B. We illustrate this idea with the problems from Sample Problem 2.12 and its associated Practice Exercise (but in reverse order).

Sample Problem 2.14 *Use Venn diagrams to prove that*

$$A\cap B\cap C\subseteq B\cap(A\cup C).$$

Solution.

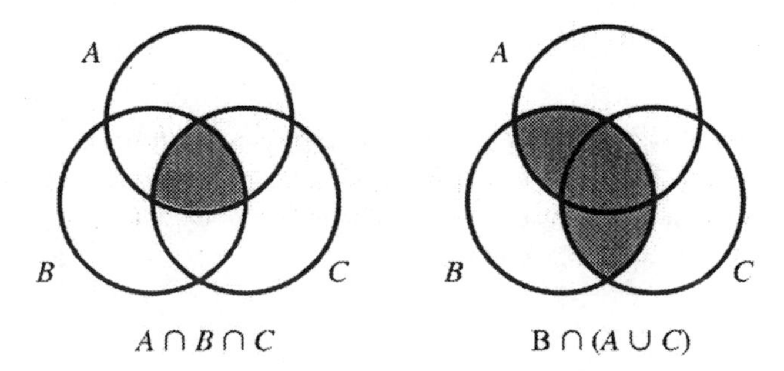

Practice Exercise. Use Venn diagrams to prove that

$$A\cap C\subseteq(\overline{A}\cap B)\cup C.$$

As before, this type of set-theoretic proof can be applied to propositions. A proof that $R \subseteq S$ serves as a proof that $r \to s$. Sample Problem 2.14 can be interpreted as a proof that

$$a \wedge b \wedge c \to b \wedge (a \vee c)$$

is true for any propositions a, b and c.

Syllogisms and Venn diagrams

Sometimes we draw a Venn diagram in order to represent some properties of sets. For example, if A and B are disjoint sets, the diagram can be drawn with A and B shown as disjoint circles. If $A \subseteq B$, the circle for A is entirely inside the circle for B.

In the classical study of logic, an argument is given in the form of a *syllogism*, a set of statements (the *premises*, or data) and a conclusion drawn from them. For example, consider the argument

> All big cities are near water; Alamagordo, NM is not near water; therefore Alamagordo is not a big city.

To examine this, suppose B is the set of all big cities, W is the set of all cities near water, and a represents Alamagordo. Since the premises tell us that $B \subseteq W$, we can draw the sets as shown. As $a \notin W$, it must lie somewhere in the outside region, so it is certainly not in B. Therefore the argument is valid.

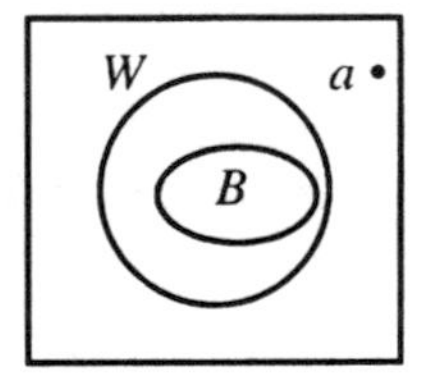

Some arguments that look logical at first sight turn out not to be valid. For example,

> All big cities are near water; Carbondale, IL is near water; therefore Carbondale is a big city.

We label the sets as before. In the diagram shown, Carbondale could be represented either by x or by y, so you can't draw any conclusion. The argument is not valid. (In fact, x is nearer the mark.)

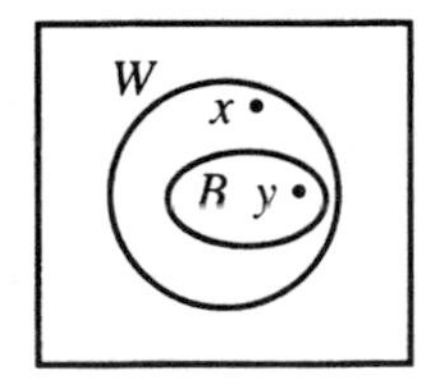

Sample Problem 2.15 *Examine the argument: no students in this class are lazy; John is a math major; all math majors are lazy; so John is not a student in this class.*

Solution. Write L, C and M for the sets of lazy students, students in this class and math majors. The premises are represented in the following diagram.

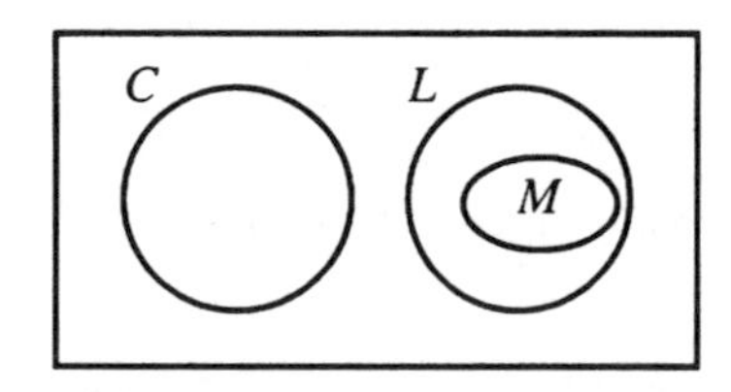

John is a member of M, which is disjoint from C. So John is not in this class, and the conclusion is valid.

Practice Exercise. Examine the argument: *some students in this class are lazy; all males are lazy; so some students in this class are males.*

Observe from the above example that the *validity* of an argument does not depend on the *truth* of the premises or conclusion. Not all math majors are lazy, and I wouldn't venture to guess how many students in your class are lazy! Another example, where the premises and conclusion are all false but the argument is valid, appears in Exercise 2.3.26.

Exercises 2.3

In Exercises 1 *to* 8, *represent the set in a Venn diagram.*

1. $R \cup S \cup T$

2. $\overline{R \cup S \cup T}$

3. $R \cup (S \cap \overline{T})$

4. $(R \cap S) \cap T$

5. $(R \backslash S) \cap T$

6. $R \cap (S \backslash T)$

7. $(R \cap S) \backslash (S \cap T)$

8. $(R \cup T) \backslash (S \cap T)$

9. Prove: $R = (\overline{\overline{R} \cup S}) \cup (R \cap S)$.

10. Find a simpler expression for $S \cup (\overline{(\overline{R} \cup S) \cap R})$.

In Exercises 11 *to* 15, *prove the rule in two ways, using truth tables and using Venn diagrams.*

11. $S \cap \overline{S} = \emptyset$.

12. $\overline{S \cup T} = \overline{S} \cap \overline{T}$.

13. $\overline{S \cap T} = \overline{S} \cup \overline{T}$.

14. $(S \cap T) \subseteq S$.

15. $S \subseteq (S \cup T)$.

16. Use truth tables to prove the commutative and associative laws for $\cup$.

17. Use Venn diagrams to prove the commutative and associative laws for $\cap$.

18. For any sets R and S, prove $R \cap (R \cup S) = R$.

19. Prove, *using Venn diagrams,* that $(R\backslash S)\backslash T = R\backslash(S\backslash T)$ does not hold for all choices of sets R, S and T.

20. (i) Prove, without using truth tables or Venn diagrams, that union is not distributive over relative difference: in other words, prove that the following statement is not always true:

$$(R\backslash S) \cup T = (R \cup T)\backslash(S \cup T).$$

(Hint: use the fact $(R\backslash S) \cup S = R \cup S$.)

(ii) Now prove this using Venn diagrams.

21. Draw Venn diagrams for use in the following circumstances:

(i) all my goldfish are tropical fish;

(ii) none of my goldfish are tropical fish.

In Exercises 22 *to* 25, *test the validity of the argument by drawing the appropriate Venn diagram.*

22. *All men are mortal; Socrates is a man. Therefore Socrates is mortal.*

23. *All my friends are students; none of my neighbors are students; Ruth is my friend. Therefore Ruth is not my neighbor.*

24. *Boston is a big city; all big cities have department stores; Shirley lives in a city with no department store. Therefore Shirley does not live in Boston.*

25. *All businessmen are wealthy; all mathematicians are cheerful; David is a businessman; no cheerful people are wealthy. Therefore David is not a mathematician.*

26. Show that the following argument is valid, although its premises and conclusion are all false: *All expensive food contains cholesterol; steak contains no cholesterol. Therefore steak is not expensive.*

27. Show that the following argument is not valid, although its premises and conclusion are all true: *Some animals walk on two legs; human beings are animals; therefore human beings walk on two legs.*

28. Consider the data: *All authors are solitary people; all physicians are rich; no solitary people are rich.* Which of the following conclusions can be drawn?

(i) No authors are rich.

(ii) All physicians are solitary.

(iii) No one can be both an author and a physician.

29. Consider the data: *I sold back all my expensive textbooks last year; all my science textbooks are green; I did not sell back any green books last year.* Which of the following conclusions can be drawn?

(i) None of my science textbooks are expensive.

(ii) All of my science textbooks were sold back last year.

(iii) Some of my science textbooks were sold back last year.

(iv) None of my science textbooks were sold back last year.

(v) No green textbooks were sold back last year.

30. Consider the data: *All topcoats are expensive; none of my clothes are expensive; all expensive clothes are well made.* Which of the following conclusions can be drawn?

(i) I do not own a topcoat.

(ii) All topcoats are well made.

(iii) None of my clothes are well made.

2.4 Some Further Set Operations

Symmetric difference

Another important set operation is *symmetric difference*. The symmetric difference of A and B is the set

$$A+B=\{x : x\in A \text{ or } x\in B \text{ but } x\notin A\cap B\}. \tag{2.10}$$

Sample Problem 2.16 *What is the truth table for $A+B$?*

Solution.

A	B	$A+B$
T	T	F
T	F	T
F	T	T
F	F	F

Practice Exercise. What is the Venn diagram for $A+B$?

This definition could be stated as

$$\begin{aligned} A+B &= (A\cup B)\backslash(A\cap B) \\ &= (A\cup B)\cap\overline{(A\cap B)} \\ &= (A\cup B)\cap(\overline{A}\cup\overline{B}), \end{aligned} \tag{2.11}$$

using (2.8). By the symmetry of the relation (2.11), it follows that

$$\overline{A}+\overline{B}=A+B.$$

From the definition, we may consider $A+B$ to be the union of the difference between A and B with the difference between B and A. This implies that

$$A+B=(A\backslash B)\cup(B\backslash A),$$

and hence that

$$A+B=(A\cap\overline{B})\cup(A\cap\overline{B}). \tag{2.12}$$

If a and b denote the propositions "$x\in A$" and "$x\in B$" respectively, then the proposition "$x\in A+B$" is denoted by $a\underline{\vee}b$. (For the definition of $\underline{\vee}$, see Exercise 2.1.29.) Then

$$a\underline{\vee}b \Leftrightarrow ((a\vee b)\vee\sim(a\wedge b)) \Leftrightarrow ((a\vee b)\wedge(\sim a\vee\sim b)) \tag{2.13}$$

is a restatement of 2.11, and similarly we see that

$$(a\underline{\vee}b)\Leftrightarrow(\sim a\underline{\vee}\sim b)$$

while (2.12) yields

$$(a\underline{\vee}b)\Leftrightarrow(a\wedge\sim b)\vee(\sim a\wedge b). \tag{2.14}$$

It is easy to see that symmetric difference satisfies the commutative law

$$A+B=B+A. \tag{2.15}$$

The associative law is also true, but it is harder to prove, so we state it as a theorem.

Theorem 2.1 *Symmetric difference satisfies the associative law*

$$A+(B+C)=(A+B)+C.$$

Proof. We start from (2.11),

$$A+B=(A\cup B)\cap(\overline{A}\cup\overline{B}).$$

Then

$$\begin{aligned} \overline{A+B} &= \overline{(A\cup B)}\cup(A\cap B) \text{ by (2.8)} \\ &= (\overline{A}\cap\overline{B})\cup(A\cap B). \end{aligned} \tag{2.16}$$

From (2.12),

$$\begin{aligned}(A+B)+C &= [(A+B)\cap\overline{C}]\cup[\overline{(A+B)}\cap C] \\ &= \{[(A\cap\overline{B})\cup(\overline{A}\cap B)]\cap\overline{C}\} \\ &\quad \cup\{[(\overline{A}\cap\overline{B})\cup(A\cap B)]\cap C\} \text{ by (2.12) and (2.16)} \\ &= \{A\cap\overline{B}\cap\overline{C}\}\cup\{\overline{A}\cap B\cap\overline{C}\} \\ &\quad \cup\{\overline{A}\cap\overline{B}\cap C\}\cup\{A\cap B\cap C\}.\end{aligned}$$

In exactly the same way we may prove that

$$\begin{aligned}(B+C)+A &= \{B\cap\overline{C}\cap\overline{A}\}\cup\{\overline{B}\cap C\cap\overline{A}\} \\ &\quad \cup\{\overline{B}\cap\overline{C}\cap A\}\cup\{B\cap C\cap A\}.\end{aligned}$$

Since union and intersection are commutative and associative operations, the right-hand sides of the last two equations are equal, so

$$(B+C)+A=(A+B)+C;$$

on applying the commutative law (2.15) to the left-hand side, we obtain

$$A+(B+C)-(A+B)+C.$$

□

In view of Theorem 2.1, we can write $A+B+C$ instead of $(A+B)+C$. In the expression

$$\{A\cap\overline{B}\cap\overline{C}\}\cup\{\overline{A}\cap B\cap\overline{C}\}\cup\{\overline{A}\cap\overline{B}\cap C\}\cup\{A\cap B\cap C\}$$

the first three terms are the sets of all elements belonging to exactly one of A,B and C, while the fourth term is the intersection of all three. So $A+B+C$ consists of all those elements which belong to an *odd* number of the sets A,B and C; see Figure 2.2, where $A+B+C$ is represented by the shaded area.

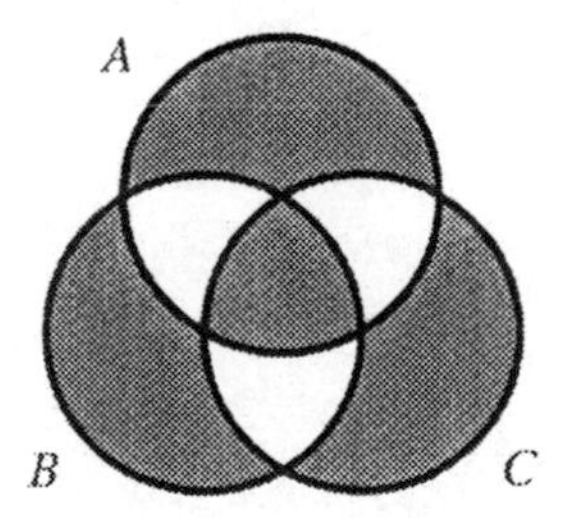

Figure 2.2: Associativity of symmetric difference

The proof of Theorem 2.1 using truth tables or Venn diagrams is left as an exercise.

Cartesian product

We define the *cartesian product* (or *cross product*) $S\times T$ of sets S and T to be the set of all ordered pairs (s,t) where $s\in S$ and $t\in T$:

$$S\times T=\{(s,t):s\in S,t\in T\}.$$

There is no requirement that S and T be disjoint; in fact, it is often useful to consider $S \times S$.

The number of elements of $S \times T$ is $|S| \times |T|$. (This is one reason why the symbol $\times$ was chosen for cartesian product.)

Sample Problem 2.17 *Suppose* $S = \{0,1\}$ *and* $T = \{1,2\}$. *What is* $S \times T$*?*

Solution. $S \times T = \{(0,1),(0,2),(1,1),(1,2)\}$, the set of all four of the possible ordered pairs.

Practice Exercise. What is $S \times T$ if $S = \{1,2\}$ and $T = \{1,4,5\}$?

The sets $(R \times S) \times T$ and $R \times (S \times T)$ are not equal; one consists of an ordered pair whose *first* element is itself an ordered pair, and the other of pairs in which the *second* is an ordered pair. So there is no associative law, and no natural meaning for $R \times S \times T$. On the other hand, it is sometimes natural to talk about ordered triples of elements, so we define

$$R \times S \times T = \{(r,s,t) : r \in R, s \in S, t \in T\}.$$

This notation can be extended to ordered sets of any length.

There are several distributive laws involving the cartesian product:

Theorem 2.2 *If* R, S *and* T *are any sets, then*

(i) $R \times (S \cup T) = (R \times S) \cup (R \times T)$;

(ii) $(R \times S) \cup T = (R \cup T) \times (S \cup T)$;

(iii) $R \times (S \cap T) = (R \times S) \cap (R \times T)$;

(iv) $(R \times S) \cap T = (R \cap T) \times (S \cap T)$.

Proof. (i) We prove that every element of $R \times (S \cup T)$ is a member of $(R \times S) \cup (R \times T)$, and conversely.

First observe that

$$\begin{aligned} R \times (S \cup T) &= \{(r,s) \mid r \in R \text{ and } s \in S \cup T\} \\ &= \{(r,s) \mid r \in R \text{ and } (s \in S \text{ or } s \in T)\}. \end{aligned}$$

On the other hand,

$$\begin{aligned} (R \times S) \cup (R \times T) &= \{(r,s) \mid (r,s) \in R \times S \text{ or } (r,s) \in R \times T \\ &= \{(r,s) \mid (r \in R \text{ and } s \in S) \text{ or } (r \in R \text{ and } s \in T)\} \\ &= \{(r,s) \mid r \in R \text{ and } (s \in S \text{ or } s \in T)\} \end{aligned}$$

where the last equality follows from the distributive law for *propositions*, applied to the propositions $r \in R$, $s \in S$ and $s \in T$.

It is now clear that $(r,s) \in R \times (S \cup T)$ and $(r,s) \in (R \times S) \cup (R \times T)$ are equivalent.

The proofs of the other parts are left as exercises. □

Exercises 2.4

1. Prove Theorem 2.1
   (i) using truth tables;
   (ii) using Venn diagrams.

2. Prove the two distributive laws

$$\begin{aligned} A \cap (B+C) &= (A \cap B)+(A \cap C) \\ (A+B) \cap C &= (A \cap C)+(B \cap C) \end{aligned}$$

   (i) using truth tables;
   (ii) using Venn diagrams.

3. Show that union is not distributive over symmetric difference. (Hint: consider the set $S \cup (S+T)$.)

4. Show that $S+T=\emptyset$ if and only if $S=T$.

In Exercises 5 *to* 14, *decide whether the given statement is true or false. Prove each of the statements that you think is true: give a counterexample for each statement that you think is false.*

5. $R+(S \cap T)=(R+S) \cap (R+T)$.

6. $R+(S \backslash T)=(R+S)\ (R+T)$.

7. $R \backslash (S \cap T)=(R \backslash S) \cup (R \backslash T)$.

8. $R \cap (S+T)=(R \cap S)+(R \cap T)$.

9. $R \cup (S \backslash T)=(R \cup S) \backslash (R \cup T)$.

10. $R+(S \cup T)=(R+S) \cup (R+T)$.

11. $R \backslash (S \cup T)=(R \backslash S) \cap (R \backslash T)$.

12. $R \backslash (S+T)=(R \backslash S)+(R \backslash T)$.

13. $R \cap (S \backslash T)=(R \cap S) \backslash (R \cap T)$.

14. $R \cup (S+T)=(R \cup S)+(R \cup T)$.

15. In each case, list all elements of $S \times T$.
   (i) $S=\{1,2,3\};T=\{1,4,5\}$.
   (ii) $S=\{x \mid x^2=1\};T=\{y \mid y^2=4\}$.
   (iii) $S=\{1,3,5,7\};T=\{1,2,3\}$.

16. In each case, list all elements of $R \times S \times T$.

(i) $R = \{1,2\}; S = \{3,4\}; T = \{5,6\}$;

(ii) $R = \{12,13,14\}; S = \{1\}; T = \{1,2,3\}$.

17. Prove the distributive law
$(R \times S) \cup T = (R \cup T) \times (S \cup T)$.

18. Prove the two distributive laws

(i) $R \times (S \cap T) = (R \times S) \cap (R \times T)$;

(ii) $(R \times S) \cap T = (R \cap T) \times (S \cap T)$.

19. (i) If $S = \emptyset, T \neq \emptyset$, what is $S \times T$?

(ii) If $S \times T = T \times S$, what can you say about S and T?

20. (i) If $A \subseteq S, B \subseteq T$, show that $A \times B \subseteq S \times T$.

(ii) Find an example of sets A, B, S, T, such that $A \times B \subseteq S \times T$ and $B \subseteq T$, but $A \nsubseteq S$.

2.5 Mathematical Induction

The principle of mathematical induction

In working with finite sets or with the sets of positive integers and of integers, one repeatedly uses a technique of proof known as the method of *mathematical induction*. The general idea is as follows: suppose we want to prove that every positive integer n has a property $P(n)$. We first prove $P(1)$ to be true. Then we prove that, for any n, the truth of P(n) implies that of P(n+1); in symbols:

$$P(n) \text{ true } \Rightarrow P(n+1) \text{ true }. \tag{2.17}$$

Intuitively, we would like to say:

$$\begin{array}{lll} & P(1) \text{ true} & \\ & P(1) \text{ true} & \Rightarrow \quad P(2) \text{ true , by 2.17,} \\ \therefore & P(2) \text{ truc;} & \\ & P(2) \text{ true} & \\ & P(2) \text{ true} & \Rightarrow \quad P(3) \text{ true , by 2.17,} \\ \therefore & P(3) \text{ true;} & \end{array}$$

and so on. There is a difficulty, however: given any positive integer k, we can select an integer n such that the proof of $P(n)$ requires at least k steps, so the proof can be arbitrarily long. As "unbounded" proofs present logical difficulties in mathematics — who could ever finish writing one down? — we need an axiom or theorem that states that induction is a valid procedure. This is the *principle of mathematical induction*, and may be stated as follows.

Principle of mathematical induction: *Suppose the proposition $P(n)$ satisfies*

(i) *$P(1)$ is true; and*

(ii) *for every positive integer n, whenever $P(n)$ is true, then $P(n+1)$ is true.*

Then $P(n)$ is true for all positive integers n.

This principle is sometimes called *weak induction*. Another form is:

Strong induction: *Suppose the proposition $P(n)$ satisfies*

(i) *$P(1)$ is true; and*

(ii) *for every positive integer n, if $P(k)$ is true whenever $1 \leq k < n$, then $P(n)$ is true.*

Then $P(n)$ is true for all positive integers n.

At first sight, the second statement looks as though we have assumed more than in the first statement. However these two forms are equivalent; a detailed proof can be found in more advanced books.

Induction can also be stated in strictly set-theoretic form. Suppose S is the set of integers n such that $P(n)$ is true. Then the principle of mathematical induction (weak induction form) is:

Let S be a subset of $\mathbb{Z}^+$ such that:

(i) *$1 \in S$; and*

(ii) *whenever $n \in S$, then $n+1 \in S$.*

Then $S = \mathbb{Z}^+$.

One could equally well state the principle in terms of non-negative integers, instead of positive integers, by changing the case $P(0)$ to $P(1)$, and converting references from "positive integers" to "non-negative integers." It can in fact be stated in terms of any starting point: if S is a set of integers for which

(i) $t \in S$,

where t is any integer, and

(ii) for every integer $n \geq t$, if $n \in S$, then $n+1 \in S$,

then the principle can be used to prove that S contains all integers equal to or greater than t. This form is sometines called "induction from t."

The well-ordering principle

Another principle that is very useful in proving results about sets of positive integers is:

The well-ordering principle: *Suppose S is any non-empty set of positive integers. Then S has a smallest member.*

Any set with the property that every non-empty subset has a least member is called being *well-ordered*; so the principle says "the positive integers are well-ordered." Many number-sets, such as the real numbers, are not well-ordered.

At first the result seems obvious. If S is not empty, then it must contain some member, x say. In order to find the smallest member of S, one need only check to see whether or not $x-1, x-2, \ldots, 1$ are members of S, and this requires only a finite number of steps. However, this "proof" contains the same problem as the "proof" of the principle of mathematical induction: the starting value, x, can be arbitrarily large.

In fact, the induction principle can be proved from the well-ordering principle. To see this, let us assume the well-ordering principle is true and suppose P is any proposition about positive integers such that $P(1)$ is true, and for every positive integer n, whenever $P(n)$ is true, then $P(n+1)$ is true. (These are the requirements for induction.) Write S for the set of positive integers n such that $P(n)$ is not true. In order to prove that induction works, we need to show that S is empty.

If S is not empty, then by well-ordering S has a smallest member, x say. So $P(x)$ is false. We know that $P(1)$ is true, so $x \neq 1$. But x is a positive integer. So $x-1$ is a positive integer, and $P(x)$ must be true — otherwise $x-1$ would be a member of S, and smaller than x. But this means $P((x-1)+1)$ is true: that is, $P(x)$ is both true and false! This can't happen, so our original assumption, that S is not empty, must have been wrong. So induction is proved.

We can also work the other way. If you assume induction is true, you can show that the positive integers are well-ordered.

However, there is no absolute proof here. It is necessary to assume that either induction or well-ordering is a property of the positive integers. So we assume these as axioms about numbers.

Some applications

We start by looking at some properties of sets and prove them by induction.

Theorem 2.3 *Let A be a set with $|A| = n$. Then $|\mathcal{P}(A)| = 2^n$.*

Proof. To apply induction, we can rephrase the statement as: *Let $P(n)$ be the statement "$|\mathcal{P}(A)| = 2^n$ for any n-element set A"; then $P(n)$ is true for all positive* *integers n*. Since the elements of A do not matter, we may as well assume $A = \{a-1, a_2, \ldots, a_n\}$.

First we consider the case $n = 1$, so $A = \{a_1\}$. Then $\mathcal{P}(A) = \{\emptyset, A\}$, so $|\mathcal{P}(A)| = 2^1$ and in this case the theorem is true. Now suppose the result has been proved for $n = k-1$ and assume $n = k$; we might as well say $A = A' \cup \{a_k\}$ where $A' = \{a-1, a_2, \ldots, a_{k-1}\}$. By the induction hypothesis $|\mathcal{P}(A')| = 2^{k-1}$. Any subset of A is either a subset of A' or a subset of A' with the element s_k adjoined to it, so to each subset of A' there correspond two subsets of A. Hence $|\mathcal{P}(A)| = 2^{k-1} \cdot 2$ and the theorem is proved. □

We consider next the cardinality of a cartesian product of two finite sets.

Theorem 2.4 *If $|S| = m$ and $|T| = n$, then $|S \times T| = mn$.*

Proof. We proceed by induction on m. If $m = 1$, then $S = \{s_1\}$ and an ordered pair with s_1 as its first element may be constructed in n ways, giving $(s_1, t_1), (s_1, t_2), \ldots, (s_1, t_n)$, so the theorem is true for $m = 1$. Now suppose the statement is true for $m = k-1$, and consider the case $m = k$.

Let $S' = \{s_1, s_2, \ldots, s_{k-1}\}$, so $|S' \times T| = (k-1)n$. Also, $|\{s_k\} \times T| = n$. But $S \times T = (S' \times T) \cup (\{s_k\} \times T)$ and since the two products on the right-hand side of this equation are disjoint, we know that $|SxT| = (k-1)n + n$, proving the theorem. □

Mathematical induction is very often used in proving general algebraic formulas.

Sample Problem 2.18 *Prove by induction that the sum of the first n positive integers is*

$$1 + 2 + \cdots + n = \tfrac{1}{2}n(n+1).$$

Solution. The case $n = 1$ is $\frac{1}{2} \cdot 1(1+1) = 1$, which is obviously true, so the formula gives the correct answer when $n = 1$. Suppose it is true when $n = k-1$; therefore

$$1 + 2 + \cdots + (k-1) = \tfrac{1}{2}(k-1)k.$$

Then

$$\begin{aligned} 1 + 2 + \cdots + (k-1) + k &= \tfrac{1}{2}(k-1)k + k \\ &= \tfrac{1}{2}(k^2 - k + 2k) \\ &= \tfrac{1}{2}k(k+1), \end{aligned}$$

and the formula is proved correct when $n = k$. So, by induction, we have the required result.

Practice Exercise. Prove that the sum of the first n odd positive integers is

$$1 + 3 + \cdots + (2n-1) = n^2.$$

The following example looks geometrical, but it also yields to induction.

Sample Problem 2.19 *Suppose n straight lines are drawn in two-dimensional space, in such a way that no three lines have a common point and no two lines are parallel. Prove that the lines divide the plane into $\frac{1}{2}(n^2+n+2)$ regions.*

Solution. We proceed by induction. If $n = 1$, the formula yields $\frac{1}{2}(1+1+2) = 2$, and one line does indeed partition the plane into two regions. Now assume that the formula works for $n = k-1$. Consider k lines drawn in the plane. Delete one line. By induction, the plane is divided into $\frac{1}{2}((k-1)^2+(k-1)+2) = \frac{1}{2}(k^2-k+2)$ regions.

Now reinsert the (deleted) k-th line. It must cross every other line exactly once, so it crosses $(k-1)$ lines, and lies in k regions. It divides each of these regions into two parts, so the k regions are replaced by $2k$ new regions; the total is

$$\begin{aligned}\tfrac{1}{2}(k^2-k+2)-k+2k &= \tfrac{1}{2}(k^2-k+2-2k+4k)\\ &= (k^2+k+2),\end{aligned}$$

and the result is true by induction.

Here is an example that uses induction from the starting point 4, rather than 0 or 1.

Sample Problem 2.20 *Prove that $n! \geq 2^n$ whenever $n \geq 4$.*

Solution. Suppose the proposition $P(n)$ means $n! \geq 2^n$. Then $P(4)$ means $4! \geq 2^4$, or $24 \geq 16$, which is true. Now suppose k is an integer greater than or equal to 4, and $P(k)$ is true: $k! \geq 2^k$. Multiplying by $k+1$, we have $(k+1)1 \geq (2^k(k+1) \geq 2^k 2 = 2^{k+1}$, so $P(k)$ implies $P(k+1)$, and the result follows by induction.

Practice Exercise. Prove that $n^2 \geq 2n+1$ whenever $n \geq 3$.

Sample Problem 2.21 *Prove by induction that $5^n - 2^n$ is divisible by 3 whenever n is a positive integer.*

Solution. Suppose $P(n)$ means 3 divides $5^n - 2^n$. Then $P(1)$ is true because $5^1 - 2^1 = 3$. Now suppose k is any positive integer, and $P(k)$ is true: say $5^k - 2^k = 3x$, where x is an integer. Then $5^{k+1} - 2^{k+1} = 5 \cdot 5^k - 2 \cdot 2^k = 3 \cdot 5^k + 2 \cdot 5^k - 2 \cdot 2^k = 3 \cdot 5^k + 2 \cdot 3x$, which is divisible by 3. So the result follows by induction.

Practice Exercise. Prove that $3^{2n} - 2^n$ is divisible by 7 whenever n is a positive integer.

The *Fibonacci numbers* $f_1, f_2, f_3, \ldots$ are defined as follows. $f_1 = f_2 = 1$, and if n is any integer greater than 2, $f_n = f_{n-1} + f_{n-2}$. This famous sequence is the solution to a problem posed by Leonardo of Pisa, or Leonardo Fibonacci (Fibonacci means *son of Bonacci*) in 1202:

> A newly born pair of rabbits of opposite sexes is placed in an enclosure at the beginning of the year. Beginning with the second month, the female gives birth to a pair of rabbits of opposite sexes each month. Each new pair also gives birth to a pair of rabbits of opposite sexes each month, beginning with their second month.

The number of pairs of rabbits in the enclosure at the beginning of month n is f_n.

Some interesting properties of the Fibonacci numbers involve the idea of *congruence* modulo a positive integer. We say a is *congruent to* b modulo n, written "$a \equiv b(\bmod n)$," if and only if a and b leave the same remainder on division by n. In other words n is a divisor of $a - b$, or in symbols $n \mid (a - b)$. This idea will be explored further in Sample Problem 4.6 and in Section 9.2.

Sample Problem 2.22 *Prove by induction that the Fibonacci number f_n is even if and only if n is divisible by* 3.

Solution. Assume n is at least 4. $f_n = f_{n-1} + f_{n-2} = (f_{n-2} + f_{n-3}) + f_{n-2} = f_{n-3} + 2f_{n-2}$, so $f_n \equiv f_{n-3}(\bmod 2)$.

We first prove that, for $k > 0, f_{3k}$ is even. Call this proposition $P(k)$. Then $P(1)$ is true because $f_3 = 3$. Now suppose k is any positive integer, and $P(k)$ is true: $f_{3k} \equiv 0(\bmod 2)$. Then (putting $n = 3k+3$) $f_{3(k+1)} \equiv f_{3k}(\bmod 2) \equiv 0(\bmod 2)$ by the induction hypothesis. So $P(k+1)$ is true; the result follows by induction. To prove that, for $k > 0, f_{3k-1}$ is odd — call this proposition $Q(k)$ — we note that $Q(1)$ is true because $f_1 = 1$ is odd, and if $Q(k)$ is true, then f_{3k-1} is odd, and $f3(k+1) - 1 \equiv f_{3k-2}(\bmod 2) \equiv 1(\bmod 2)$. We have $Q(k+1)$ and again the result follows by induction. The proof for $k \equiv 1(\bmod 3)$ is similar.

Practice Exercise. Prove by induction that the f_n is divisible by 3 if and only if n is divisible by 4.

Some further properties of Fibonacci numbers appear among the Exercises.

Exercises 2.5

In Exercises 1 *to* 6, *prove the given proposition by induction.*

1. $\sum_{r=1}^{n} r^2 = \frac{1}{6}n(n+1)(2n+1)$.

2. $\sum_{k=1}^{n} k^3 = \left[\sum_{k=1}^{n} k\right]^2 = \frac{1}{4}n^2(n+1)^2.$

3. $1+4+7+\cdots+(3n-2) = \frac{1}{2}n(3n-1).$

4. $2+6+12+\cdots+n(n+1) = \sum_{k=1}^{n} k(k+1) = \frac{1}{3}n(n+1)(n+2).$

5. $\frac{1}{1\cdot 3}+\frac{1}{3\cdot 5}+\frac{1}{5\cdot 7}+\cdots+\frac{1}{(2n-1)(2n+1)} = \frac{n}{2n+1}.$

6. $1+3+3^2+\cdots+3^n = \frac{1}{2}(3^{n+1}-1).$

7. Write down the first twelve Fibonacci numbers.

In Exercises 8 *to* 11, *prove the given result about the Fibonacci numbers, for all positive integers* n.

8. f_n is divisible by 4 if and only if n is divisible by 6.

9. $f_1+f_2+\cdots+f_n = f_{n+2}-1.$

10. $f_1+f_3+\cdots+f_{2n-1} = f_{2n}.$

11. $f_n^2+f_{n+1}^2 = f_{2n+1}.$

12. The numbers $a_0, a_1, a_2, \ldots$ are defined by $a_0 = \frac{1}{4}$ and

$$a_{n+1} = 2a_n(1-a_n) \text{ when } n > 0.$$

Prove that

$$a_n = \frac{1}{2}\left(1-\frac{1}{2^{2^n}}\right).$$

13. The numbers $a_0, a_1, a_2, \ldots$ are defined by $a_0 = 3$ and

$$a_{n+1} = 2a_n - a_n^2 \text{ when } n > 0.$$

Prove that $a_n = 1-2^{2^n}$ when $n > 0$, although this formula does not apply when $n = 0$.

14. Show by induction that $2^n \geq n^2$ for $n \geq 4$.

In Exercises 15 *to* 18, *prove that the divisibility result holds for all positive integers* n.

15. 2 divides $3^n - 1$.

16. 6 divides $n^3 - n$.

17. 5 divides $2^{2n-1} + 3^{2n-1}$.

18. 24 divides $n^4 - 6n^3 + 23n^2 - 18n$.

19. Prove that the sum of the cubes of any three consecutive integers is a multiple of 9.

20. The numbers $x_1, x_2, \ldots$ are defined as follows. $x_1 = 1$, $x_2 = 1$, and if $n \geq 2$ then $x_{n+1} = x_n + 2x_{n-1}$. Prove that x_n is divisible by 3 if and only if n is divisible by 3.

21. Prove by induction: if n people stand in line at a counter, and if the person at the front is a woman and the person at the back is a man, then somewhere in the line there is a man standing directly behind a woman.

22. Assume that the sum of the angles of a triangle is π radians. Prove by induction that the sum of the angles of a convex polygon with n sides is $(n-2)\pi$ radians when $n \geq 3$.

23. Consider the set of real numbers: $\{x : x^2 < 1\}$. Show that this set has no least member. Use this to prove that the real numbers are not well-ordered.

24. Let $\mathbb{R}^0$ be the set of non-negative real numbers, $\{x : x \in \mathbb{R}, x \geq 0\}$. Is $\mathbb{R}^0$ well-ordered?

25. Assuming the principle of mathematical induction, show that the positive integers are well-ordered.

26. Find the errors in the following "proofs":

(i) **Theorem** *All computer programs contain the same number of bugs.*

Proof. If we show that in any set of n programs, all the programs contain the same number of bugs, then we have proved the theorem. We proceed by induction on n.

First, let $n = 1$. Certainly, in a set consisting of one program, all the programs contain the same number of bugs, so the statement is true for $n = 1$.

Now suppose that for every set containing fewer than n programs, all the programs in the set contain the same number of bugs, and consider a set D of n programs, $D = \{p_1, p_2, \ldots, p_n\}$. Remove the first program and consider $D_1 = \{p_2, p_3, \ldots, p_n\}$, a set of $n-1$ programs. By the induction hypothesis, all the programs $p_2, \ldots, p_n$ contain the same number of bugs. Now replace the first program and remove the last, forming $D_2 = \{p_1, p_2, \ldots, p_{n-1}\}$, another set of $n-1$ programs. By the induction hypothesis, all of $p_1, \ldots, p_{n-1}$ have the same number of bugs. Hence p_1 and p_n each have the same number of bugs as the other programs in the set, so all the programs contain the same number of bugs. The theorem follows by induction.

(ii) **Theorem** *All computer programs contain the same number of bugs.*

Proof. Any program must have a non-negative number of bugs, so the possible numbers are $\{0, 1, \ldots, N\}$ for some large (but finite) number N. Choose any two programs and compare the number of bugs, say r and s, contained in them. If $\max\{r, s\} = 0$, then $r = s = 0$. Now suppose that if $\max\{r, s\} \leq n - 1$, then $r = s$, and consider the case where max $\{r, s\} = n$. This implies that $\max\{r - 1, s - 1\} = n - 1$ and hence $r - 1 = s - 1$ by the induction hypothesis. Hence $r = s$. Since any two programs have the same number of bugs, all programs must have the same number of bugs, and the theorem is proved.

3

Boolean Algebras and Circuits

Observe the similarities between the algebra of sets and the algebra of propositions: the universal set plays the same role as T, the empty set $\emptyset$ corresponds to F, $\cup$ is like $\vee$ and $\cap$ is like $\wedge$. This similarity is exploited by defining an object, called a *Boolean algebra*, such that the algebra of all subsets of a fixed set is a Boolean algebra, and a well-defined set of propositions also forms a Boolean algebra.

Boolean algebras are defined from a set of axioms. We show how to derive a number of properties directly from these axioms, and investigate other ways to prove results about Boolean algebras.

A Boolean algebra can be used to analyze electrical circuits. We shall describe how this is done, and introduce the circuits used to perform arithmetic in computers.

3.1 Boolean Algebra

Boolean algebras defined

By a *binary operation* on a set S we mean a way of combining an ordered pair of elements of S to give a uniquely defined element of the set. There are many familiar examples of binary operations in arithmetic: for example, addition is a binary operation on $\mathbb{Z}$, or on $\mathbb{Z}^+$, or on other number sets. Subtraction is a binary operation on $\mathbb{Z}$, but it is not a binary operation on $\mathbb{Z}^+$, because some pairs do not give a member of the set: for example, 1 and 2 are members of $\mathbb{Z}^+$, but $1-2$ is not in $\mathbb{Z}^+$. This property of belonging to the original set is called *closure*. Similarly, a *unary operation* is a rule that associates with each member of S some uniquely defined member of S; for example, the multiplicative inverse operation is a unary operation on $\mathbb{R}^*$, but not on $\mathbb{R}$ (because 0 has no inverse).

Suppose $\mathcal{B}$ is a set containing two distinct special elements called 0 and 1: suppose there are defined on $\mathcal{B}$ two binary operations, which we shall denote by $+$ and $\times$, and a unary operation, denoted by $'$. (We write $x+y$ for the result of applying the operation $+$ to (x,y), and x' for the result of applying $'$ to x.) Then we call $\mathcal{B}$ a *Boolean algebra* if the following five pairs of laws hold for all members x, y and z in $\mathcal{B}$.

B1. *Commutative laws*:
(a) $x+y=y+x$,
(b) $x\times y=y\times x$,

B2. *Associative Laws*:
(a) $(x+y)+z=x+(y+z)$,
(b) $(x\times y)\times z=x\times(y\times z)$,

B3. *Distributive Laws*:
(a) $x+(y\times z)=(x+y)\times(x+z)$,
(b) $x\times(y+z)=(x\times y)+(x\times z)$,

B4. *Identity Laws*:
(a) $x+0=x$,
(b) $x\times 1=x$,

B5. *Complement Laws*:
(a) $x+x'=1$,
(b) $x\times x'=0$.

We shall call these laws the *Axioms of Boolean algebra*.

The elements 0 and 1 are usually called the "zero" and "unity" elements of the Boolean algebra. Similarly, the operations $+$ and $\times$ are often called "sum" and "product," but they are not the same as the ordinary sum and product of numbers. In particular, notice that the first distributive law, B3(a), is not true in ordinary arithmetic. The element x' is called the *complement* of x.

Sample Problem 3.1 *Let U be any set. Show that the power set $\mathcal{P}(U)$ is a Boolean algebra with the roles of $0,1,+,\times$ and $'$ taken by $\emptyset,U,\cup,\cap$ and $^-$ respectively.*

Solution. Rules B1, B2 and B3 are the commutative and associative laws for union and intersection and the distributive laws (2.2) and (2.3). The identity laws simply state $S\cup\emptyset=S$ and $S\cap U=S$, and the complement laws were proved in Exercises 2.2.7 and 2.2.11.

Practice Exercise. Let $\mathcal{B}$ be the set of all propositions that can be expressed in English. Show that $\mathcal{B}$ is a Boolean algebra with the roles of $0, 1, +, \times$ and $'$ taken by $F, T, \vee, \wedge$ and $\sim$ respectively.

We now introduce another important Boolean algebra, the two-element algebra $\mathcal{B}_2$. This algebra is used to analyze electrical circuits. $\mathcal{B}_2$ has two elements, called 0 and 1. The operations on $\mathcal{B}_2$ can be described in ordinary arithmetic as follows: $x+y$ equals the greater of x and y, $x \times y$ equals the lesser of x and y, and $x' = 1-x$. In other words, the operations are given by the tables

$+$	0	1
0	0	1
1	1	1

$\times$	0	1
0	0	0
1	0	1

x	x'
0	1
1	0

For brevity we adopt three notational conventions that are like those used in ordinary arithmetic. First, we say that *$'$ takes precedence over $\times$ and $\times$ takes precedence over $+$*. For example, the expression $x \times y + z$ means $(x \times y) + z$, *not* $x \times (y+z)$, and $x \times y'$ means $x \times (y')$, *not* $(x \times y)'$. Second, we omit the $\times$ sign when no confusion arises, so that xy means the same as $x \times y$. And finally, we omit brackets in the expression $x + (y+z)$ and $(x+y)+z$: the associative law tells us that these two expressions are equal, so we simply write $x+y+z$. We do the same for sums of more than three terms, and also for extended products.

Sample Problem 3.2 *Write the following expressions with as few symbols as possible, using the conventions:* $x + ((y \times (z')) + t)$, $x \times ((x + (y')) + z)$.

Solution. $x + yz' + t$, $x(x+y'+z)$.

Practice Exercise. Do the same for the expression $(x \times (y')) \times ((z') \times x)$

Some theorems about Boolean algebras

Suppose you are given any formula in a Boolean algebra. You could derive another formula by interchanging $+$ and $\times$ throughout, and also interchanging 0 and 1. This new formula is called the *dual* of the original. For example, the dual of $x + y' = 0$ is $x \times y' = 1$.

Observe that each of the axioms B1 through B5 consists of a statement and its dual. So, if we can use the axioms to prove any theorem about Boolean algebras, then we could also use them to prove the dual of the theorem. We state this result formally as follows.

Principle of Duality: *The dual of every theorem in Boolean algebra is also a theorem.*

The effect of this is to halve the amount of work we need to do, in the sense that every proof gives both a theorem and its dual theorem.

We shall list several theorems that hold in any Boolean algebra. Some proofs will be given as examples; others are left as exercises. In each case, the theorems come in pairs, one statement being the dual of the other.

Theorem 3.1 (Further Distributive Laws) *In any Boolean algebra, all elements x, y and z satisfy* (a) $(x+y)z = xz + yz$ *and* (b) $xy + z = (x+z)(y+z)$.

Theorem 3.2 (Idempotent Laws) *In any Boolean algebra, every element x satisfies* (a) $x + x = x$ *and* (b) $xx = x$.

Sample Problem 3.3 *Prove the idempotent laws.*

Solution. We prove the statement (a) using the axioms. Statement (b) is then true because it is the dual of (a).

$$\begin{aligned} x &= x+0 && \text{by B4(a)} \\ &= x+xx' && \text{by B5(b)} \\ &= (x+x)(x+x') && \text{by B3(a)} \\ &= (x+x)1 && \text{by B5(a)} \\ &= (x+x) && \text{by B4(b).} \end{aligned}$$

Practice Exercise. Prove the second idempotent law *from the axioms.*

Theorem 3.3 (Absorption Laws) *In any Boolean algebra, all elements x and y satisfy* (a) $x + xy = x$ *and* (b) $x \times (x+y) = x$.

Theorem 3.4 (Nullity Laws) *In any Boolean algebra, every element x satisfies* (a) $x + 1 = 1$ *and* (b) $x \times 0 = 0$.

The nullity laws are so named because 0 is a null element for product, just as the number 0 is for ordinary multiplication. They are also called *boundedness laws.*

Sample Problem 3.4 *Prove the boundedness laws from the axioms.*

Solution. Again we prove only one statement using the axioms — in this case the statement (b).

$$\begin{aligned} x0 &= x0+0 && \text{by B4(a)} \\ &= x0+xx' && \text{by B5(b)} \\ &= x(0+x') && \text{by B3(b)} \\ &= x(x'+0) && \text{by B1(a)} \\ &= xx' && \text{by B4(a)} \\ &= 0 && \text{by B5(b).} \end{aligned}$$

Practice Exercise. Prove the absorption laws from the axioms.

Although we write the laws in terms of single letters, they apply to any expressions involving elements of a Boolean algebra.

Sample Problem 3.5 *Prove that* $zt + ytz = tz$.

Solution. From the commutative laws, $zt = tz$, and $ytz = tzy$, so $zt + ytz = tz + tzy$. Define $x = tz$. From the absorption law, $x + xy = x$, so $tz + tzy = tz$.

Practice Exercise. Prove that $yz + yz = yz$.

The complement of a set is defined in a natural way, and obviously no set can have two complements. Similarly, the negation of a proposition is a unique proposition. This uniqueness is actually a property of all Boolean algebras:

Theorem 3.5 (Uniqueness of Complement) *If* x *and* y *are any elements of any Boolean algebra, and if* $x + y = 1$ *and* $xy = 0$, *then* $y = x'$.

Proof.

$$\begin{aligned}
x' &= x' + 0 && \text{by B4(a)} \\
&= x' + xy && \text{from the data} \\
&= (x' + x)(x' + y) && \text{by B3(a)} \\
&= 1(x' + y) && \text{by B5(a)} \\
&= (x + y)(x' + y) && \text{from the data} \\
&= (y + x)(y + x') && \text{from the data} \\
&= y + xx' && \text{by B3(b)} \\
&= y0 && \text{by B5(a)} \\
&= y && \text{by B4(b).}
\end{aligned}$$

□

This enables us to prove some further theorems about complements.

Theorem 3.6 (Involution Law) $(x')' = x$.

Theorem 3.7 $0' = 1$ *and* $1' = 0$.

Theorem 3.8 (de Morgan's Laws) *In any Boolean algebra, all elements* x *and* y *satisfy* (a) $(x + y)' = x'y'$ *and* (b) $(xy)' = x' + y'$.

Exercises 3.1

1. Let $\mathcal{B}$ be the set of all positive integer divisors of 15, that is $\mathcal{B} = \{1, 3, 5, 15\}$. Prove that $\mathcal{B}$ forms a Boolean algebra with zero element 1 and unity element 15, provided the operations are defined as follows: $x + y$ is the least common multiple of x and y, xy is the greatest common divisor, and x' is the ordinary arithmetical quotient $15/x$.

2. Show that the procedure outlined in Exercise 1 also produces a Boolean algebra when 15 is replaced by

(i) 30; (ii) 42.

3. Show that the procedure outlined in Exercise 1 does *not* produce a Boolean algebra when 15 is replaced by 12.

In Exercises 4 *to* 11, *give the dual of the Boolean expression.*

4. $xy+1$

5. $xy+z'$

6. $yz+zx$

7. $x'yz$

8. $x+y'$

9. $x(y+z')y$

10. $x(xy+z)$

11. $xy+yz+zx$

12. $\mathcal{B}$ consists of the eight elements $\{0,1,a,b,c,d,e,f\}$ with the following operations.

$+$	0	a	b	c	d	e	f	1
0	0	a	b	c	d	e	f	1
a	a	a	c	c	e	e	1	1
b	b	c	b	c	f	1	f	1
c	c	c	c	c	1	1	1	1
d	d	e	f	1	d	e	f	1
e	e	e	1	1	e	e	1	1
f	f	1	f	1	f	1	f	1
1	1	1	1	1	1	1	1	1

$\times$	0	a	b	c	d	e	f	1
0	0	0	0	0	0	0	0	0
a	0	a	0	a	0	a	0	a
b	0	0	b	b	0	0	b	b
c	0	a	b	c	0	a	b	c
d	0	0	0	0	d	d	d	d
e	0	a	0	a	d	e	d	e
f	0	0	b	b	d	d	f	f
1	0	a	b	c	d	e	f	1

x	x'
0	1
a	f
b	e
c	d
d	c
e	b
f	a
1	0

Prove that $\mathcal{B}$ is a Boolean algebra.

13. Use the axioms to prove the third and fourth distributive laws (Theorem 3.1).

14. Use the axioms to prove de Morgan's Laws (Theorem 3.8).

In Exercises 15 *to* 18, *prove that the results hold in any Boolean algebra.*

15. $x+x'y=x+y$ for any x and y.

16. $x+xy+y=(x+y)(y+1)$ for any x and y.

17. $(x+y')z'=((x'+z)(y+z))'$ for any x,y and z.

18. $x+y(x+z)=(x+y)(x+z)$ for any x,y and z.

19. Write down the duals of the equations in Exercises 15 to 18.

20. Prove that, in any Boolean algebra, $x+y+x'y'z = x+y+z$, for any x,y and z.

21. Suppose the elements x, y, and z in a Boolean algebra satisfy $x = yz' + y'z$. Prove that $x' = yz + y'z'$.

22. In ordinary arithmetic, the *cancellation law for addition* is the law that says: if the integers x,y and z satisfy $x+y = x+z$, then $y = z$. Prove that there is no cancellation law for $=$ or for $\times$ in Boolean algebras. (It is sufficient to show that, in any Boolean algebra, there are elements x,y and z such that $x+y = x+z$, but $y \neq z$.)

23. Show that in any Boolean algebra, if the elements x,y and z satisfy $x+y = x+z$ and $x'+y = x'+z$, then $y = z$. (Compare this with the preceding exercise.)

24. Define a structure $\mathcal{B}$ as follows. The elements of $\mathcal{B}$ are the integers $\mathbb{Z}$. However, $x+y$ means the greater of x and y, and xy means the lesser of x and y.

(i) Show that $\mathcal{B}$ satisfies the associative, commutative and distributive laws B1, B2 and B3.

(ii) Show that, no matter which element of $\mathbb{Z}$ is chosen to be 0, the identity law $x+0 = x$ does not hold.

(iii) Is this true if the elements of $\mathcal{B}$ are taken to be the non-negative integers?

25. Suppose $\mathcal{B}$ is any Boolean algebra. Define $\mathcal{A}$ to be the set $\mathcal{B} \times \mathcal{B}$, consisting of all the ordered pairs (x,y) with x and y in $\mathcal{B}$. Define operations sum, product and complement on $\mathcal{A}$ by

$$\begin{aligned} (x,y)+(z,t) &= (xy, z+t), \\ (x,y)(z,t) &= (x+y, zt), \\ (x,y)' &= (x', y'). \end{aligned}$$

Prove that $\mathcal{A}$ is a Boolean algebra. What are its zero and unity?

3.2 Boolean Forms

Disjunctive forms

By a *Boolean expression* in the variables $\{x, y, \ldots\}$ we mean any expression built up from the variables using the variables, the operators $+$, $\times$ and $'$, and parentheses. Examples of Boolean expressions in $\{x,y,z\}$ include x, y', $xy + yzx'$ and $(x+yz)(x'+yz)$. In particular, Boolean expressions consisting of a single variable or its complement, like x and y', are called *literals*.

Sample Problem 3.6 *Consider the Boolean expression* $\alpha(x,y,z) = xyz' + x(y+z) + (x+z)'$. *Find the values of* $\alpha(1,1,1), \alpha(1,0,0)$ *and* $\alpha(0,1,1)$.

Solution. $\alpha(1,1,1) = 111' + 1(1+1) + (1+1)' = 11' + 10 + 0' = 0 + 0 + 1 = 1$.
$\alpha(1,0,0) = 100' + 1(0+0) + (1+0)' = 0 + 10 + 1' = 0 + 0 + 0 = 0$.
$\alpha(0,1,1) = 011' + 0(1+1) + (1+1)' = 0 + 0 + 0' = 0 + 0 + 1 = 1$.

Practice Exercise. If $\beta(x,y,z) = xy + yzx'$, evaluate $\beta(1,0,1)$, $\beta(1,1,0)$ and $\beta(0,1,1)$.

Intuition suggests that a Boolean expression is less complicated when it involves fewer symbols. For this reason, we prefer forms in which the number of repeated symbols is smaller. We define a *fundamental product* to be a literal or a product of literals that contains no repeated variable. The products $x, xy, x'zy$ and y' are fundamental products, but xyx' and $yxyz$ are not. Any product is equal to a fundamental product or to 0: if it contains two copies of the same literal x, we can use the idempotence law (Theorem 3.2(b)) to replace xx by x, and if it contains both x and x' we can use the complement law (Axiom B5(b)) to replace their product, and in fact the whole product, by 0.

Sample Problem 3.7 *Reduce the following products to a fundamental product or to* 0: $zxyxz, x'yyx'$.

Solution. $zxyxz = xxyyz$ (repeated use of the commutative law) $= xyz$ (two uses of the idempotent law). $x'yyx' = x'x'yy$ (commutative law) $= 0y = 0$ (idempotent law, nullity law).

Practice Exercise. Reduce the following products to a fundamental product or 0: $xy'x'z', x'yz'y$.

If α and β are two fundamental products, we say that β *includes* α and write $\alpha \leq \beta$ if every literal in α is also a literal in β. A literal and its complement are treated as different. So $xz \leq xyz$ is true, but $xz' \leq xyz$ is false. If $\alpha \leq \beta$, then $\alpha + \beta = \alpha$, by the absorption law: for example, $xz + xyz = xz + (xz)y = xz$.

We say that a Boolean expression is a *disjunctive form* if it equals a sum of fundamental products and none of the products includes any other product. For example, $xy + xz + yz$ is in disjunctive form, but $xy + xz + xyz$ is not. However, $xy + xz + xyz = xy + xz$, so $xy + xz + xyz$ equals a disjunctive form. The fundamental products are called the *terms* of the form. (This is just the way "term" is used in ordinary arithmetic.)

Theorem 3.9 *Any Boolean expression is equal to a disjunctive form.*

Proof. We present an algorithm, in four steps, to reduce any normal form to a disjunctive form.

Step 1. By repeated application of de Morgan's laws, move all complement signs inside parentheses. Then use the involution law to reduce any strings of complementation to one or no complements ($x''' = x', x'''' = x$, and so on.) Then the form will be comprised entirely of literals, parentheses, and $+$ and $\times$ operations.

Step 2. We may assume that the form is something like $\alpha + \beta + \cdots$, where each of $\alpha, \beta, \ldots$ is a product. If any of these terms is a product of literals, leave it unchanged. Otherwise, if (say) α is a product in which one or more of the terms is a product of sums, transform it into a sum of products using the distributive law B3(a).

Step 3. Use the commutative, idempotent and complement laws to transform each product into 0 or a fundamental product, as we did in Sample Problem 3.7.

Step 4. Any zero term can be deleted. If any term is included in any other, use the absorption law to delete one term. □

Sample Problem 3.8 *Illustrate Theorem 3.9 by reducing* $(xy')'(x+x'z)'$ *to disjunctive form.*

Solution. Step 1: de Morgan's laws give $(xy')' = x' + y'' = x' + y$ and $(x + x'z)' = x'(x'z)' = x'(x'' + z') = x'(x + z')$. So the original form equals $(x' + y)(x'(x + z'))$.

Step 2: $(x' + y)(x'(x + z')) = (x' + y)(x'x + x'z') = (x' + y)x'x + (x' + y)x'z' = x'x'x + yx'x + x'x'z' + yx'z'$.

Step 3: $x'x'x + yx'x + x'x'z' + yx'z' = x'x + yx'x + x'z' + yx'z' = 0 + y0 + x'z' + yx'z' = x'z' + yx'z'$.

Step 4: $x'z' + yx'z' = x'z' + x'z'y = x'z'$.

Practice Exercise. Illustrate Theorem 3.9 by reducing $(x'y)'(x' + xyz')$ to disjunctive form.

Minimal forms and prime implicants

Suppose α and β are two disjunctive forms. We shall say that α is *simpler* than β if either α contains fewer terms than β and contains no more literals than β, or α contains fewer literals than β and contains no more terms than β. We say the disjunctive form α is *minimal* if no disjunctive form equal to α is simpler than α.

A fundamental product π is called a *prime implicant* for a Boolean expression α if $\pi + \alpha = \alpha$ and if no fundamental product included in π has this property.

To show that π is a prime implicant for a Boolean expression α involves two kinds of proof. We usually prove that $\pi + \alpha = \alpha$ by manipulating the formula for $\pi + \alpha$. For every fundamental product σ included in π we are then required to prove that $\sigma + \alpha \neq \alpha$. This is normally done by showing that there is some assignment of values to the variables that gives $\sigma + \alpha$ and α different values.

Sample Problem 3.9 *Show that xz is a prime implicant of $\alpha = xy + xy'z + x'yz$.*

Solution.
$$\begin{aligned} xz+\alpha &= xz+xy+xy'z+x'yz \\ &= x(y+y')z+xy+xy'z+x'yz \\ &= xyz+xy'z+xy+xy'z+x'yz \\ &= (xy+xyz)+(xy'z+xy'z)+x'yz \\ &= xy+xy'z+x'yz \\ &= \alpha. \end{aligned}$$

The only fundamental products included in xz are x and z. To show that $x+\alpha \neq \alpha$, it is sufficient to observe that if $x = 1$ and $y = z = 0$, we have $x+\alpha = 1, \alpha = 0$. For the form z, the substitution $z = 1, x = y = 0$ gives $z+\alpha = 1, \alpha = 0$.

Practice Exercise. Show that xy is a prime implicant of $\beta = xyz + xy'z + xyz' + x'y'z'$.

The importance of prime implicants is explained by the following theorem.

Theorem 3.10 *If α is a minimal disjunctive form, then the terms of α are prime implicants of α.*

Proof. Suppose α is a minimal disjunctive form. We show that every term of α is a prime implicant. Let π be such a term. Then we can write

$$\alpha = \pi + \gamma,$$

where γ is the sum of the other terms of α. Then

$$\pi + \alpha = \pi + \pi + \gamma = \pi + \gamma = \alpha.$$

On the other hand, if π included any product β that satisfied $\beta + \alpha = \alpha$, then

$$\alpha = \beta + \alpha = (\beta + \pi) + \gamma = \beta + \gamma.$$

This new form is simpler than the original, since it contains as many terms as the original and contains fewer literals (β contains fewer literals than π). But this is impossible, as α is minimal. □

It follows that a minimal disjunctive form is the sum of its prime implicants, so we can construct the minimal disjunctive form of a Boolean expression by finding its prime implicants. For this reason we shall present (in the next section) a technique for finding all prime implicants of a Boolean expression.

The disjunctive normal form

A Boolean expression in the set of variables S is said to be in *disjunctive normal form* (or *complete*) if it is in disjunctive form and every member of S appears in every term. For example, $x'y+xy'$ is in disjunctive normal form over $\{x,y\}$ but $x'y+x$ is not.

The set S is important in this definition. If S is $\{x,y,z\}$, then $xy+x'y'$ is *not* complete over S. To avoid the need to specify S every time, we usually assume that S consists of precisely the variables appearing in the expression, so that if we write $xy+x'y'$ we assume that $S=\{x,y\}$ unless told otherwise.

Every Boolean expression can be put into disjunctive normal form using the complement law. For example $x=x1=x(y+y')=xy+xy'$, so $x'y+x=y+xy+xy'$, which is complete.

Sample Problem 3.10 *Put into disjunctive normal form:* $xyz+xz'$, $xy+xz$.

Solution. $xyz+xz'=xyz+x(y+y')z'=xyz+xyz'+xy'z'$, $xy+xz=xy(z+z')+x(y+y')z=xyz+xyz'+xyz+xy'z$.

Practice Exercise. Put $xy'+xz'+yz$ into disjunctive normal form.

Disjunctive normal forms are useful because two disjunctive normal forms in the same set of variables are equal if and only if they contain precisely the same terms. Surprisingly enough, we shall also use them in the next section as a tool for finding prime implicants, and consequently minimal forms.

Exercises 3.2

1. The Boolean expression $\alpha(x,y)=(xy+xy')(x'+y)$. Evaluate

(i) $\alpha(0,0)$; (ii) $\alpha(0,1)$; (iii) $\alpha(1,0)$; (iv) $\alpha(1,1)$.

2. The Boolean expression $\beta(x,y)=(x+y+xy)(x+y')$. Evaluate

(i) $\beta(0,0)$; (ii) $\beta(0,1)$; (iii) $\beta(1,0)$; (iv) $\beta(1,1)$.

3. The Boolean expression $\gamma(x,y,z)=(x+y+z)(x'+y'+z')$. Evaluate

(i) $\gamma(0,0,1)$; (ii) $\gamma(0,1,0)$; (iii) $\gamma(1,1,0)$; (iv) $\gamma(1,1,1)$.

4. The Boolean expression $\delta(x,y,z)=(x+y'z)(y+z')$. Evaluate

(i) $\delta(1,0,0)$; (ii) $\delta(0,0,1)$; (iii) $\delta(1,1,0)$; (iv) $\delta(1,1,1)$.

In Exercises 5 *to* 20, *the Boolean expressions* α, β, γ *and* δ *are defined as in Exercises* 1 *to* 4. *Evaluate the given expression over the eight-element Boolean algebra* $\mathcal{B}$ *defined in Exercise* 3.1.12.

5. $\alpha(1,e)$

6. $\alpha(a,b)$

7. $\alpha(f,1)$

8. $\beta(c,e)$

9. $\beta(d,d)$

10. $\beta(b,a)$

11. $\alpha(e,1)\beta(a,c)$

12. $\alpha(a,b)+\beta(e,f)$

13. $\gamma(1,e,a)$

14. $\gamma(a,b,e)$

15. $\gamma(1,f,0)$

16. $\delta(e,c,1)$

17. $\delta(f,d,d)$

18. $\delta(1,b,a)$

19. $\gamma(e,a,1)\delta(b,b,c)$

20. $\gamma(0,a,b)+\delta(e,d,1)$

In Exercises 21 *to* 24, *reduce each product to a fundamental product or zero.*

21. $xyy'xy$

22. $xyxzyzx$

23. $xy'zx$

24. $yyxyyz$

In Exercises 25 *to* 32, *reduce the Boolean expression to disjunctive form.*

25. $x(xy'+x'y)$

26. $x(x'+y')'$

27. $(x+y)(x'+z)(y+z)$

28. $(x+y)'(x'y)'$

29. $(x+y)(x+z)(y+z)$

30. $(x+y)xyz$

31. $(x+y)(x'+y')$

32. $(x+y)(x+z')(y'+z)$

In Exercises 33 *to* 38, *find a simpler form for the Boolean expression.*

33. $x+x'y$

34. $xy'x'+xyx'$

35. $yx+yz'+y'xz+z$

36. $x+yz+x'y+y'xx$

37. $xy'z'+xyz'$

38. $xy+x+y$

In each of Exercises 39 *to* 42, *is* π *a prime implicant of* α?

39. $\pi = xz', \alpha = xy'+xyz'+x'yz'.$

40. $\pi = xy', \alpha = xy+xy'z'+x'y'z.$

41. $\pi = xy, \alpha = z+xyz+xyz'.$

42. $\pi = xyz, \alpha = x'yz+xy'z+xyz.$

In Exercises 43 *to* 46, *write the Boolean expression in disjunctive normal form.*

43. $(x' + y)' + yz'$

44. $(x + y + z)(x' + y + z)$

45. $(x + y' + z)(x' + y + z)$

46. $y(x + yz')$

3.3 Finding Minimal Disjunctive Forms

Karnaugh maps

In this section we outline a technique to find a minimal disjunctive form for any Boolean expression. We first convert the expression into a disjunctive form. If we find all prime implicants of this form, then the expression is equal to the sum of those prime implicants.

Interestingly, the method we use to find prime implicants operates on a disjunctive normal form. Paradoxical as it may seem, the first step in finding the *simplest* form is to convert to a form that will usually be *less* simple.

The *Karnaugh map* of a disjunctive normal form is a diagram in which the fundamental products are represented by small square cells. It is something like a stylized map of a country, where the cells represent states. We shall call two cells *adjacent* if they have a common border line (not just a single point). So two cells are adjacent if they differ only because one of the literals is complemented in one case and not complemented in the other. By looking for these adjacent pairs of cells you can reduce the number of terms in the form: for example, if both xyz and xyz' are represented, then the form contains $xyz + xyz'$, which can be replaced by xy. Karnaugh maps are useful in expressions involving as many as four variables.

Two-variable Karnaugh maps

Two-variable Karnaugh maps are not particularly useful, but the ideas we see here will be used in three- and four-variable cases.

A two-variable Karnaugh map is drawn in a 2×2 array of squares. The two horizontals, or *rows*, represent one variable, while the verticals, or *columns*, represent the other. We shall assume the variables are x and y, and associate x with rows and y with columns; the first (upper) row corresponds to x and the second to x', while the first (left) column corresponds to y and the second to y'. Each cell represents the corresponding fundamental product. So the correspondences are as follows.

	y	y'
x	xy	xy'
x'	$x'y$	$x'y'$

For example, the form $xy+x'y'$ corresponds to a map with the upper left and lower right cells. For our purposes, rather than deleting the missing cells, it will be more convenient to draw the whole diagram, with the required cells occupied by a 1:

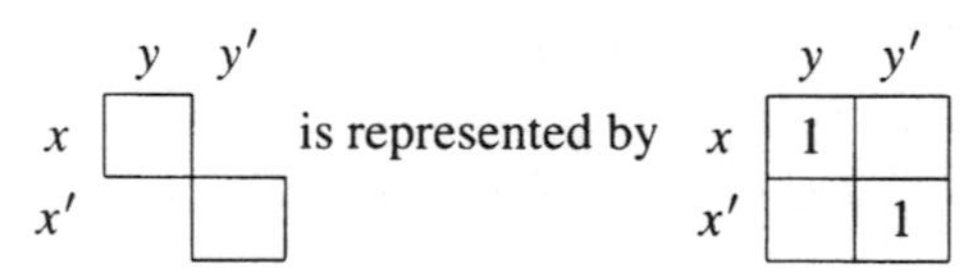

Sample Problem 3.11 *Draw Karnaugh maps for $xy+x'y$ and $x'y+x'y'$.*

Solution.

	y	y'
x	1	
x'	1	

	y	y'
x		
x'	1	1

Practice Exercise. Draw Karnaugh maps for xy and $x'y+xy'$.

Once the Karnaugh map is drawn, it is easy to simplify the expression. Look for pairs of adjacent occupied cells. Each pair corresponds to a sum (such as $xy+x'y$) that can be reduced to a simpler single term (y in the example).

If the whole 2×2 rectangle contains all 1's, then it will be called a *basic rectangle*. Each adjacent pair that both contain 1's will also be called basic, unless the whole rectangle is basic, and any cell with a 1 that has no 1 adjacent to it is also called basic. In other words, a rectangular area in which all cells contain 1 will be called basic unless it is contained in some larger rectangle of all 1's. Then each prime implicant is the sum of the elements in a basic rectangle. For example, in Sample Problem 3.11, the basic rectangle joining $x'y$ and $x'y'$ gives prime implicant $x'y+x'y'=x'$.

To find a minimal form of a Boolean expression: first put the expression in disjunctive normal form and draw the Karnaugh map. Identify the basic rectangles. Among the basic rectangles, find a set that between them cover every 1, with no unnecessary rectangles; this is called a *minimal cover*. The minimal form is the sum of the basic rectangles in a minimal cover.

The "no unnecessary rectangles" condition is not important in the two-variable case, but we shall return to it later. In cases with more variables, there can possibly be more than one minimal form.

Sample Problem 3.12 *Use Karnaugh maps to find minimal forms for $xy+x'y$, $xy+x'y+xy'$, and $xy+x'y+xy'+x'y'$.*

Solution. The maps for the three forms are

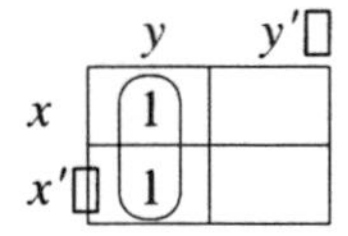

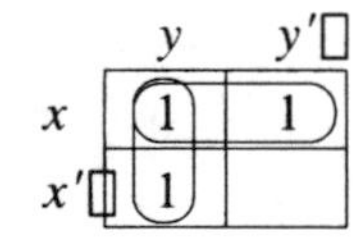

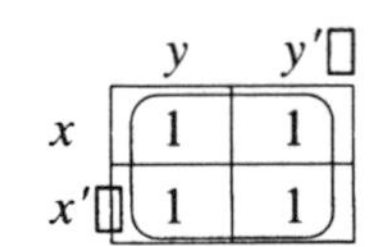

respectively. The basic rectangles are marked. (Notice that two rectangles overlap in the second diagram.) These give the expressions $xy + x'y = y$, $xy + x'y + xy' = x + y$, and $xy + x'y + xy' + x'y' = 1$.

Practice Exercise. Find minimal forms for $xy + x'y'$ and $xy + xy'$.

In the diagram for $xy + x'y + xy'$ we did not circle the single 1 corresponding to xy', because that 1×1 rectangle is enclosed in a 2×2 basic rectangle.

Three-variable and four-variable maps

Three-variable maps use the eight-cell pattern

	yz	yz'	$y'z'$	$y'z$
x				
x'				

They are analyzed similarly to two-variable maps, with the additional convention that cells in the first and last column are treated as adjacent if they are in the same row ("rows wrap around"). Because of the larger size, larger basic rectangles are possible: sizes 1×4 and 2×4 can occur. But 1×3 rectangles are not considered basic, because they do not lead to simplification. A 1×3 rectangle gives two overlapping 1×2 rectangles. For example, the map for $xyz + xyz' + xy'z'$ is

	yz	yz'	$y'z'$	$y'z$
x	1	1	1	
x'				

and the minimal form is $xy + xz'$.

Four-variable maps use the sixteen-cell pattern

	zt	zt'	$z't'$	$z't$
xy				
xy'				
$x'y'$				
$x'y$				

The wrap-around property is applied to columns as well as rows, and 4×1, 4×2 and 4×4 basic rectangles are possible.

Sample Problem 3.13 *Use Karnaugh maps to find minimal forms for* $xyz + x'yz' + xyz' + x'y'z$, $xyz + xy'z + xy'z' + x'yz + x'y'z$, $xyz + xyz' + xy'z' + x'y'z + x'y'z'$.

Solution. The maps for the first two forms are

	yz	yz'	$y'z'$	$y'z$
x	1	1		
x'		1		1

	yz	yz'	$y'z'$	$y'z$
x	1		1	1
x'	1			1

respectively. The basic rectangles are marked. Notice that two 1×2 rectangles and two 2×1 rectangles are not marked in the second diagram. They are not basic, because they are in the 2×2 rectangle. In the first case the prime implicants are $xyz + xyz' = xy$, $xyz' + x'yz' = yz'$ and $x'y'z$. So the minimal form is

$$xyz + x'yz' + xyz' + x'y'z = xy + yz' + x'y'z.$$

In the second case the prime implicants are $xy'z + xy'z' = xy'$ and $(xyz + xy'z) + (x'yz + x'y'z) = xz + x'z = z$ and the minimal form is

$$xyz + xy'z + xy'z' + x'yz + x'y'z = xy' + z.$$

In the third example we obtain

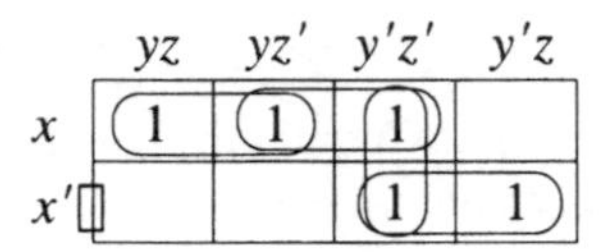

Not all of these basic rectangles are necessary. We can omit either the rectangle joining xyz' to $xy'z'$, or the one joining $xy'z'$ to $x'y'z'$, and all 1's will still be covered. So we can use either of the diagrams

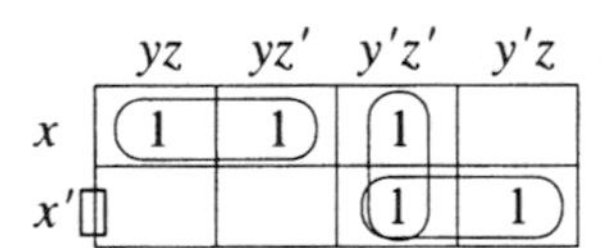

	yz	yz'	$y'z'$	$y'z$
x	1	1	1	
x'			1	1

and obtain two different minimal forms:

$$xyz + xyz' + xy'z' + x'y'z + x'y'z' = xy + y'z' + x'y',$$
$$xyz + xyz' + xy'z' + x'y'z + x'y'z' = xy + xz' + x'y'.$$

Sample Problem 3.14 *Use Karnaugh maps to find minimal forms for* $\alpha = xyz't + xyz't' + xy'zt + xy'zt' + x'yz't + x'yz't' + x'y'zt + x'y'zt'$ *and* $\beta = xyzt + xyz't + xy'zt + xy'z't + xyzt' + xyz't' + xy'zt' + xy'z't' + x'yz't'$.

Solution. The maps for the two forms are

	zt	zt'	$z't'$	$z't$
xy			1	1
xy'	1	1		
$x'y'$	1	1		
$x'y$			1	1

	zt	zt'	$z't'$	$z't$
xy	1	1	1	1
xy'	1	1	1	1
$x'y'$			1	
$x'y$				

respectively. The basic rectangles are marked. We get $\alpha = yz' + y'z$ and $\beta = x + y'z't'$.

Sometimes there can be a large number of basic rectangles, and care must be taken to find a minimal cover.

Sample Problem 3.15 *Use Karnaugh maps to find a minimal form for* $xyzt + xyzt' + xyz't' + xy'z't' + x'y'zt + x'y'zt' + x'y'z't' + x'y'z't + x'yzt + x'yzt'$.

Solution. The map is shown in the left-hand part of the following diagram, and six basic rectangles are marked. However, on examination we see that three basic rectangles cover all the 1's, as shown in the right-hand figure.

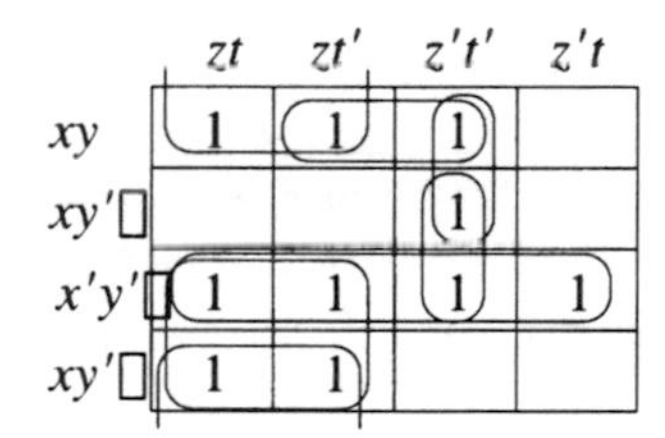

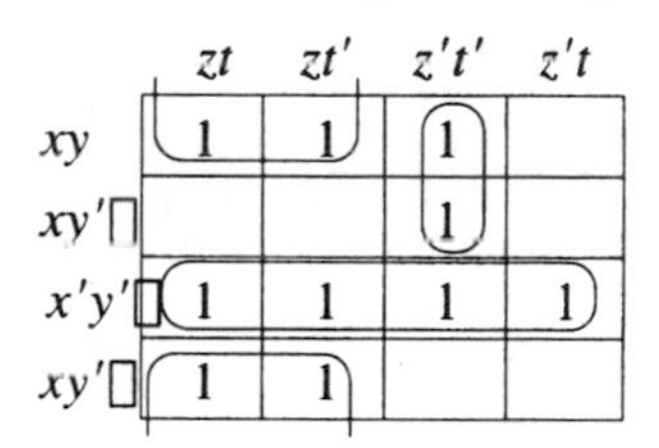

So the minimal form is $xz't' + x'y' + yz$.

Exercises 3.3

Use Karnaugh maps to find minimal forms for the two-variable expressions in Exercises 1 *to* 6.

1. $xy + xy'$

2. $xy' + x'y + x'y'$

3. $xy + x'y + x'y'$

4. $x'y + xy'$

5. $xy' + x'y'$

6. $xy + xy' + x'y$

7. What is the fundamental product represented by each of the following Karnaugh maps?

(i)

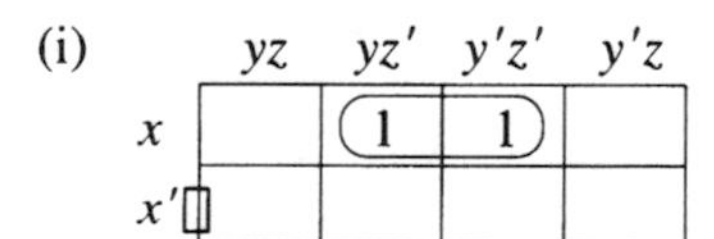

(ii)

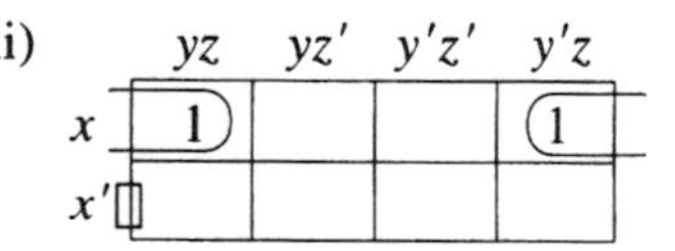

8. What is the fundamental product represented by each of the following Karnaugh maps?

(i)

(ii)

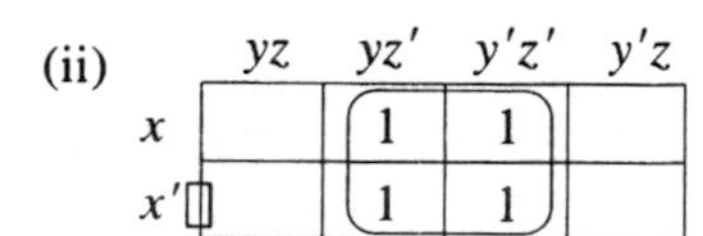

Use Karnaugh maps to find minimal forms for the three-variable expressions in Exercises 9 *to* 20.

9. $xyz + xyz' + xy'z$

10. $xyz + xyz' + xy'z' + x'y'z$

11. $xyz' + xy'z + xy'z' + x'yz + x'yz' + x'y'z$

12. $xyz' + xy'z' + x'yz + x'y'z$

13. $xyz + xyz' + x'yz + x'y'z$

14. $xyz' + xy'z + x'yz + x'yz'$

15. $xyz + xyz' + x'yz'$

16. $xyz' + xy'z + xy'z' + x'yz' + x'y'z$

17. $xyz + xyz' + xy'z + xy'z' + x'y'z$

18. $xyz + xyz' + xy'z + xy'z' + x'yz + x'yz' + x'y'z + x'y'z'$

19. $xyz' + xy'z + xy'z' + x'yz + x'y'z$

20. $xy'z + xy'z' + x'yz + x'y'z'$

21. What is the fundamental product represented by each of the following Karnaugh maps?

(i)

	zt	zt'	$z't'$	$z't$
xy				
xy'			1	1
$x'y'$				
$x'y$				

(ii)

	zt	zt'	$z't'$	$z't$
xy	1			1
xy'	1			1
$x'y'$	1			1
$x'y$	1			1

22. What is the fundamental product represented by each of the following Karnaugh maps?

(i)

	zt	zt'	$z't'$	$z't$
xy		1	1	
xy'				
$x'y'$				
$x'y$		1	1	

(ii)

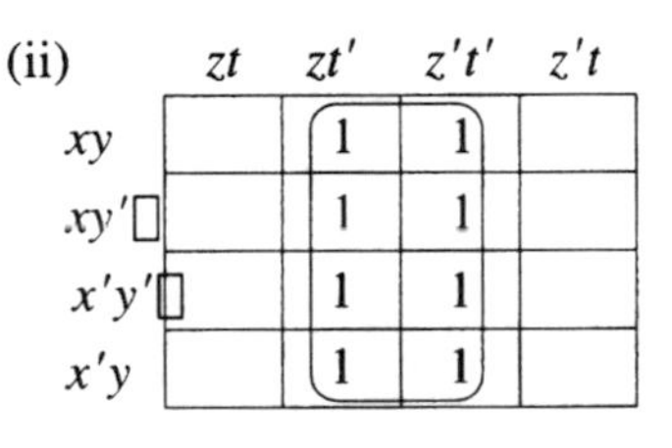

Use Karnaugh maps to find minimal forms for the four-variable expressions in Exercises 23 *to* 30. *(There will sometimes be more than one correct answer.)*

23. $xyzt + xyz't + xyzt' + x'yzt + x'y'zt + x'yzt'$

24. $xyz't + xy'zt + xy'z't + x'yzt + x'yz't + x'yzt' + x'y'z't'$

25. $xy'z't + xy'zt' + xy'z't' + x'y'zt + x'y'z't + x'y'zt' + x'y'z't'$

26. $xyz't' + xy'zt' + xy'z't' + x'yzt + x'y'zt + x'yzt' + x'yz't'$

27. $xyzt + xy'z't + xyzt' + xy'z't' + x'yzt + x'yz't + x'yz't' + x'y'zt'$

28. $xy'z't + x'y'zt + x'y'z't + x'yzt' + x'yz't' + x'y'zt'$

29. $x'yzt + x'yz't + x'y'zt + x'y'z't + x'yzt' + x'yz't' + x'y'zt' + x'y'z't'$

30. $xyz't + xyzt' + xy'zt' + xy'z't' + x'yzt' + x'yz't'$

31. In each case, write down the disjunctive form of the Boolean expression represented by the given Karnaugh map, and write down a minimal form for it.

(i)

	zt	zt'	$z't'$	$z't$
xy				1
xy'		1	1	
$x'y'$	1	1		
$x'y$	1			

(ii)

	zt	zt'	$z't'$	$z't$
xy			1	1
xy'	1	1	1	1
$x'y'$		1	1	1
$x'y$			1	

32. In each case, write down the disjunctive form of the Boolean expression represented by the given Karnaugh map, and write down a minimal form for it.

(i)

	zt	zt'	$z't'$	$z't$
xy		1		
xy'		1	1	1
$x'y'$	1	1	1	
$x'y$	1	1	1	

(ii)

	zt	zt'	$z't'$	$z't$
xy		1		
xy'	1	1	1	
$x'y'$		1	1	1
$x'y$			1	

33. In each case, write down the disjunctive form of the Boolean expression represented by the given Karnaugh map, and write down a minimal form for it.

(i)

	zt	zt'	$z't'$	$z't$
xy			1	1
xy'	1	1		
$x'y'$	1	1		
$x'y$			1	

(ii)

	zt	zt'	$z't'$	$z't$
xy	1	1		
xy'	1	1	1	1
$x'y'$			1	1
$x'y$	1	1		

34. In each case, write down the disjunctive form of the Boolean expression represented by the given Karnaugh map, and write down a minimal form for it.

(i)

	zt	zt'	$z't'$	$z't$
xy	1			1
xy'		1	1	
$x'y'$			1	
$x'y$	1			1

(ii)

	zt	zt'	$z't'$	$z't$
xy			1	1
xy'	1	1	1	1
$x'y'$	1			1
$x'y$	1		1	

In Exercises 35 *to* 40, *convert the Boolean expression to a disjunctive normal form and find a minimal form.*

35. $xyz + xyz' + x'z' + x'y'z + x'yz'$

36. $xy' + xyz' + x'yz'$

37. $(x+y)'(z'+t')'$

38. $xy + x'(y'+z') + yz'$

39. $x'(yz + zt + ty)$,

40. $xyz't + xy'zt' + xyt' + zt't + xz'$

3.4 Digital Circuits

Doing arithmetic with circuits

Computers use electricity, and as such they are essentially *binary*: either a switch is open or it is closed; either there is current in a wire or there is not. As we pointed out in Section 1.5, this means that computers use binary arithmetic. So it is necessary to build electric circuits that carry out pieces of arithmetic.

The typical circuit consists of a number of inputs, a processor and a number of outputs. For example, suppose you have a circuit that carries out addition. Specifically, suppose it takes two variables x and y, each of which can equal 0 or 1, and outputs the value $z = x + y$. As z could have two binary digits, we shall write it as t_1t_0, the usual binary representation of the decimal number $2t_1 + t_0$. Now all the quantities involved have value 0 or 1, so each can be represented by a single electrical entity, such as a wire (carrying current or not) or a switch (open or closed).

We represent each variable as a line; you may wish to think of it as a physical wire. Then the addition process looks like

The x and y wires either contain current (variable $= 1$) or not. The current in the t_0 and t_1 wires will depend on the outcome of the addition. The question is, what goes on in the gray box?

Suppose x, y, t_0 and t_1 are treated as Boolean variables. Then t_1 will be 1 if both x and y are 1, and 0 otherwise. So $t_1 = xy$. Similarly, t_0 will be 1 if $x = 1, y = 0$ or $x = 0, y = 1$ and 0 in the other cases. After a little thought we see that $t_0 = x'y + xy'$. So the arithmetical processes can be represented by Boolean operations. If we use the Boolean algebra $\mathcal{B}_2$, defined in Section 3.1, then no values other than 0 and 1 ever appear.

The building blocks

The circuits we consider are called *digital* circuits, because they deal only with inputs that are single binary digits. They are also called *logic circuits*.

The simplest component circuits are called *gates*. All digital circuits can be built up using three fundamental building blocks, the AND, OR and NOT gates. The standard symbols for these gates are shown in Figure 3.1.

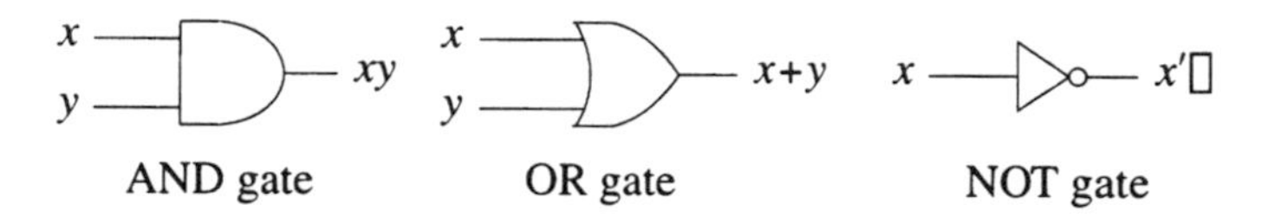

Figure 3.1: The three basic gates

The AND gate takes as its input two variables, x and y, and outputs the value xy. This is easily realized physically by a series circuit. The AND gate has input x, y and output $x + y$, while the NOT gate (or *inverter*) has input x and output x'.

Larger circuits are formed by putting these components together. For example, the following circuits produce $x'y$ and xy':

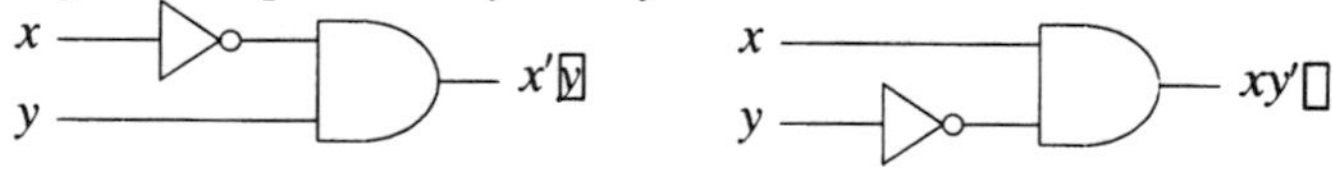

These can be combined as follows to give a circuit to produce $x'y + xy'$.

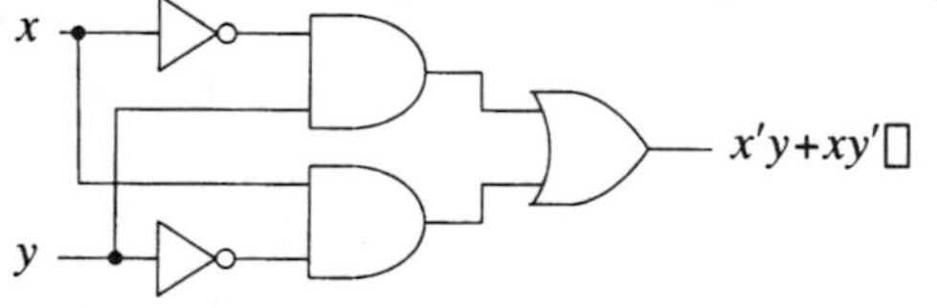

Observe the use of a dot (*splitter*) to show that two lines carry the same charge, and note that we allow two lines to cross without interacting (just as two insulated wires could cross, one on top of the other).

Sample Problem 3.16 *Construct a circuit to implement* $xy+z$*.*

Solution. An appropriate circuit is

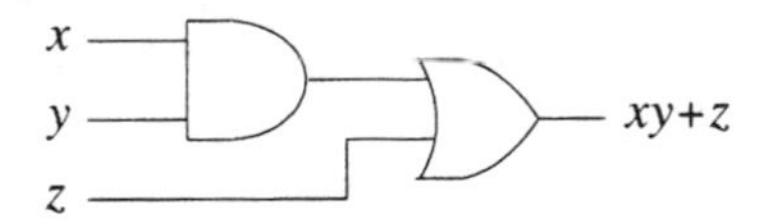

Practice Exercise. Construct a circuit to implement $x'+yz$.

Half adders

We continue with the example of addition. A circuit that carries out the addition of two one-bit variables is called a *half adder*. Following from what we have done so far, it is easy to construct a half adder: take the circuit for $x'y+xy'$ and attach to it a circuit to produce xy:

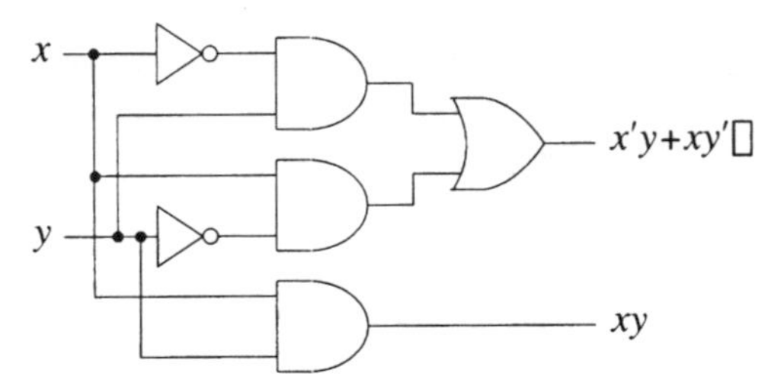

However, a simpler version is available. The given half adder involves six gates (three AND, one OR and two NOT gates). But

$$\begin{aligned} x'y+xy' &= x'y+xx'+xy' \text{ (as } xx'=0) \\ &= x'y+x(x'+y') \\ &= x(x'+y')+yx' \text{ (reordering)} \\ &= x(x'+y')+y(x'+y') \text{ (adding 0 again)} \\ &= (x+y)(x'+y') \\ &= (x+y)(xy)' \text{ (de Morgan's laws).} \end{aligned}$$

This expression is simpler to evaluate. First we find $(x+y)$ (one OR gate) and $(xy)'$ (an AND and a NOT). Then we find their product (another AND). Then we are finished, because the product xy was already computed, earlier in the process. So the number of gates is reduced to four. The corresponding circuit is

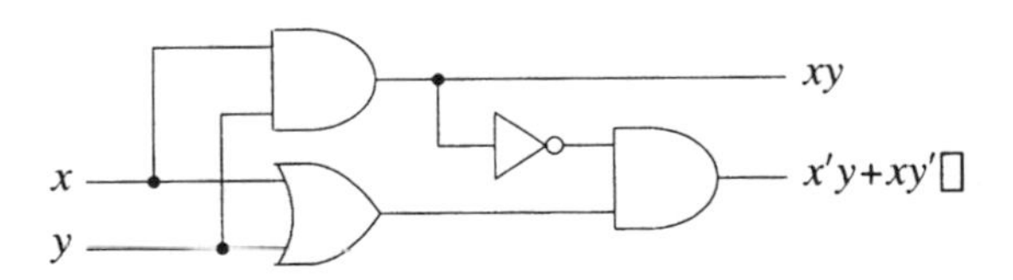

Full adders

Of course, addition in a computer involves summing numbers with more than one binary digit. The way we do addition by hand is: add the units, write down the unit digit, move the carry digit to the tens column; add the tens (there are now three figures to add), write down the new tens digit, move the carry to the hundreds column; and so on. Computers add in the same way. So we need a circuit to add *three* binary digits. Such a circuit is called a *full adder*.

A full adder could be constructed from half adders, but the circuit is cumbersome. Suppose we want to find $x+y+z$. We could find $x+y$, getting the binary number t_1t_0 (this takes one half adder, or four gates), then find $z+t_0$, getting say w_1w_0 (another half adder), and finally add t_1 to w_1. This latter is a little simpler, because t_1 and w_1 cannot both be 1, but still requires an OR gate. So nine gates are required.

There is a simpler way to find $x+y+z$ in binary notation. The units digit will be 1 if exactly one of x, y and z equals 1, or if all three are 1. The "twos" digit will be 1 when two or three of the summands is 1. So the units digit is

$$t_0 = x'y'z + x'yz' + xy'z' + xyz$$

and the twos digit is

$$t_1 = x'yz + xy'z + xyz' + xyz.$$

These two expressions can be simplified as follows. First we observe that

$$t_0 = x'(y'z + yz') + x(y'z' + yz)$$

and

$$t_1 = yz(x + x') + x(y'z + yz') = yz + x(y'z + yz').$$

The expression $\alpha = y'z + yz'$ appears in both equations. Moreover it is not hard to see that $\alpha' = y'z' + yz$. So

$$t_0 = x'\alpha + x\alpha', \; t_1 = yz + x\alpha'.$$

So we first form the sum of y and z with a half adder. The units term of this is α, and the twos term is yz. If we now add this to x, using a half adder, the units term is t_0 and the twos term is $x\alpha$. Finally, t_1 is produced by putting this value $x\alpha$ and the value yz through an AND gate. The circuit is

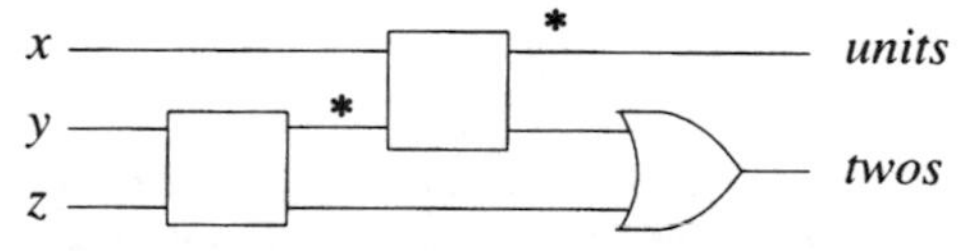

where the square is shorthand for a half adder whose units output is marked by an asterisk. This involves a total of nine gates.

A computer can carry out extended addition by the repeated use of full adders.

General circuits

In general, it is more desirable to build a circuit with fewer gates. To implement any Boolean expression, a first step is to replace the expression by its minimal form. Then, if possible, make any other transformation to reduce the number of gates — usually by using de Morgan's laws to reduce the number of NOT gates.

Sample Problem 3.17 *Find a circuit to implement* $xyzt + xyz't + xy'zt + xy'z't + xyzt' + xyz't' + xy'zt' + xy'z't' + x'yz't'$.

Solution. In Sample Problem 3.14 we saw that the given expression has minimal form $x + y'z't'$. To cut down the number of gates, we note that $y'z't' = (x+y+z)'$. So a suitable circuit is

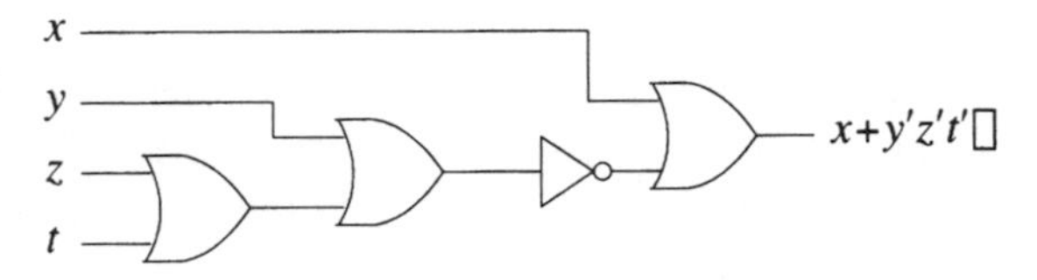

Practice Exercise.

Find a circuit to implement

$$xyz't + xyz't' + xy'zt + xy'zt' + x'yz't + x'yz't' + x'y'zt + x'y'zt'$$

(see Sample Problem 3.14 again).

Exercises 3.4

1. Verify that if $\alpha = y'z + yz'$, then $\alpha' = y'z' + yz$.

In Exercises 2 to 8, *draw circuits to implement the given expression.*

2. $x + y + z$
3. xyz
4. xyz'
5. $x + yz'$
6. $xy + z'$
7. $(x + yz)' + y$
8. $(xy)' + (x + z)'$

In Exercises 9 *to* 12, *draw a circuit to implement the given expression exactly as it is written. Then try to find a simpler circuit to implement the expression.*

9. $(xy')'z$
10. $xy' + (x + z)'$
11. $xy' + (xy)'$
12. $xy'z + x'yz$

13. Find a Boolean expression corresponding to the circuit

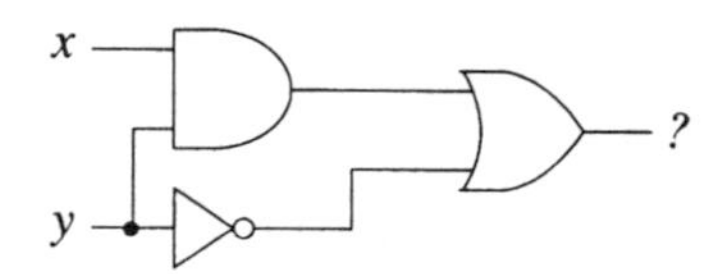

14. Find a Boolean expression corresponding to the circuit

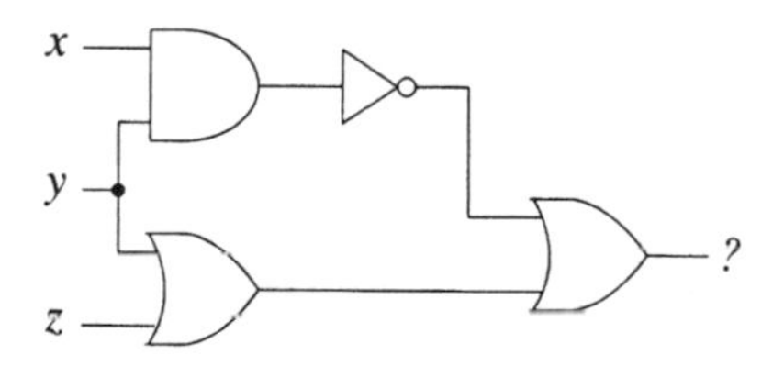

15. Find a Boolean expression corresponding to the circuit

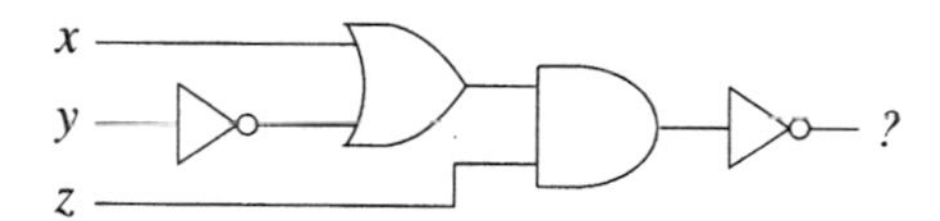

16. Find a Boolean expression corresponding to the circuit

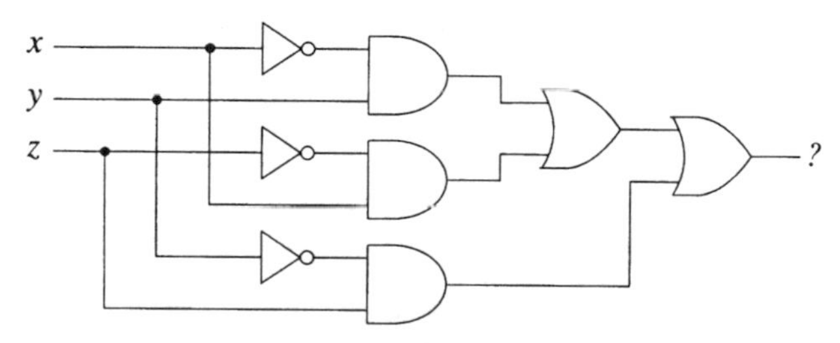

17. Find a Boolean expression corresponding to the following circuit. Simplify the expression, and find a simpler circuit that achieves the same result.

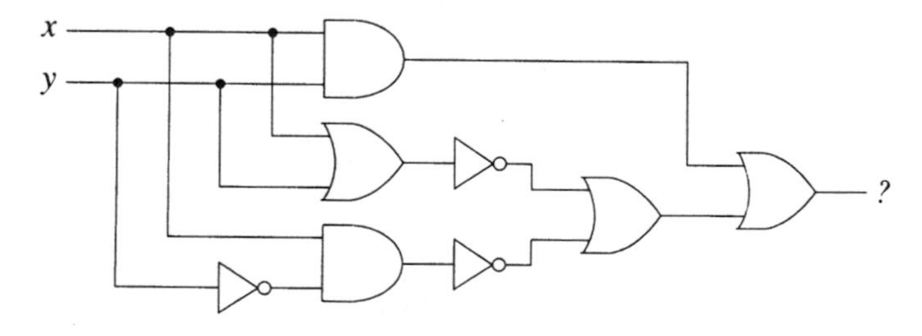

4

Relations and Functions

Functions are important in many areas of mathematics. Elementary algebra starts to differ from arithmetic when the concept of a function is developed. Calculus can be viewed as being the study of functions, and of certain ways of associating new functions with a given one.

We start by discussing binary relations. These can be seen as a generalization of functions, and we lead into functions by looking at special relations.

The reader who has already started to study calculus will find that we focus on slightly different aspects of functions. Our viewpoint may be less algebraic than what you have seen before.

4.1 Relations

First definitions

A (*binary*) *relation* ρ from a set S to a set T is a rule that stipulates, given any element s of S and any element t of T, whether s bears a certain relationship to t (written $s \rho t$) or not (written $s \not\rho t$). For example, if S is the set of living males and T is the set of living females, the relation ρ might be "is the son of"; if s denotes a certain man and t denotes a certain woman, we write $s \rho t$ if s is the son of t, and $s \not\rho t$ otherwise.

Binary relations are very common in mathematics. We have aleady seen relations such as $\leq$, $=$ and $<$, and all of these are relations from $\mathbb{R}$ to $\mathbb{R}$ (or from $\mathbb{Z}$ to $\mathbb{Z}$, or ...). The relation $|$, or "divides," is a relation from $\mathbb{Z}^+$ to $\mathbb{Z}^+$, or from $\mathbb{Z}^+$ to $\mathbb{Z}$. If $\mathcal{P}$ and $\mathcal{Q}$ are any collections of sets, then $\subseteq$ and $\supseteq$ are relations from $\mathcal{P}$ to $\mathcal{Q}$.

Alternatively, we can define a binary relation ρ from the set S to the set T as a set ρ of ordered pairs (s,t), where s belongs to S and t belongs to T, with the notation that $s \rho t$ when $(s,t) \in \rho$ and $s \not\rho t$ otherwise. This means that, formally, a binary relation from S to T can be defined as a subset of $S \times T$, the Cartesian product of S and T defined in Section 2.4.

Sample Problem 4.1 *Suppose $S = \{1,2,3\}$ and $T = \{1,2,3,4\}$. Find the sets corresponding to $s < t$ and $s^2 = t$, in the usual arithmetical sense.*

Solution. The relation $s < t$ corresponds to the set L, defined by

$$L = \{(1,2),(1,3),(1,4),(2,3),(2,4),(3,4)\}.$$

$s^2 = t$ has

$$R = \{(1,1),(2,4)\}$$

as its corresponding set.

Practice Exercise. What are the sets corresponding to $s \leq t$ and $s \geq t$?

New relations from old

Sometimes it is useful to be able to combine relations. If we have two binary relations ρ and σ, each from S to T, we can define two further binary relations, their union $\rho \cup \sigma$ and their intersection $\rho \cap \sigma$, from S to T in the following way:

$s(\rho \cup \sigma)t$ if and only if $s \rho t$ or $s \sigma t$ or both;
$s(\rho \cap \sigma)t$ if and only if both $s \rho t$ and $s \sigma t$.

In other words, $(s,t) \in \rho \cup \sigma$ if and only if $(s,t) \in \rho$ or $(s,t) \in \sigma$ or both, and $(s,t) \in \rho \cap \sigma$ if and only if $(s,t) \in \rho$ and $(s,t) \in \sigma$, so that the notation agrees exactly with the usual set-theoretic notation.

We can also construct new relations from old ones by *composition*. If τ is a binary relation from R to S, and σ a binary relation from S to T, then the composition $\tau\sigma$ is a binary relation from R to T, where $r(\tau\sigma)t$ if and only if there exists at least one element $s \in S$ such that $r \tau s$ and $s \sigma t$.

If τ is a binary relation from R to S, and ν a binary relation from P to T, then the composition $\tau\nu$ is not defined unless $P = S$.

Sample Problem 4.2 *Suppose $R = \{1,2\}$, $S = \{3,4,5\}$, and $T = \{6,7\}$. The binary relations τ from R to S, ρ from S to T and σ from S to T are* *defined by*

$$\begin{aligned}\tau &= \{(1,3),(2,4),(2,5)\},\\ \rho &= \{(3,6),(3,7),(5,7)\},\\ \sigma &= \{(3,6),(4,6),(4,7)\}.\end{aligned}$$

What are the binary relations $\tau\rho$, $\rho\sigma$ *and* $\rho\cap\sigma$*?*

Solution.

$$\begin{aligned}\tau\rho &= \{(1,6),(1,7)\},\\ \rho\sigma &\text{ is not defined as } T \text{ is not equal to } S\\ \rho\cap\sigma &= \{(3,6)\}.\end{aligned}$$

Practice Exercise. What are the binary relations $\rho\tau$, $\tau\sigma$ and $\rho\cup\sigma$?

Theorem 4.1 *Composition of binary relations is associative: if* α, β *and* γ *are given binary relations and if either of the compositions* $\alpha(\beta\gamma)$ *and* $(\alpha\beta)\gamma$ *exists, then so does the other, and* $\alpha(\beta\gamma) = (\alpha\beta)\gamma$.

Proof. Let Q,R,S,T be sets where a $\alpha \subseteq Q\times R$, $\beta \subseteq R\times S$ and $\gamma \subseteq S\times T$. Suppose there are some elements $q \in Q$ and $t \in T$ that satisfy $q\,(\alpha\beta)\,\gamma t$. Then by the definition of composition, there exists an $s \in S$ such that $q\alpha\beta s$ and $s\gamma t$. Again by the definition, there exists $r \in R$ such that $q\alpha r$ and $r\beta s$. But now since $r\beta s$ and $s\gamma t$, it follows that $r\beta\gamma t$, and since $q\alpha r$, we have $q\alpha(\beta\gamma)t$. Hence $(\alpha\beta)\gamma \subseteq \alpha(\beta\gamma)$. A similar argument shows that $\alpha(\beta\gamma) \subseteq (\alpha\beta)\gamma$ and therefore the two compositions are equal. □

Every binary relation has an *inverse relation*: if α is a binary relation from S to T, so that $\alpha \subseteq S\times T$, then α^{-1} is a binary relation from T to S, or $\alpha^{-1} \subseteq T\times S$, and α is defined by

$$\alpha^{-1} = \{(t,s) : (s,t) \in \alpha\}.$$

In other words, $t\alpha^{-1}s$ if and only if $s\alpha t$.

Sample Problem 4.3 *What is the inverse relation of* τ*, in the preceding Sample Problem?*

Solution. $\tau^{-1} = \{(3,1),(4,2),(5,2)\}$.

Practice Exercise. What is the inverse relation of ρ, in the preceding Sample Problem?

The inverse of a composition may be expressed as a composition of inverses.

Theorem 4.2 *If* α *and* β *are binary relations from* R *to* S *and from* S *to* T *respectively, then* $(\alpha\beta)^{-1} = \beta^{-1}\alpha^{-1}$.

Proof. $\alpha\beta$ is a binary relation from R to T, and hence $(\alpha\beta)^{-1}$ is a binary relation from T to R.

Suppose $(t,r) \in (\alpha\beta)^{-1}$. Then $(r,t) \in \alpha\beta$. Hence for some $s \in S$, we have $(r,s) \in \alpha$ and $(s,t) \in \beta$, so $(t,s) \in \beta^{-1}$ and $(s,r) \in \alpha^{-1}$. So $(t,r) \in \beta^{-1}\alpha^{-1}$, and therefore $(\alpha\beta)^{-1} \subseteq \beta^{-1}\alpha^{-1}$.

Reversing the argument at each step shows that $\beta^{-1}\alpha^{-1} \subseteq (\alpha\beta)^{-1}$, so the two relations are equal. □

Relations on a set

Suppose α is a binary relation from a set A to A itself or, in other words, $\alpha \subseteq A \times A$. Then we say that α is a binary relation *on* A. One special relation on a set is the *identity* relation: ι_A is the identity relation on A if $a\,\iota_A\,b$ is true precisely when $b = a$, that is, $\iota_A = \{(a,a) : a \in A\}$.

There are several important properties that a relation on a set may (or may not) have. Suppose α is a relation on A.

(i) α is called *reflexive* if and only if $a\alpha a$ for every $a \in A$, or, as we would usually say, α *is always true*. α is *irreflexive* if and only if $a\alpha a$ is *never* true for any element a of A.

(ii) α is called *symmetric* if and only if $a\alpha b$ implies $b\alpha a$ or, in other words, if $(a,b) \in \alpha$ implies $(b,a) \in \alpha$. Since, by the definition of inverse relation, we know that $a\alpha b$ if and only if $b\alpha^{-1}a$, we could say that α is symmetric if and only if $\alpha = \alpha^{-1}$. α is called *antisymmetric* if and only if $a\alpha b$ and $b\alpha a$ together imply $a = b$ or, in other words, $\alpha \cap \alpha^{-1} \subseteq \iota_A$. If $a\alpha b$ and $b\alpha a$ can never both be true, then α is called *asymmetric*, or sometimes *skew*. (If $a\alpha b$ and $b\alpha a$ are never both true, then $(a\alpha b \wedge b\alpha a \rightarrow a\alpha a$ is vacuously true, so all asymmetric relations are antisymmetric, but this gives us no information.)

(iii) α is called *transitive* if and only if $a\alpha b$ and $b\alpha c$ together imply $a\alpha c$, and *atransitive* if and only if, whenever $a\alpha b$ and $b\alpha c$, $a \not\alpha c$, or in other words if and only if $\alpha \cap \alpha\alpha = \emptyset$.

> **Sample Problem 4.4** *What are the reflexive, symmetric and transitive properties of $<$ and $\leq$ on the set $\mathbb{Z}$?*
>
> **Solution.** $<$ is irreflexive ($a < a$ is never true) while $>$ is reflexive ($a \leq a$ is always true). Both relations are antisymmetric, but $<$ is in fact asymmetric. Both are transitive.
>
> **Practice Exercise.** What are the reflexive, symmetric and transitive properties of $=$ and $\geq$ on $\mathbb{Z}$?

The relation of *adjacency* on the set of integers, defined by $i\alpha j$ if and only if $|i-j| = 1$ for i and j positive integers, is irreflexive (since $|i-i| = 0$), symmetric (since $|i-j| = |j-i|$) and not transitive (since $1\alpha 2$ and $2\alpha 3$ but $1 \not\alpha 3$). It is in fact atransitive, for $i \alpha j$ and $j \alpha k$ together imply $i = j \pm 1$ and $k = j \pm 1$, so $i = k$ or $i = k \pm 2$, but i cannot equal $c \pm 1$. So $a \not\alpha c$.

The properties of relations can be exhibited even in quite small examples. For example, say $A = \{1,2,3\}$ and consider the relation β on A, defined by

$$\beta\{(1,1),(2,2)\}.$$

Since $\iota_A \not\subseteq \beta$, it is not reflexive, but since $\iota_A \cap \beta \neq \emptyset$, it is not irreflexive either. Since $\beta = \beta^{-1}$, it is symmetric, and since $\beta \cap \beta^{-1} \subseteq \iota A$, it is also antisymmetric. Since $\beta\beta = \beta$, it is transitive.

If α and β are relations on A and $\beta \subseteq a$, then we say α *covers* β. Some of the properties we have defined can be described in terms of covering. For example, if α is a relation on A, then α is *reflexive* if and only if it covers ι_A. By the definition of composition we know that $a\alpha b$ and $b\alpha c$ together imply that $a\alpha\alpha c$, so α is transitive if and only if it covers $\alpha\alpha$.

Sample Problem 4.5 *Let $A = \{1,2,3\}$ and define the relation γ on A by*

$$\gamma = \{(1,1),(2,2),(3,3),(1,2),(2,3)\}.$$

What are the reflexive, symmetric and transitive properties of γ?

Solution. Since $\iota_A \subseteq \gamma$, γ covers ι_A, so it is reflexive; since $\gamma^{-1} = \{(1,1),(2,2),(3,3),(2,1),(3,2)\}$ and $\gamma^{-1} \cap \gamma = \iota_A$, it is antisymmetric; since $1\gamma 2$ and $2\gamma 3$ but $1 \not\gamma 3$, it is not transitive, but since $\gamma \subseteq \gamma\gamma$, it is not atransitive either.

Practice Exercise. Let $A = \mathbb{Z} \cup \{\sqrt{2}\}$ and define the relation δ on A by $a\delta b$ if and only if $a+b \in \mathbb{Z}$. Show that $\delta = \mathbb{Z} \times \mathbb{Z}$. What are the reflexive, symmetric and transitive properties of δ?

Exercises 4.1

In Exercises 1 *to* 6, *a binary relation from $S = \{1,2,3,4,5,6,7,8,9\}$ to itself is defined. For each of these relations, what is the corresponding subset of $S \times S$? For each relation, what is the inverse relation?*

1. $x\alpha y$ means $x = y^2$.

2. $x\beta y$ means $x+y = 9$.

3. $x\gamma y$ means x divides y.

4. $x\delta y$ means $2x < y$.

5. $x\varepsilon y$ means $x^2 + y = 12$.

6. $x\phi y$ means $x - y = 0$.

7. Relations α and β are binary relations from $S = \{1,2,3\}$ to $T = \{1,2,3,4\}$ defined by

$x \alpha y$ means $x = y^2$;

$x \beta y$ means $x = y - 1$.

For each of these relations, what is the corresponding subset of $S \times T$?

8. Suppose ρ and σ are relations defined on $S = \{1,2,3,4\}$ by the sets

$\rho = \{(2,2),(3,2),(3,4),(4,1),(4,3)\}$;

$\sigma = \{(2,3),(1,2),(3,1),(3,2),(4,4)\}$.

Find the set form of the composition $\rho\sigma$.

9. Given ρ and σ as in Exercise 8, what is the set form of the relation $\sigma\rho$?

10. Given ρ and σ as in Exercise 8, what are the set forms of the relations $\rho\rho$ and $\sigma\sigma$?

11. Given ρ as in Exercise 8, what is the set form of the relation $\rho\rho\rho$?

12. Let μ and ν be two binary relations from the set $\mathbb{Z}$ of integers to itself, defined by

$$\mu = \{(m,2m) \mid m \in \mathbb{Z}\}$$

and

$$\nu = \{(n,3n) \mid \cap \in \mathbb{Z}\}.$$

What is the composite relation $\mu\nu$?

13. Let α and β be any two given binary relations from A to B such that $\alpha \subseteq \beta$ or, in other words, such that $a \alpha b$ implies $a \beta b$.

(i) Show that if γ is a binary relation from B to C, then $\alpha\gamma \subseteq \beta\gamma$.

(ii) Show that if δ is a binary relation from D to A, then $\delta\alpha \subseteq \delta\beta$.

In Exercises 14 *to* 20, *prove that the indicated identity holds for any binary relations* α *and* β *from* A *to* B *and* γ *from* C *to* A.

14. $(\alpha^{-1})^{-1} = \alpha$.

15. $\alpha^{-1} = \beta^{-1}$ if and only if $\alpha = \beta$.

16. $\alpha^{-1} \subseteq \beta^{-1}$ if and only if $\alpha \subseteq \beta$.

17. $(\alpha \cap \beta)^{-1} = \alpha^{-1} \cap \beta^{-1}$.

18. $(\alpha \cup \beta)^{-1} = \alpha^{-1} \cup \beta^{-1}$.

19. $\gamma(\alpha \cup \beta) = \gamma\alpha \cup \gamma\beta$.

20. $\gamma(\alpha \cap \beta) \subseteq \gamma\alpha \cap \gamma\beta$.

21. If we try to prove equality in Exercise 20, where does the proof break down? (Hint: consider the sets $A = \{2,3\}, B = \{4\}, C = \{1\}$, and the relations $\gamma = \{(1,2),(1,3)\}$, $\alpha = \{(2,4)\}$, $\beta = \{(3,4)\}$.)

22. The relation ρ is defined on $\{a,b,c,d\}$ by

$$\rho = \{(a,a),(b,b),(b,c),(c,c),(c,d),(d,c)\}.$$

Prove that ρ has none of the properties: reflexive, transitive, symmetric, antisymmetric.

23. If α is a relation from A to B, then we can treat $a \not\alpha b$ as defining a new binary relation $\not\alpha$, called the *complement* of α, from A to B. In terms of sets, $\not\alpha = (A \times B)\backslash\alpha$. Show that $\not\alpha^{-1}$ is the complement of α^{-1}.

24. Consider the relations α and β of Exercise 7. What are the sets $\not\alpha$ and $\not\beta$?

25. Define the relations α, β and γ on the set $\{1,2,3\}$ by the sets

$$\alpha = \{(1,1),(2,2),(3,3)\};$$
$$\beta = \{(1,3),(2,2),(3,1)\};$$
$$\gamma = \{(2,3),(1,2),(1,3),(3,2)\}.$$

Which of these relations are

(i) reflexive;
(ii) irreflexive;
(iii) symmetric;
(iv) antisymmetric;
(v) asymmetric;
(vi) transitive?

Exercises 26 to 33 show relations from the set $\mathbb{Z}^+$ of positive integers to itself. Determine whether each relation has each of the following properties: reflexivity, symmetry, antisymmetry, transitivity.

26. m is divisible by n.

27. $m+n \geq 50$.

28. $m+n$ is even.

29. $m+n$ is odd.

30. mn is even.

31. mn is odd.

32. $3|(m+n)$.

33. m equals some positive integer power of n.

34. A group of people, denoted by $A, B, \ldots$, are at a party. Consider the following relations: is each reflexive, symmetric, antisymmetric, asymmetric or transitive?

(i) $A\,\alpha\,B$ means A and B spoke to each other at the party.
(ii) $A\,\beta\,B$ means A spoke to B at the party.
(iii) $A\,\gamma\,B$ means A and B come from the same city.
(iv) $A\,\delta\,B$ means A and B met for the first time at the party.

(v) $A \varepsilon B$ means A is the father of B.

35. Find the error in the following " proof."

Theorem. *A symmetric and transitive relation must be reflexive.*

Proof. Let α be a symmetric and transitive relation on a set A. If $a \alpha b$, then $b \alpha a$ because a is symmetric. But $a \alpha b$ and $b \alpha a$ together imply $a \alpha a$, since α is transitive. Therefore α is reflexive.

36. Show that if α is reflexive (respectively irreflexive, symmetric, antisymmetric, transitive, atransitive), then so is α^{-1}.

37. Let α and β be relations on a set A. Show that

(i) if α and β are reflexive, so is $\alpha\beta$;

(ii) if α and β are symmetric, then $\alpha\beta$ is symmetric if and only if $\alpha\beta = \beta\alpha$;

(iii) if α and β are antisymmetric, $\alpha\beta$ is not in general antisymmetric;

(iv) if α and β are transitive, $\alpha\beta$ is transitive if $\alpha\beta = \beta\alpha$.

4.2 Some Special Kinds of Relations

Equivalence relations

Let α be a relation on a set A. Then α is said to be an *equivalence relation* on A if and only if it is reflexive, symmetric, and transitive.

The obvious equivalence relation is equality, on any set. In sets other than number sets, equal objects are often called "equal in all respects." More generally, an equivalence relation can be considered as a statement that two objects are "equal in some (specified) respects." One example, on the integers, is the relation α, where $a \alpha b$ is true if and only if $a = \pm b$ — a and b have the same absolute value. Another, which you have probably seen in high school geometry, is congruence, on the set of all plane triangles.

The word "congruence" is also used for another important equivalence relation.

Sample Problem 4.6 *The relation "congruence modulo n" for a positive integer n is defined on the set $\mathbb{Z}$ of integers by specifying that a is congruent to b modulo n, written "$a \equiv b \pmod{n}$," if and only if $n \mid (a-b)$. Show that congruence modulo n is an equivalence relation.*

Solution. Since $n \mid (a-a)$, the relation is reflexive; if $n \mid (a-b)$ then $n \mid (b-a)$, so the relation is symmetric; if $n \mid (a-b)$ and $n \mid (b-c)$ then we have $a-b = kn$, and $b-c = ln$, for some $k, l \in \mathbb{Z}$, hence $a-c = (k+l)n$ and $n \mid (a-c)$, which shows that the relation is transitive. So congruence modulo n is an equivalence relation on the integers.

Practice Exercise. Let $\mathcal{S}$ be the set of all geometrical solids. Write $a\sigma b$ to mean that a and b have the same volume. Show that σ is an equivalence relation on $\mathcal{S}$.

The relation of congruence modulo a positive integer will be discussed at length in Section 9.2.

There is a very important connection between equivalence relations and partitions of a set. Suppose S is a set and α is an equivalence relation on S. Given $s \in S$, we define

$$[s] = \{t : t \in S, t\alpha s\}.$$

Then $[s]$ is called the *α-class* of s or the *equivalence class* of s (under α). By the reflexive law, $[s]$ is never empty: it always contains s.

If two or more equivalence relations are being discussed, we write $[s]_\alpha$ to denote the equivalence class under the relation α.

Sample Problem 4.7 *What are the equivalence classes of congruence mod 5 on $\mathbb{Z}$?*

Solution. $a \equiv b$ means $5 \mid (a-b)$, so two numbers are in the same equivalence class if and only if they leave the same remainder on division by 5. So

$$[a] = \{\dots, a-10, a-5, a, a+5, a+10, \dots\},$$

which is sometimes abbreviated to $a + 5\mathbb{Z}$. There are five classes, namely $[0]$, $[1]$, $[2]$, $[3]$, $[4]$.

Practice Exercise. What are the equivalence classes of the relation $a = \pm b$ on $\mathbb{Z}$?

Theorem 4.3 *Suppose α is an equivalence relation on S. Then*

(i) *$s\alpha t$ if and only if $[s] = [t]$;*

(ii) *$t \in [s]$ if and only if $[t] = [s]$;*

(iii) *given $s, t \in S$, either $[s] \cap [t] = \emptyset$ or $[s] = [t]$;*

(iv) *the set of disjoint equivalence classes partitions the set S.*

Proof. (i) Suppose $s\alpha t$. Let $x \in [s]$. By the definition of equivalence class $x\alpha s$, and hence by transitivity $x\alpha t$. Again by the definition of equivalence class, $x \in [t]$. Hence $[s] \subseteq [t]$. But by symmetry, $s\alpha t$ implies $t\alpha s$, so a repetition of the above argument shows that $[t] \subseteq [s]$. Hence $[s] = [t]$.

Conversely, let $[s] = [t]$. By reflexivity $s \in [s]$ so $s \in [t]$ and $s\alpha t$.

(ii) Let $t \in [s]$. This means that $t \subseteq s$ and by (i) $[t] = [s]$. Conversely, if $[t] = [s]$, then by (i) $t \subseteq s$ and we have $t \in [s]$.

(iii) Given $s, t \in S$, either $[s] \cap [t] = \emptyset$ or $[s] \cap [t] \neq \emptyset$. In the former case, the theorem is satisfied; in the latter case, there must exist an element $x \in S$ such that

$x \in [s] \cap [t]$. By (iii), since $x \in [s]$ we have $[x] = [s]$. Similarly $[x] = [t]$. Hence $[s] = [t]$.

(iv) For every $s \in S$, we have $s \in [s]$. Hence the union of all the equivalence classes contains the whole of S. But by (iii), distinct equivalence classes are disjoint, which proves the theorem. □

Theorem 4.3 shows that every equivalence relation on a set induces a partition of that set. Next we show that the converse of this theorem is also true.

Theorem 4.4 *Consider a partition of the set S given by*

$$S = S_1 \cup \cdots \cup S_n.$$

Define a relation α on the set S as follows: $s\alpha t$ if and only if s and t both belong to the same set S_i of the partition. Then α is an equivalence relation on S.

Proof. We must verify the three properties of an equivalence relation. First, since any element s belongs to exactly one set of the partition, we have $s\alpha s$ and the relation is reflexive. Next, if $s\alpha t$, then s and t belong to the same set of the partition, so that we also have $t\alpha s$ and the relation is symmetric. Finally, if $r\alpha s$ and $s\alpha t$, then r, s and t must all belong to the same set of the partition, so that $r\alpha t$ and the relation is transitive. □

Order relations

A relation α on a set A is called a *partial order relation* on A if and only if it is antisymmetric and transitive. If α is also reflexive, it is said to be a *weak* partial order; if it is irreflexive, it is called a *strong* partial order.

We say two elements of a set A are *comparable* under α if either $a\alpha b$ or $b\alpha a$ is true. A partial order relation for which every pair of elements is comparable is called a *total* order.

Familiar examples of order relations include the relations $<$ and $\leq$ on the set $\mathbb{Z}$ of integers (or on other sets of real numbers). Both relations are transitive; $<$ is asymmetric and $\leq$ is antisymmetric. So $<$ is a strong partial order and $\leq$ is a weak partial order. If a and b are real numbers and $a \neq b$, then either $a < b$ or $b < a$, and similarly $a \leq b$ or $b \leq a$, so both are total orders.

Another important example is the relation of containment, denoted by $\subseteq$, on the power set $\mathcal{P}(S)$ of a set S. If $A \subseteq B$ and $B \subseteq A$, then $A = B$, so $\subseteq$ is antisymmetric; if $A \subseteq B$ and $B \subseteq C$, then $A \subseteq C$, so $\subseteq$ is transitive. So containment is a partial order. Since $A \subseteq A$, containment is a weak partial order. If S contains at least two elements, say s_1 and s_2, then $\{s_1\} \not\subseteq \{s_2\}$ and $\{s_2\} \not\subseteq \{s_1\}$, so the order is not total. (However if S is a one-element set, $S = \{s\}$, so that $\mathcal{P}(S) = \{\emptyset, S\}$, then the order is total.)

Sample Problem 4.8 *Consider the relation* $|$ *on the set* $\mathbb{Z}^+$ *of positive integers:* $a \mid b$ *means that for some* $n \in \mathbb{Z}^+, b = na$. *Show that* $|$ *is a weak partial order, and is not total.*

Solution. If $a \mid b$ and $b \mid a$, then $a = b$, so $|$ is antisymmetric. If $a \mid b$ and $b \mid c$, say $b = n_1 a, c = n_2 b$ for some positive integers n_1 and n_2, then $b = n_1 a, c = n_1 n_2 b$ and $a \mid c$, so $|$ is transitive and is a partial order relation. Since $a \mid a$, it is weak. Since 2 does not divide 3 and 3 does not divide 2, the integers 2 and 3 are not comparable, it is not a total order.

Practice Exercise. Write $a \wr b$ to mean "$a \mid b$ but $a \neq b$." Show that $\wr$ is a partial order on $\mathbb{Z}^+$. Is it weak or strong? Is it total?

Suppose α is a relation on A and suppose $a_1 \alpha a_2, a_2 \alpha a_3, \ldots, a_{k-1} \alpha a_k$ and $a_k \alpha a_1$ are all true, where the elements $a_1, a_2, a_3, \ldots, a_k$ are all different. Then we say α contains a *cycle of length k*.

Theorem 4.5 *No order relation contains a cycle of length greater than* 1.

Proof. Let α be an order relation on the set A. We may have $a \alpha a$ for some $a \in A$, and this is a cycle of length 1. There cannot be a cycle of length 2, for that would mean that $a \alpha b$ and $b \alpha a$ for some $a \neq b$, contradicting the antisymmetry of α.

Consider the set L of all lengths, other than 1, of cycles in α. Suppose L is not empty. Then we are led to a contradiction as follows. There must be a least element $k \in L$, and because $2 \notin L$, we know $k \geq 3$. So for some elements $a_1, a_2, \ldots, a_k$ of A, $a_1 \alpha a_2, a_2 \alpha a_3, \ldots, a_{k-1} \alpha a_k$, and $a_k \alpha a_1$, where $k \geq 3$ and the only shorter cycles have length 1. But α is transitive, so a $a_{k-1} \alpha a_k$ and $a_k \alpha a_1$ together imply $a_{k-1} \alpha a_1$. So α contains a cycle of length $k-1$, and $2 \leq k-1 < k$. This contradicts the choice of k. So $L = \emptyset$, and no cycles of length 2 or longer exist. □

Predecessor-successor relations

Relations that are irreflexive, asymmetric and atransitive are called *predecessor-successor relations*. An example is the relation "is the daughter of" on the set of all women. No one is her own daughter, so the relation is irreflexive; if Amy is the daughter of Belle, then Belle can never be the daughter of Amy, so the relation is asymmetric; if Belle is the daughter of Claire, and Amy is the daughter of Belle, then Amy is not the daughter of Claire, so the relation is atransitive.

Exercises 4.2

In Exercises 1 *to* 4, *is the relation* α *an equivalence relation on the set* S?

1. $S = \mathcal{R}$, $a \alpha b$ means $ab \geq 0$.

2. $S = \mathcal{R}$, $a \alpha b$ means $ab > 0$.

3. S is the set of all non-zero real numbers, $a \alpha b$ means $ab > 0$.

4. $S = \mathbb{Z}^+$, $a \alpha b$ means a and b have greatest common divisor 16.

In Exercises 5 *to* 7, *is the relation* β *a partial order on the set* S? *For each partial order state whether it is weak or strong, total or not total.*

5. $S = \mathbb{Z}$; $a \beta b$ means $a \mid b$.

6. $S = \mathbb{Z}$; $a \beta b$ means $a \wr b$ (as defined in Sample Problem 4.8).

7. S is the English alphabet; $a \beta b$ means a comes before b is the usual alphabetical order. (Don't be confused; a and b do not necessarily mean the letters "a" and "b" here.)

8. Which, if any, of the relations in Exercises 4.1.26 to 4.1.34 are equivalence relations?

9. Suppose A is the set $\mathbb{Z} \times \mathbb{Z}$. Prove that the relation α on A, defined by

$$(x, y)\, \rho\, (z, t) \text{ if and only if } x + t = y + z,$$

is an equivalence relation.

10. Prove that the relation β on $\mathbb{Z}^0 \times \mathbb{Z}^0$, defined by

$$(x, y)\, \beta\, (z, t) \text{ if and only if } xt = yz,$$

is an equivalence relation. Would this remain true if $\mathbb{Z}^0$ were replaced by $\mathbb{Z}$?

11. A relation ρ on a set S is said to be *circular* if and only if for every $x, y, z \in S$, $x \rho y$ and $y \rho z$ together imply that $z \rho x$. Show that a reflexive circular relation is an equivalence relation. Give an example of a circular relation that is not an equivalence relation.

12. Suppose ρ is a reflexive and transitive relation on a set S. Define a new relation σ as follows: $a \sigma b$ is true if and only if both $a \rho b$ and $a \rho a$ are true. Show that σ is an equivalence relation.

13. Consider the set $\mathbb{Z}$ of integers and its power set $P(\mathbb{Z})$. Define a relation ρ on $P(\mathbb{Z})$ by

$$A \rho B \text{ if and only if } A + B \text{ is finite,}$$

where $A, B \subseteq \mathbb{Z}$ and $+$ is the symmetric difference of sets (defined in Section 2.4). Show that ρ is an equivalence relation.

14. Let A be a finite set. Define a relation ρ on $P(A)$ by $B_i \rho B_j$ if and only if there exists a function $f_{ij} : B_i \to B_j$ such that f is one-one and onto, where $B_i, B_j \subseteq A$. Show that ρ is an equivalence relation on $P(A)$.

15. Let α and β be equivalence relations on a set A.

(i) Show that $\alpha\beta$ is an equivalence relation if and only if $\alpha\beta = \beta\alpha$.

(ii) Show that $\rho = \alpha \cap \beta$ is an equivalence relation.

(iii) Is $\alpha \cup \beta$ necessarily an equivalence relation?

(iv) Let α and β be congruence modulo m and n respectively, on $\mathbb{Z}$. Let $\rho = \alpha \cap \beta$. Find the smallest positive integer equivalent to 0 under ρ.

16. Let α and β be equivalence relations on A and let $\rho = \alpha \cap \beta$. By Exercise 15(ii), ρ is an equivalence relation. Describe the partition induced by ρ in terms of those induced by α and β.

17. (i) Consider the relation "congruence modulo 7" (written $a \equiv b(\text{mod } 7)$) defined on $\mathbb{Z}$. Into how many classes does this relation partition $\mathbb{Z}$?

(ii) Now define a relation ρ on $\mathbb{Z}$ by $a \rho b$ if and only if $a^2 \equiv b^2(\text{mod } 7)$. Show that ρ is an equivalence relation. Into how many classes does ρ partition $\mathbb{Z}$?

18. Let Π_1 and Π_2 be two partitions of a given set U. We say Π_1 *is finer than* Π_2, and write $\Pi_1 \prec \Pi_2$, if and only if $X \in \Pi_1$ implies $X \subseteq Y$ for some $Y \in \Pi_2$.

(i) Show that $\prec$ is a weak partial ordering on the set of partitions of U for any set U.

(ii) Show that $\prec$ cannot be a total ordering when $|U| \geq 3$. What is the situation when $|U| = 2$?

19. Suppose ρ is a partial order on a set A. Show that its inverse relation ρ^{-1} is also a partial order on A.

20. Show that the relation $\triangleleft$, where $a \triangleleft b$ means $b = a + 1$, is a predecessor-successor relation on $\mathbb{Z}^0$.

4.3 Functions

Functions arise in all parts of mathematics. The most familiar ones are simple algebraic functions such as linear and quadratic functions, exponentials and logarithms. Other number-based functions arise in trigonometry and calculus. We need to discuss functions in a more general way.

Definition of a function

A *function*, or *mapping*, f from a set S to a set T, is a rule that associates with each member of S exactly one member of T. In other words, a function is a special kind of relation from S to T, one in which each element of S occurs as the first element of *precisely one* ordered pair. This uniqueness property is the main point of the idea of a function.

The set S is called the *domain* of the function, and the set T its *codomain*. If we used notation consistent with that of the last section, we would write $f \subseteq S \times T$, but more often when dealing with functions we write $f : S \longrightarrow T$; the set form is then called the *defining set* or *associated set* of f. To indicate the correspondence between elements, we could again write $(s,t) \in f$, but more often we write $f(s) = t$, as for example in familiar algebraic functions like $f(x) = x^3 - 3x + 1$. This is referred to as *left notation*. In algebraic and set-theoretic contexts it is also common to use *right notation*, and we would write xf instead of $f(x)$; you will find both notations in books.

The *image* of x under f is the value $f(x)$; if R is any subset of S, then $f(R) = \{f(r) : r \in R\}$ is the set of images of elements of R and is called the *image* of R. In particular $f(S)$, the image of S, is called the *range* of f and necessarily $f(S) \subseteq T$. The range and domain of f will be denoted $\mathcal{R}(f)$ and $\mathcal{D}(f)$ respectively. To avoid trivial cases, it is usual to require that $\mathcal{D}(f)$ should not be empty.

The *composition* of two functions is defined just as the composition of relations is. In functions, it can be more simply described by saying "first perform one function, then the other" (where *performing* f means replacing the value x by $f(x)$). The composition "first f, then g" is written $g(f(x))$ in left notation, and this can be interpreted as defining a new function $g(f)$, by the natural rule

$$g(f)(x) = g(f(x)).$$

In order for $g(f)(x)$ to be defined, it is necessary that every value of $f(x)$ be in the domain of g. This will not necessarily be true for all x in $\mathcal{D}(x)$; in fact, the domain of $g(f)$ is

$$\mathcal{D}(g(f)) = \{x : x \in \mathcal{D}(f), f(x) \in \mathcal{D}(g)\}.$$

This method of combining functions is *not* commutative; quite possibly $g(f)$ and $f(g)$ will be different functions. In fact, one might be defined and the other not. However, composition is associative, because it is just a special case of composition of relations.

In right notation, we would write xfg instead of $g(f)(x)$; in either notation, the function to be applied first is written closest to the variable.

Sample Problem 4.9 *S is the set of all real numbers other than 0 and 1. Consider the following functions from S to S:*

$$f_1(x) = x; \qquad f_2(x) = \frac{1}{x}; \qquad f_3(x) = 1 - x;$$
$$f_4(x) = \frac{1}{1-x}; \quad f_5(x) = \frac{x}{x-1}; \quad f_6(x) = \frac{x-1}{x}.$$

Find formulae for $f_2(f_2)$, $f_3(f_5)$ and $f_5(f_3)$, and verify that they are all members of the original set of six functions. Use these to verify that composition is not commutative.

Solution.

$$\begin{aligned} f_2(f_2)(x) &= f_2(\frac{1}{x}) = \frac{1}{1/x} = x = f_1(x); \\ f_3(f_5)(x) &= f_3(\frac{x}{x-1}) = 1 - \frac{x}{x-1} = \frac{1}{1-x} = f_4(x); \\ f_5(f_3)(x) &= f_5(1-x) = \frac{1-x}{1-x-1} = \frac{x-1}{x} = f_6(x). \end{aligned}$$

So $f_2(f_2) = f_1$, $f_3(f_5) = f_4$ and $f_5(f_3) = f_6$.

Observe that $f_5(f_3) \neq f_3(f_5)$, an example of non-commutativity.

Practice Exercise. Find formulae for $f_1(f_1)$, $f_2(f_4)$ and $f_5(f_6)$, and verify that they are all members of the original set of six functions.

One-one and onto functions

Two properties of functions are particularly important: we say that a function f is *one-to-one* if $s_1 f = f(s_2)$ always implies $s_1 = s_2$ for any $s_1, s_2 \in \mathcal{D}(f)$; we say that f is *onto* if $f(S) = T$ or, in other words, if for every $t \in T$, there exists an $s \in S$ such that $f(s) = t$.

These properties depend on the domain of f. For example, consider the function f defined by $f(x) = x^2$ or $f = \{(x, x^2) : x \in \mathbb{R}\}$, which associates its square with each real number. This satisfies the definition of function, for if we choose x, then x^2 is uniquely determined.

(i) If we define the domain of f to be the set $\mathbb{R}$ of all real numbers, and the codomain to be $\mathbb{R}$ also, then f is neither one-to-one (since $f(x)f = f(-x)$) nor onto (since $f(\mathbb{R}) = \{y \in \mathbb{R} : y \geq 0\}$).

(ii) If instead we define the domain of f to be the set of all non-negative real numbers, but let the codomain still be the whole of $\mathbb{R}$, then f is one-to-one but is not onto.

(iii) If we let the domain of f be R but restrict the codomain to be the set of all non-negative real numbers, then f is onto, but is not one-to-one.

(iv) If both the domain and the codomain are defined to be the set of non-negative reals, then f is both one-to-one and onto.

Inverse functions

Given any set S we can define the identity function $\iota_S : S \to S$ by $\iota_S(s) = s$ for every $sE \in S$; in other words,

$$\iota_S = \{(s,s) : s \in S\}.$$

For a function, as for any relation, we can define an inverse relation. If the function is, say $f(x) = x^2$, and if we let the domain and the codomain equal $\mathbb{R}$, then the inverse relation f^{-1} contains the pairs $(4,2)$ and $(4,-2)$, so it is not a function: $f^{-1}(x)$ would not be uniquely defined. This raises an important question: when is the inverse relation of a function also a function? The following theorem gives the necessary and sufficient conditions.

Notice that the function f_1 in Sample Problem 4.9 is the identity function on the set S, and in that problem we showed that f_2 is its own inverse.

Sample Problem 4.10 *What is the inverse of the function f_4 from Sample Problem* 4.9*?*

Solution. $f_4(x) = 1/(1-x)$. Say $y = 1/(1-x)$. Then $1-x = 1/y$ so $x = 1-1/y = (y-1)/y$. So $f_4^{-1}(y) = (y-1)/y$. So $f_4^{-1} = f_6$.

Practice Exercise. What is the inverse of the function f_3 from Sample Problem 4.9?

Theorem 4.6 *Let f be a function, $f : S \to T$. Then f is one-to-one and onto if and only if its inverse relation f^{-1} is a function from T to S.*

Proof. (i) Suppose f is one-to-one and onto. Since f is one-to-one each element of T occurs as the second element of at most one ordered pair in f; since f is onto, each element of T occurs as the second element of at least one ordered pair in f. Hence if $t \in T$, exactly one ordered pair $(s,t) \in f$.

Now consider the inverse relation: $(t,s) \in f^{-1}$ exactly when $(s,t) \in f$. Hence for every $t \in T$, exactly one ordered pair $(t,s) \in f^{-1}$, so f^{-1} is a function from T to S.

(ii) Suppose conversely that f^{-1} is a function from T to S. Then for every $t \in T$, exactly one ordered pair $(t,s) \in f^{-1}$ and hence exactly one ordered pair $(s,t) \in f$. Since at least one ordered pair has t as its second element for each $t \in T$, we see that f is onto; since at most one ordered pair has t as its second element for each $t \in T$, we see that f is one-to-one. □

The following theorem gives us some information about the behavior of identity and inverse functions.

Theorem 4.7 *Suppose f is any function from S to T.*

(i) $f(\iota_S) = f = (\iota_T)f$.

(ii) *If f is one-to-one and onto, then $f^{-1}(f) = \iota_S$ and $f(f^{-1}) = \iota_T$.*

(iii) *If both f and $g : T \to U$ are one-to-one and onto functions, so that f^{-1} and g^{-1} are functions, then the composition $g(f)$ is defined, $(g(f))^{-1}$ is a function and $(g(f))^{-1} = f^{-1}(g^{-1})$.*

Proof. (i) All three functions map T into S. For all $t \in T$,

$$(f(\iota_S))(s) = f(\iota_S(s)) = f(s) = \iota_T(f(s)) = S(\iota_T(f))(s),$$

which proves the statement.

(ii) Since f is one-to-one and onto, f^{-1} is a function. Moreover, both $f^{-1}(f)$ and ι_S map S to S; both $f(f^{-1})$ and ι_T map T to T. For every $s \in S$, exactly one ordered pair (s,t) belongs to f; the corresponding pair $(t,s) \in f^{-1}$ and is the only ordered pair in f^{-1} with t as its first element. Hence $(s,s) \in f^{-1}(f)$ for every $s \in S$, and no element (s_1,s_2), with $s_1 \neq s_2$, belongs to $f^{-1}(f)$. So $f^{-1}(f) = \iota_S$. Similarly $f(f^{-1}) = \iota_T$.

(iii) Since the image of f equals the domain of g, the composition $g(f)$ is defined. Suppose $g(f)(s_1) = g(f)(s_2)$. Since g is one-to-one, $f(s_1) = f(s_2)$; since f is one-to-one, $s_1 = s_2$. So fg is one-to-one. Now select $u \in U$. Since g is onto, there exists an element $t \in T$ such that $g(t) = u$; since f is onto, there exists an element $s \in S$ such that $f(s) = t$. So $g(f)(s) = g(t) = u$ and $g(f)$ is onto. Since fg is one-to-one and onto, $(fg)^{-1}$ is defined.

Now consider $u \in U$ and let s be the unique element of S such that $g(f)(s) = u$ and, equivalently, $(g(f)^{-1})(u) = s$. Then

$$\begin{aligned} f^{-1}(g^{-1})(u) &= f^{-1}(g^{-1}(u)) \\ &= f^{-1}(t) \text{ where } g(t) = u \\ &= s \text{ where } f(s) = t, \end{aligned}$$

so $(g(f))^{-1}$ and $f^{-1}(g^{-1})$ have exactly the same action on u. But since u was an arbitrary element of U, this means that $(g(f))^{-1} = f^{-1}(g^{-1})$. □

Exercises 4.3

In Exercises 1 *to* 8, $S = \{1,2,3,4\}$. *Is the given set a function from S to S?*

1. $\{(1,4),(2,3),(3,2),(2,1),(4,4)\}$

2. $\{(1,1),(3,1),(4,3)\}$

3. $\{(2,1),(3,4),(1,4),(2,1),(4,2)\}$

4. $\{(1,4),(2,4),(3,3),(4,2)\}$

5. $\{(1,4),(2,1),(3,4),(4,1),(3,4)\}$

6. $\{(1,1),(3,1),(1,2),(4,2)\}$

7. $\{(1,2),(2,2),(3,1),(4,2)\}$

8. $\{(2,2),(1,2),(2,1),(3,4),(4,2)\}$

9. Prove that ordinary addition of integers is a function $+ : (\mathbb{Z}\times\mathbb{Z}) \to \mathbb{Z}$. What is the associated set?

10. Define $A = \{a,b,c,d\}$. Does the following set define a function?

$$\{(a,(b,c),(a,(c,b)),(b,(c,c))\}$$

11. Prove that $\{(x^2,x+1) : x \in \mathbb{R}\}$ is not the defining set of a function. Does $\{(x^2,x) : x \in \mathbb{R}^*\}$ represent a function, where $\mathbb{R}^*$ is the set of the non-negative real numbers?

In Exercises 12 *to* 14, *is the function one-to-one? Is it onto?*

12. $f : K \to K$, where $K = \{1,2,\ldots,k\}$ and $f(j) = j+1$, for $j = 1,2,\ldots,k-1, f(k) = 1$.

13. $f : \mathbb{Z}^+ \to \{0,1,2,3\}$, where $\mathbb{Z}^+$ is the set of positive integers, and

$$f(j) = \begin{cases} 0 \text{ if } 7 \nmid j, 3 \mid j, \\ 1 \text{ if } 7 \mid j, 3 \nmid j, \\ 2 \text{ if } 21 \mid j, \\ 3 \text{ otherwise.} \end{cases}$$

14. $f : \mathbb{Q} \to \mathbb{Q}$, where $f(q) = q^3$ for $q \in \mathbb{Q}$. ($\mathbb{Q}$ denotes the set of rational numbers.)

In Exercises 15 *to* 18, *find the inverse of the given function.*

15. $f_6 : S \to S$ from Sample Problem 4.9, where S is the set of all real numbers other than 0 and 1.

16. $f_5 : S \to S$ from Sample Problem 4.9 (same S as the preceding question).

17. $f : \mathbb{R} \to \mathbb{R}^+$, where $\mathbb{R}$ is the set of real numbers, $\mathbb{R}^+$ is the set of positive real numbers and

$$f(x) = \begin{cases} 2-x \text{ if } x \le 1, \\ 1/x \text{ if } x > 1. \end{cases}$$

18. $g : \mathbb{R} \to \mathbb{R}$, where $\mathbb{R}$ is the set of real numbers and

$$g(x) = \begin{cases} 2x \text{ if } x < 0, \\ 3x \text{ if } x \ge 0. \end{cases}$$

19. Let $f : A \to B$ be a given function. Suppose $A' \subseteq A$ and $B' \subseteq B$. Show that
(i) $A' \subseteq (f^{-1}(f))(A')$ and
(ii) $(f(f^{-1})(B') \subseteq B'$

where $f^{-1}(B') = \{a \in A : f(a) \in B'\}$.

(Note: f^{-1} is the inverse *binary relation* of f *considered as a binary relation*; f^{-1} need not be a *function*.)

20. Let $f : A \to B$ be a function. Show that it can be written as a composition $f = h(g)$, where $g : A \to C$ is an onto function and $h : C \to B$ is a one-to-one function, for some appropriate set C.

21. Let f, g, h be functions such that f is one-to-one, $g(f)$ and $h(f)$ are defined, and $g(f) = h(f)$. Show that $g = h$.

22. Let $f : S \to S$ be one-to-one onto, and let $g : S \to S$. Show that

(i) g is one-to-one if and only if $f(g)$ is one-to-one;

(ii) g is onto if and only if $f(g)$ is onto.

5

The Theory of Counting

Enumeration answers the question "how many?" In a practical case, we find the number of elements in a set by counting. So the theoretical counterpart is often called the *theory of counting*.

A typical practical counting question is *how many subsets of size* 2 *are there in the set* $\{a,b,c\}$*?* This can be answered by listing the subsets — $\{a,b\}$, $\{a,c\}$ and $\{b,c\}$ — and then counting that there are three of them. A more general question is *how many subsets of size 2 are there in a* 3*-element set?* A little thought convinces us that the number is the same for any 3-element set, so the answer is again three. But consider the question *how many subsets of size* 2 *are there in an* n*-element set?* This is truly theoretical, and to achieve the answer $\frac{1}{2}n(n-1)$ is a piece of enumeration.

In this chapter we introduce the idea of an "event," which enables us to clarify the sets we are counting and lays the foundation for the application of enumeration in studying probability, later. We then study the basic tools for theoretical counting, including one-to-one correspondence, selections, arrangements, and the binomial theorem.

5.1 Events

Experiments and events

In many cases we want to know how many ways an event can occur, or in how many ways a task can be performed. For this it is convenient to use the language of experimentation.

An *experiment* is defined to be any activity with well-defined, observable outcomes or results. For example, a coin flip has the outcomes "head" and "tail." The act of looking out the window to check on the weather has the possible results "it is sunny," "it is cloudy," "it is raining," "it is snowing," and so on. Both of these activities fit our definition of an "experiment."

The different possible outcomes of an experiment will be called *sample points* of the experiment. The set of all possible outcomes is the *sample space*. Each subset of the sample space is called an *event*. So discussions of experiments and their outcomes are in fact discussions of sets.

This terminology is used because counting results, and the corresponding results on probability that will be developed in the next chapter, are often used by statisticians in analyzing experiments and testing conclusions from them. However, we shall also use the same words in other situations, such as the weather observations listed in the preceding paragraph, or listing the possible results in games.

Many examples will involve ordinary dice, such as are used in games like Trivial Pursuit. Each die has six faces, with the numbers 1 through 6 on them. The sample space is the set $S = \{1,2,3,4,5,6\}$. The event "an odd number is rolled" is $\{1,3,5\}$.

Sample Problem 5.1 *Betty and John roll a die. If the result is a* 1 *or* 2*, John wins; otherwise Betty wins. What is the event* John wins*?*

Solution. $\{1,2\}$.

Practice Exercise. In the above game, what is the event *Betty wins*?

Sample Problem 5.2 *How many events are there associated with the roll of one die?*

Solution. Each event corresponds to a subset of $\{1,2,3,4,5,6\}$. There are 2^6 subsets, so there are 2^6 possible events.

Practice Exercise. How many events are there associated with the flip of one coin?

Sample Problem 5.3 *In an experiment three identical coins are flipped simultaneously and the results are recorded. What is the sample space?*

Solution. $\{HHH, HHT, HTT, TTT\}$. Notice that the order is not relevant here, so that for example, the events HHT, HTH and THH of Sample Problem 5.5 are all the same event in this case.

Practice Exercise. An experiment consists of flipping a quarter and noting the result, then flipping two pennies and noting the number of heads. What is the sample space?

Set language

Since events are sets, we can use the language of set theory in describing them. We define the union and intersection of two events to be the events whose sets are the union and intersection of the sets corresponding to the two events. For example, in Sample Problem 5.1, the event "either John wins or the roll is odd" is the set $\{1,2\} \cup \{1,3,5\} = \{1,2,3,5\}$. (For events, just as for sets, "or" carries the understood meaning, "or both.") The complement of an event is defined to have associated with it the complement of the original set, so the complement of A is "A does not happen." Venn diagrams can represent events, just as they can represent sets.

The language of events is different from the language of sets in a few cases. If S is the sample space, then the events S and $\emptyset$ are called "certain" and "impossible." If U and V have empty intersection, they are disjoint sets, but we call them *mutually exclusive* events.

Sample Problem 5.4 *A die is thrown. Represent the following events in a Venn diagram:*

A: an odd number is thrown;

B: a number less than 5 is thrown;

C: a number divisible by 3 is thrown.

Solution.

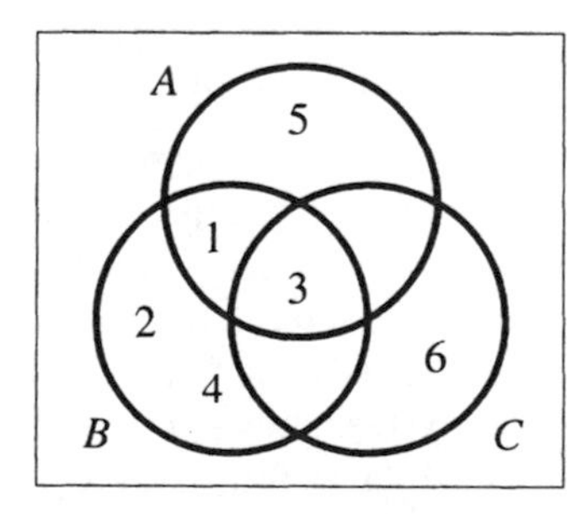

Practice Exercise. Repeat the above example for the events:

A: an even number is thrown;

B: John wins (in the game of Sample Problem 5.1);

C: Betty wins.

The multiplication principle. Tree diagrams

It is sometimes useful to break an experiment down into several parts, forming what we shall call a *compound experiment*. For example, suppose you are planning a trip from Los Angeles to Paris, with a stopover in New York. You have two options for the flight to New York: a direct flight with United or an American flight that stops in Chicago. For the second leg, you consider the direct flight with Air France, a British Airways flight through London, and Lufthansa stopping in Frankfurt. There are two ways to make the first flight and three to make the second, for a total of six combinations.

In general, suppose the event A can occur in a ways, and the event B can occur in b ways. Then the combination of events A and B can occur in ab ways. This very obvious principle is sometimes called the *multiplication principle* or *rule of product*. In set-theoretic terminology, the combined event is $A \times B$. The principle only works when the events are performed independently — if the result of A is somehow used to affect the performance of B, some combined results may be impossible.

Sample Problem 5.5 (i) *A coin is flipped three times; each time the result is recorded. What is the sample space? How many elements does it contain?* (ii) *Repeat this example in the case where you stop flipping as soon as a head is obtained.*

Solution. (i) The sample space has eight elements,

$$\{HHH, HHT, HTH, HTT, THH, THT, TTH, TTT\}.$$

(ii) The sample space has only four elements:

$$\{H, TH, TTH, TTT\}.$$

Notice that the multiplication principle says there will be $2 \times 2 \times 2 = 8$ outcomes. It applies in the first problem, but not in the second.

The possible outcomes of a compound experiment can be shown in a diagram called a *tree diagram*. A special point (called a *vertex*) is drawn to represent the start of the experiment, and lines (*branches*) are drawn from it to further vertices representing the outcomes, which are called the *first generation*. The outcome is shown as a label, either on its vertex or on the branch leading to it. If the experiment has two or more stages, the second stage is drawn onto the outcome vertex of the first stage, and all these form the *second generation*, and so on.

Sample Problem 5.6 *Draw tree diagrams for the experiments in Sample Problem 5.5.*

Solution. The tree diagrams are (a) and (b) respectively.

(a) (b)

Practice Exercise.

(i) Draw a tree diagram for the experiment described in Sample Problem 5.1, with outcomes "Betty wins" and "John wins."

(ii) Betty and John play their game twice. Draw a tree diagram for this new experiment.

Even when the two parts of a compound experiment are performed simultaneously, we can use the multiplication principle and draw a tree diagram. For example, if you simultaneously flip a nickel and a dime there are four possibilities:

head on nickel, head on dime;
head on nickel, tail on dime;
tail on nickel, head on dime;
tail on nickel, tail on dime.

Diagram (a) is the tree diagram for a coin flip, and diagram (b) describes the flipping of a nickel and a dime.

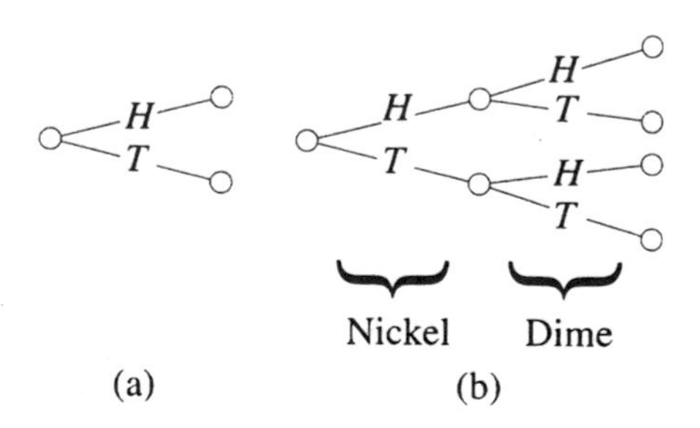

(a) (b)

Exercises 5.1

1. A student takes courses in mathematics and computer science. Define events:

E : She passes mathematics

F : She passes computer science.

Write symbolic expressions (using $E, F, \cap, \cup, ^-, \backslash, \times, +$) for the events that the student

(i) fails mathematics;

(ii) passes both subjects;

(iii) passes exactly one subject;

(iv) passes at least one subject;

(v) fails both subjects;

(vi) passes mathematics but fails computer science.

2. There are three roads from town X to town Y, four roads from town Y to town Z, and two roads from town X to town Z.

(i) How many routes are there from town X to town Z with a stopover in town Y?

(ii) How many routes are there in total from town X to town Z?

(Assume that no road is traveled twice.)

3. An experiment consists of studing families with three children. B represents "boy," G represents "girl," and for example BGG will represent a family where the oldest child is a boy and the young children are both girls. For convenience, assume there are no twins (so the three children have different ages).

(i) What is the sample space for this experiment?

(ii) We define the following events.

E : The oldest child is a boy.

F : There are exactly two boys.

(a) What are the members of E and F?

(b) Describe in words the event $E \cap F$.

(iii) Draw a Venn diagram for this experiment, and show the events E and F on it.

4. A die is rolled twice, and the results are recorded as an ordered pair.

(i) How many outcomes are there in the sample space?

(ii) Consider the events:

E : The sum of the throws is 4.

F : Both throws are even.

G : The first throw was 3.

H : The two throws were identical.

(a) Write down the elements of the events $E, F, G, H, E \cup F, G \cap H, \overline{F} \cap H$.

(b) Write descriptions, in words, of the sets $E \cup F, G \cap H, \overline{F} \cap H$.

(c) Are any pair of the events E, F, G, H mutually exclusive?

(iii) Draw a tree diagram for this experiment.

(iv) Draw a Venn diagram that shows the outcomes of the experiment. Show the events F and G in the diagram.

5. An experimenter tosses four coins — two quarters and two nickels — and records the number of heads. For example, two heads on the quarters and one on the nickels is recorded as 21.

(i) Write down all members of the sample space.

(ii) We define the events:

E : There are more heads on the quarters than on the nickels.

F : There are exactly two heads in total.

G : The number of heads is even.

(a) List the members of events $E, F, G, E \cup F, E \cap F, E \cap G, \overline{F} \cap G$.

(b) Write descriptions in words of the events $E \cap F, \overline{F} \cap G$.

(c) Are any two of the events E, F, G mutually exclusive?

(iii) Draw a tree diagram for this experiment.

(iv) Represent the outcomes of the experiment in a Venn diagram. Show the events E, F and G in the diagram.

6. Your doctor tests your cholesterol level each month. She makes three tests, and records whether the reading is higher than (H), the same as (S) or lower than (L) the corresponding test from the preceding month.

(i) What are the possible outcomes of this experiment? List all members of the sample space.

(ii) Consider the events:

E : The cholesterol level never decreases.

F : The cholesterol level decreases at least twice.

(a) Write down the elements of E, F and $\overline{F}$.

(b) Write a description in words of the sets $\overline{E}, \overline{F}$.

(c) Are E and F mutually exclusive?

(iii) Draw a tree diagram for this experiment.

(iv) Draw a Venn diagram to represent the outcomes of the experiment, and show the events E and F on it.

7. A bag contains two red, two yellow and three blue balls. In an experiment, one ball is drawn from the bag and its color is noted, and then a second ball is drawn and its color noted.

(i) Draw a tree diagram for this experiment.

(ii) How many outcomes are there in the sample space?

(iii) What is the event: "two balls of the same color are selected"?

8. Workers in a factory are classified by gender, experience and union membership. Events are defined as follows:

E : The worker is male.

F : The worker has one year's experience at least.

G : The worker belongs to the union.

(i) Write descriptions in words of the following events: $\overline{E}, \overline{F}, \overline{G}$, linebreak $E \cap G, E \cap \overline{E}, E \cup \overline{E}, G \cap \overline{F}, \overline{E} \cap F, E \cap F \cap G$.

(ii) Write expressions in symbols (using $E, F, G, \cap, \cup$ and $^-$) for the events:

(a) The worker is a woman who belongs to the union.

(b) The worker has less than one year's experience and is a man.

(c) The worker has at least one year's experience and is a woman union member.

(iii) Represent the set of all workers on a Venn diagram, with the sets E, F and G shown. Shade in the set defined in (ii)(c) above.

In Exercises 9 *to* 16 *two events are shown. Are they mutually exclusive?*

9. A die shows a 4.
A die shows an odd number.

10. A die shows a 4.
A die shows a perfect square.

11. A person is male.
The same person is a store clerk.

12. A person is a college freshman.
The same person is a college sophomore.

13. A coin is tossed and falls heads.
The coin falls tails.

14. A student wears a watch.
The student wears sneakers.

15. Joe is a freshman.
Joe is a sophomore.

16. A die is rolled and gives an odd number.
The roll is greater than 4.

17. The events E and F are mutually exclusive. Which of the following are mutually exclusive pairs? Draw a Venn diagram in each case.

(i) E and $\overline{F}$

(ii) E and $\overline{E}$
(iii) $\overline{E}$ and $\overline{F}$

18. If E and F are two events, write symbolic expressions (using the symbols $E, F, \cap, \cup$ and $^{-}$) for the events:
(i) E occurs but F does not.
(ii) E and F both occur.
(iii) E does not occur.
(iv) Exactly one of the two occurs.

19. Suppose E and F are two events. Must E and $\overline{(E \cup F)}$ be mutually exclusive?

5.2 Unions of Events

The rule of sum

Suppose S and T are any two sets, and you want to list all members of $S \cup T$. If you list all members of S, then list all the members of T, you will cover all members of $S \cup T$, but those in $S \cap T$ will be listed twice. To count all members of $S \cup T$, you could count all members of both lists, then subtract the number of duplicates. In other words,

$$|S \cup T| = |S| + |T| - |S \cap T|. \tag{5.1}$$

If S is the set of all the objects in some set that have some property A and T is the set of all the objects with property B, then (5.1) expresses the way to count the objects that either have property A or property B:

(i) count the objects with property A;
(ii) count the objects with property B;
(iii) count the objects with both properties;
(iv) subtract the third answer from the sum of the other two.

Another rule, sometimes called the *rule of sum*, can be simply expressed by saying "the number of objects with property A equals the number that have both property A and property B, plus the number that have property A but not property B"; in terms of the sets S and T, this is

$$|S| = |S \cap T| + |S \backslash T|. \tag{5.2}$$

The two rules (5.1) and (5.2) can be combined to give

$$|S \cup T| = |T| + |S \backslash T|. \tag{5.3}$$

Sample Problem 5.7 *There are* 40 *students in Dr. Brown's discrete mathematics course and* 50 *in his calculus section. If these are his only classes, and if 20 of the students are taking both subjects, how many students does he have altogether?*

Solution. Using the notations D for the set of students in the discrete class and C for calculus,

$$|D \cup C| = |D| + |C| - |D \cap C| = 40 + 50 - 20 = 70,$$

so he has 70 students.

Practice Exercise. A survey shows that twelve newspaper readers buy the morning edition and seven buy the evening edition; three of these also bought the morning paper. How many readers were surveyed?

Using Venn diagrams

The kind of data we have been discussing is very easily represented in a Venn diagram. The number of elements in a set is written in the area corresponding to that set.

Sample Problem 5.8 *Represent the data concerning Dr. Brown's students (from Sample Problem* 5.7*) in a Venn diagram.*

Solution. There are four areas in the diagram: the set of students in both classes (the center area) has 20 members; the set of students in discrete mathematics only (the left-hand enclosed area) has 20 members — subtract 20 from 40; the area corresponding to calculus-only students, with $50 - 20 = 30$ members, and the outside area, which has no members (only Dr. Brown's students are being considered). The diagram is

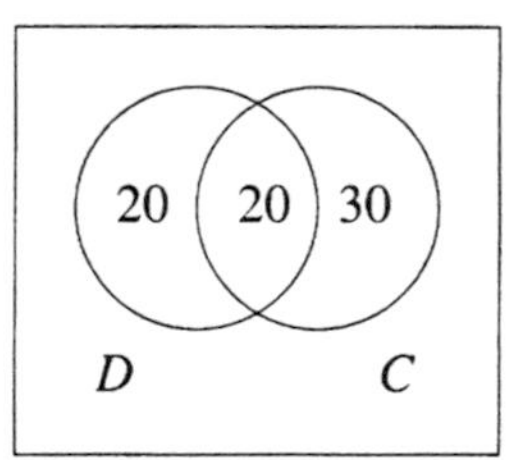

Practice Exercise. Represent the data concerning newspaper readers in a Venn diagram.

This method can be used to find the number of elements in the set represented by any of the areas or unions of areas, provided there is enough information.

Sample Problem 5.9 100 *people were surveyed to find out whether newspaper advertisements or flyers were more efficacious in advertising supermarket specials.* 20 *of them said they pay no attention to either medium.* 50

said they read the flyers, and 15 *of those said they also check the newspapers. How many use the newspaper ads, in total?*

Solution. This can be done in various ways. In the obvious notation, $|F| = 50$ and $|N \cap F| = 15$. Moreover, since 20 of the 100 read neither, $|N \cup F| = 80$. Using (5.1), $|N \cup F| = |N| + |F| - |N \cap F|$ yields

$$|N| = |N \cup F| - |F| + |N \cap F| = 80 - 50 + 15 = 45.$$

Alternatively, we could observe that $|F \backslash N| = 35$, so (5.3) is $|N \cup F| = |N| + |F \backslash N|$, so

$$|N| = |N \cup F| - |F \backslash N| = 80 - 35 = 45.$$

Or we could draw the Venn diagram

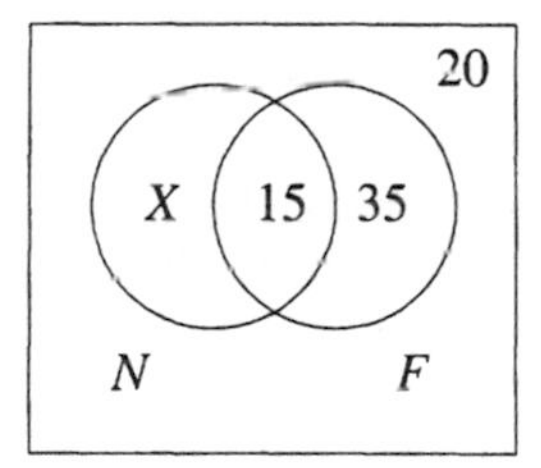

and observe that in order for the total to add to 100, $X = 30$, so $X + 15 = 45$.

These methods can be applied to three or more sets.

Sample Problem 5.10 500 *people were asked about their morning vitamin intake. It was found that* 150 *take vitamin B,* 200 *take vitamin C,* 165 *take vitamin E,* 57 *take both B and C,* 125 *take both B and E,* 82 *take both C and E, and* 52 *take all three vitamins. How many take both B and E but do not take C? How many take none of these vitamins?*

Solution. We start with the diagram

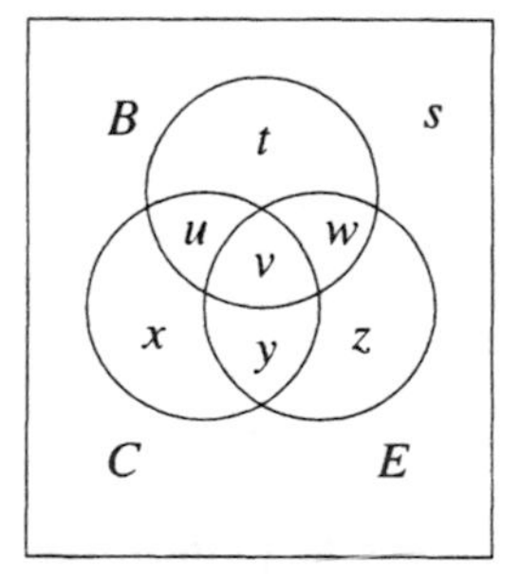

From the data, $v = 52$, and $u + v = 57$, so $u = 5$. Similarly, $w = 73$ and $y = 30$. Now $t + u + v + w = 150$, so $t = 20$. The other sizes are calculated similarly, and we get the diagram

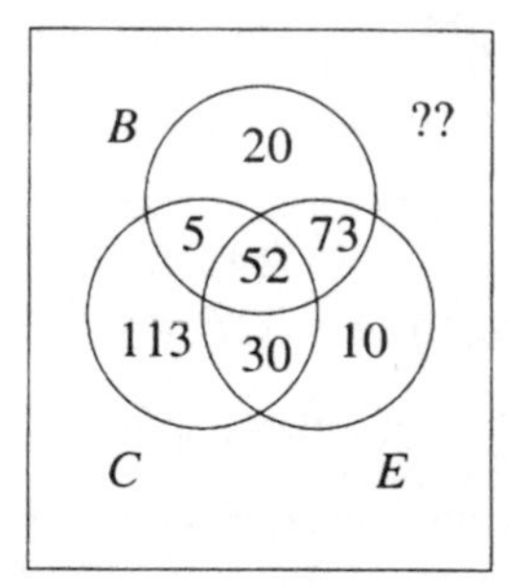

The cell corresponding to "B and E but not C" has 73 elements. There are 303 elements in total, so there are 197 people who take none.

Practice Exercise. In a survey of 150 moviegoers, it was found that 80 like horror movies, 75 like police procedurals, and 60 like romances. In total, 35 like both horror and procedurals, 25 like horror and romance, and 30 like procedurals and romance. 15 like all three types of movie. Represent these data in a Venn diagram. How many like horror and police procedural movies, but do not like romances? How many like romances only? How many like none of these three types?

Inclusion and exclusion

Equation (5.1), with S replaced by X, is

$$|X \cup T| = |X| + |T| - |X \cap T|.$$

Now suppose T is itself a union, $Y \cup Z$. Then

$$|X \cup Y \cup Z| = |X| + |Y \cup X| - |X \cap (Y \cup Z)|.$$

The term $|Y \cup Z|$ can be replaced by

$$|Y| + |Z| - |Y \cap Z|;$$

as a consequence of the distributive laws for intersections and unions of sets, $X \cap (Y \cup Z) = (X \cap Y) \cup (X \cap Z)$, and the last term can be written as

$$|X \cap Y| + |X \cap Z| - |X \cap Y \cap Z|.$$

So the whole expression is

$$\begin{aligned} |X \cup Y \cup Z| &= |X| + |Y| + |Z| \\ &\quad -|X \cap Y| - |Y \cap X| - |X \cap Z| \\ &\quad +|X \cap Y \cap Z|. \end{aligned}$$

It is instructive to look at the Venn diagram in Figure 5.1. We want to count $X \cup Y \cup Z$. If we add $|X|$ to $|Y|$ and $|Z|$, we count all points lying in the lightly shaded area twice, and all points in the heavily shaded area three times. Subtracting $|X \cap Y|$, $|X \cap Z|$ and $|Y \cap Z|$ takes away 1 for each point in the shaded area,

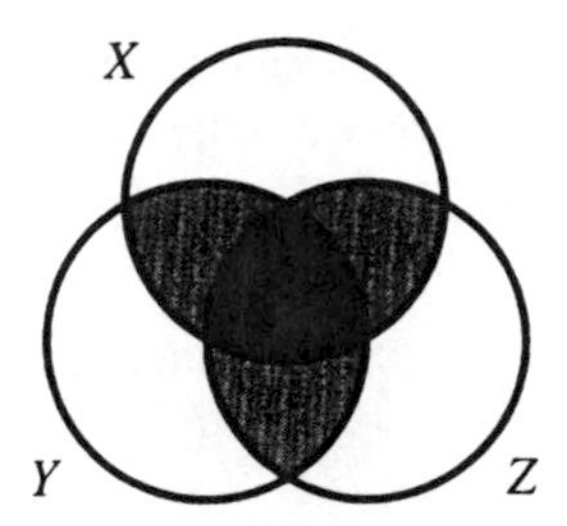

Figure 5.1: Enumerating $X \cup Y \cup Z$

but takes away 3 for every heavily shaded point. So we must add the number of heavily shaded points, which equals $|X \cap Y \cap Z|$.

This formula can be generalized. The general form is called the *principle of inclusion and exclusion.*

Theorem 5.1 *The number of elements in* $X_1 \cup X_2 \cup \cdots \cup X_n$ *is*

$$\begin{aligned}&\textstyle\sum_{i_1=1}^{n}|X_{i_1}| - \sum_{i_1=1}^{n}\sum_{i_2=i_1+1}^{n}|X_{i_1} \cap X_{i_2}| + \cdots \\ &+(-1)^{k-1}\textstyle\sum_{i_1=1}^{n}\sum_{i_2=i_1+1}^{n}\cdots\sum_{i_k=i_{k-1}+1}^{n}|X_{i_1} \cap X_{i_2} \cap \cdots \cap X_{i_k}| \\ &+\cdots+(-1)^{n-1}|X_{i_1} \cap X_{i_2} \cap \cdots \cap X_{i_n}|.\end{aligned}$$

We shall prove this theorem in Section 5.7, below.

Exercises 5.2

1. 1865 voters were surveyed about a new highway. Of them, 805 were in favor of financing it with a new tax, 627 favored tolls on the highway, while 438 said they would vote in favor of either measure.

(i) How many favor taxes but not tolls?

(ii) How many favor tolls but not taxes?

(iii) How many would vote against either form of funding?

2. Among 1000 personal computer users it was found that 375 have a scanner and 450 have a DVD player attached to their computer. Moreover 150 had both devices.

(i) How many had either a scanner or a DVD player?

(ii) How many had neither device?

3. Some shoppers buy bread, butter and coffee. Ten buy bread, 15 buy butter and 14 buy coffee. The number buying both bread and butter, both bread and coffee and both butter and coffee are four, five and eight respectively. Three buy all three products. How many shoppers were there (assuming all bought at least one of the three)?

4. A survey was carried out. It was found that ten of the people surveyed were drinkers, while five were smokers; only two both drank and smoked, but seven did neither. Use formula (5.1) to find out how many people were interviewed.

5. Sixty people were asked about news magazines. It was found that 32 read *Newsweek* regularly, 25 read *Time* and 20 read *U.S. News and World Report*. Nine read both *Time* and *U.S. News and World Report*, 11 read both *Newsweek* and *Time*, and eight read both *Newsweek* and *U.S. News and World Report*. Eight of the people do not read any of the magazines.

(i) How many read all three magazines?

(ii) Represent the data in a Venn diagram.

(iii) How many people read exactly one of the three magazines?

6. Of 100 high-school students, 45 were currently enrolled in mathematics courses, 38 were enrolled in science, and 21 were enrolled in geography. There were 18 in both math and science and nine in both math and geography, while 12 were taking geography but neither of the other subjects. There were 23 not currently enrolled in any of the three subjects, and five enrolled in all three. Represent these figures in a Venn diagram. How many students were taking precisely one of the three subjects? How many were taking precisely two?

7. Eighteen people are interviewed. Of these, seven dislike the Republican party, ten dislike the Democrats and eleven dislike the Reform Party. Moreover, five dislike both Democrats and Republicans, five dislike both Republicans and the Reform Party, while six dislike both Democrats and the Reform Party. And four dislike all three types of politicians, on principle. How many *like* all three?

8. Of 100 personal computer users surveyed, 27 use Dell, 35 use Gateway, and 35 use Hewlett-Packard. Ten of them use both Dell and Gateway, eight use both Dell and Hewlett-Packard, and twelve use both Gateway and Hewlett-Packard. Four use all three.

(i) How many use exactly one of these brands?

(ii) How many only use other brands?

9. A shopping mall contains K-Mart, Walmart and Target stores. In a survey of 100 shoppers, it is found that

(a) 47 shopped at Walmart;

(b) 61 shopped at K-Mart;

(c) 52 shopped at Target;

(d) 32 shopped at both Walmart and K-Mart;

(e) 35 shopped at both K-Mart and Target;

(f) 22 shopped at both Walmart and Target;

(g) 12 shopped at all three stores.

(i) How many people shopped only at Walmart?

(ii) How many people shopped at Walmart and K-Mart, but not at Target?

(iii) How many people shopped at exactly one of the stores?

(iv) How many people shopped at none of the three stores?

10. A survey of students found that

(a) 62 were enrolled in calculus;

(b) 71 were enrolled in algebra;

(c) 67 were enrolled in discrete mathematics;

(d) 37 were enrolled in both calculus and algebra;

(e) 32 were enrolled in both calculus and discrete mathematics;

(f) 40 were enrolled in both algebra and discrete mathematics;

(g) 12 were enrolled in all three subjects;

(h) 44 were enrolled in none of the three.

(i) How many were enrolled in both calculus and algebra but not in discrete mathematics?

(ii) How many were enrolled in exactly one of the three subjects?

(iii) How many students were surveyed?

11. Four hundred researchers at IBM were surveyed about their qualifications in computer science. It was found that 300 have at least one degree. 214 have a Bachelor's degree, 123 have a Master's degree, and 99 have a Ph.D. It was further found that 57 have both Bachelor's and Master's degrees, 74 have both Master's and Ph.D., and 22 have both Bachelor's and Ph.D. degrees.

(i) How many of the researchers have all three degrees?

(ii) How many came into computer science from another field: that is, they have either a Master's or a Ph.D., but no Bachelor's degree?

12. Let A be a set with n elements. How many relations can be defined on A? How many of the relations on A are

(i) reflexive?

(ii) irreflexive?

(iii) symmetric?

(iv) antisymmetric?

(v) symmetric and antisymmetric?

(vi) reflexive and symmetric?

13. Derive the formula

$$\begin{aligned} |A \cup B \cup C| &= |A| + |B \cap C| + |\overline{B} \cap C| \\ &\quad - |A \cap B \cap C| - |A \cap \overline{B} \cap C|. \end{aligned}$$

5.3 One-to-one Correspondences and Infinite Sets

One-to-one correspondence

Two sets, S and T, are said to be *in one-to-one correspondence* if there is a one-to-one function from S onto T. We then write $S \equiv T$.

Suppose f is such a one-to-one, onto function. We say the correspondence is given by f. For every element s of S, there is exactly one corresponding element of T, namely $f(s)$. Moreover f has an inverse, f^{-1}, and for every element t of T, there is exactly one corresponding element of S, namely tf^{-1}. So f^{-1} gives a one-to-one correspondence from T to S. In fact we can say more.

Theorem 5.2 *Suppose U is any set. Then $\equiv$ is an equivalence relation on the power set $\mathcal{P}(U)$.*

Proof. It is necessary to show that $\equiv$ is reflexive, symmetric and transitive.

The reflexive property is easy: for any set S, the identity function ι_S is a one-to-one map from S to itself.

If $S \equiv T$, with the correspondence given by $f : S \to T$, then f^{-1}gives maps $T \equiv S$. Therefore $\equiv$ is symmetric.

Suppose $R \equiv S$ and $S \equiv T$: say the two correspondences are given by $f : R \to S$ and $g : S \to T$. Then $g(f)$ is a one-to-one function from R onto T, so $R \equiv T$. Therefore $\equiv$ is transitive. $\square$

Sample Problem 5.11 *Show that the set $\mathbb{O}$ of all odd integers is in one-to-one correspondence with the set $\mathbb{E}$ of all even integers.*

Solution. We use the function $f(n) = n + 1$. If n is odd then $n + 1$ is defined and is even, so f is a function from $\mathbb{O}$ to $\mathbb{E}$. To see that f is one-to-one, we observe that if $m + 1 \neq n + 1$, then subtracting 1 from each side gives $m \neq n$. Every even integer arises from an odd integer: m is $f(n - 1)$; so f is onto. (We could have shown that f is onto by noting that the inverse function f^{-1} is defined by $f^{-1}(n) = n - 1$ and may be applied to any even integer.)

Practice Exercise. Show that the set of all positive integers less than 27 is in one-to-one correspondence with the set of letters of the English alphabet.

The importance of one-to-one correspondences in the theory of enumeration comes from the following theorem:

Theorem 5.3 *Suppose S is a set with n elements, and suppose there is a one-to-one correspondence $f : S \to T$. Then T has n elements.*

Proof. For convenience, let us write $S = \{a_1, a_2, \ldots, a_n\}$. Then $T = f(S) = \{f(a_1), f(a_2), \ldots, f(a_n)\}$, because T is onto. As we have written only n elements $f(a_i)$, T can have at most n elements. On the other hand, if T has fewer than n elements, then $f(a_i) = f(a_j)$ for some i and j, $i \neq j$. But this is impossible, because f is one-to-one. So T has precisely n elements. □

The usual way in which Theorem 5.3 is used is as follows. If it is required to find the number of elements of a set S, find some set T that is in one-to-one correspondence with S, such that $|T|$ is already known, or is easy to find. Then $|S| = |T|$.

Finite and infinite sets

The idea of a one-to-one onto correspondence is useful in the formal discussion of finite and infinite sets. Suppose S is any finite set, and T is a subset of S. If there is a one-to-one function from S onto T, then Theorem 5.3 says that T has $|S|$ elements, so $T = S$. On the other hand, if S is infinite, it is possible to find a proper subset T of S and a one-to-one correspondence taking S to T. The simplest example is the function $f(x) = x + 1$, which maps $\mathbb{Z}^*$ onto $\mathbb{Z}^+$, proving that $\mathbb{Z}^* \equiv \mathbb{Z}^+$, even though $\mathbb{Z}^+$ is a proper subset of $\mathbb{Z}^*$. The inverse function, $f^{-1}(x) = x - 1$, is a one-to-one correspondence from $\mathbb{Z}^+$ to $\mathbb{Z}^*$.

Sample Problem 5.12 *Consider the function $f(z) = z^2$. Show that f is a one-to-one correspondence from $\mathbb{Z}^*$ to the set of integer squares, $S = \{0, 1, 4, 9, 16, \ldots\}$.*

Solution. Every perfect square arises from squaring an integer, so f is onto. If $f(z) = f(x)$, then $z^2 = x^2$, and (as negatives have been excluded) $z = x$. So f is one-to-one.

Practice Exercise. Show that the function $g : g(x) = 2x$ is a one-to-one correspondence from $\mathbb{Z}$ to the set $\mathbb{E}$ of even integers.

Theorem 5.4 *Let A be a finite set, and let f be a function from A to itself.*

(i) *If f is one-to-one, then f is onto.*

(ii) *If f is onto, then f is one-to-one.*

Proof. (i) We leave the proof of this part as an exercise.

(ii) Let A be $\{a_1, a_2, \ldots, a_n\}$. Since f is onto, $f(A) = A$, so

$$\{f(a_1), f(a_2), \ldots, f(a_n)\} = A.$$

Suppose $f(a_i) = f(a_j)$ for some $i \neq j, 1 \leq i, j \leq n$. Then $f(A)$ has at most $n - 1$ distinct elements, contradicting the fact that A must have exactly n distinct elements. □

Countable sets

Among the infinite sets, those that can be put into one-to-one correspondence with $\mathbb{Z}^+$, the set of positive integers, are called *countably infinite*, and the others are *uncountable*. Many (but not all) writers also say that finite sets are also countable.

Sample Problem 5.13 *Prove that the set* Π *of prime numbers is countably infinite.*

Solution. We know from Theorem 1.2 that Π is infinite. The smallest prime number is 2. Write $p(1) = 2$. Similarly, define $p(n) = 3$. In general, define $p(n)$ to equal the n-th prime number. This is well defined even though we know of no formula for $p(n)$: once the value $p(n-1)$ is known, it is at worst necessary to test all the integers from $p(n-1)+1$ to $p(1) \times p(2) \times \cdots \times p(n-1)$ for primacy (see the proof of Theorem 1.2). If $i \neq j$ then $p(i) \neq p(j)$. The function p is a one-to-one correspondence from $\mathbb{Z}^+$ to Π.

Practice Exercise. Prove that the set of odd integers is countably infinite.

Theorem 5.5 *Every countably infinite set can be put in one-to-one correspondence with a proper subset of itself.*

Proof. Suppose S is a countable set, and suppose f is a one-to-one and onto function from S to $\mathbb{Z}^+$. There will be a unique element of S that is mapped to 1 by f; let us call this element z. Now consider the function g, defined by $g(x) = f^{-1}(1 + f(x))$. As f, f^{-1} and $x \to x+1$ are all one-to-one and onto functions, it follows that g is one-to-one and onto. It carries S to $T = S \backslash \{z\}$, the set of all elements of S other than z. So S is in one-to-one correspondence with its proper subset T. □

It may be shown that every infinite set contains a countably infinite subset. To prove this we need to use some advanced set theory, including the *Axiom of Choice*, which lies outside the scope of this book. The argument goes roughly as follows.

Suppose A is any infinite set. Select an element — call it a_1 — of S. Write S_1 to denote the set $\{a_1\}$, and write A_1 for the result of deleting a_1 from S: $A_1 = S \backslash \{a_1\}$. A_1 is again infinite. Now select an element a_2 in A_1, and write $S_2 = \{a_1, a_2\}$ and $A_2 = A_1 \backslash \{a_2\}$. Proceed in this way: when S_n and A_n have been determined, then select a_{n+1} in A_n, and write $S_{n+1} = S_n \cup \{a_{n+1}\} = \{a_1, a_2, \ldots, a_{n+1}\}$ and $A_{n+1} = A_n \backslash \{a_{n+1}\} = A \backslash S_{n+1}$. It can be seen that this process defines an infinite subset

$$S = \{a_1, a_2, \ldots\}$$

of A, and S is clearly countably infinite because the function $f : f(i) = a_i$ is a one-to-one correspondence between $\mathbb{Z}^+$ and S.

(The problem, and the need for an axiom, comes because the definition of S requires an infinite number of steps. The situation is rather like the one for which the Principle of Mathematical Induction had to be introduced.)

Now suppose A is an infinite set and S is a countably infinite subset of A. Select a proper subset T of S that is in one-to-one correspondence with S; say the function $g : S \to T$ is one-to-one and onto. Define $B = (A \backslash S) \cup T$, the set obtained by deleting from A all those elements of S that are not in T. Then B is a proper subset of A, and the function f, defined by

$$f(s) = \begin{cases} g(s) & \text{if } s \in S, \\ s & \text{if } s \in a \backslash S, \end{cases}$$

is a one-to-one correspondence from A to B.

So any infinite set is in one-to-one correspondence to a proper subset of itself.

This in fact characterizes infinite sets. We could say that a set is infinite if and only if it is in one-to-one correspondence with some proper subset of itself.

Exercises 5.3

1. Suppose $|A| = m$ and $|B| = n$, and suppose $C \subseteq A \times B$. What is the greatest value possible for $|C|$ if C is the set corresponding to a function?
2. If $f : A \to B$ and $|A| < |B|$, can f be onto B?
3. Let $|S| = 2$. How many functions are there from S to S? How many onto functions? How many one-to-one functions?
4. Repeat the preceding exercise in the case where $|S| = n$.
5. Show that the elements of $A \times B$ are in one-to-one correspondence with the elements of $B \times A$.
6. Prove that $f : x \to 2x$ is a one-to-one correspondence from $\{x : 0 < x < 1\}$ to $\{x : 0 < x < 2\}$.
7. Find a one-to-one correspondence between the set of all positive real numbers and the set of all real numbers.
8. Prove Theorem 5.4(i).
9. Say $|A| = m, |B| = n$. What can you say about the following orders?

   (i) $|A \cup B|$ (iii) $|A \backslash B|$;
   (ii) $|A \cap B|$; (iv) $|A + B|$

10. Suppose S is a countable set and T is a finite set. Prove that $S \cup T$ is countable.

11. Suppose S and T are countable sets. Prove that $S \cup T$ is countable.

12. If A and B are uncountable sets, is $A \cup B$ always uncountable?

13. If A and B are uncountable sets, is $A \cap B$ always uncountable?

14. Consider the set S whose elements are all the sets T of integers such that both T and $\overline{T}$ (its complement in the universal set $\mathbb{Z}$) are finite sets. Suppose A and B are members of S. Prove that the following sets are members of S.

(i) $\overline{A}$

(ii) $A \cup B$

(iii) $A \cap B$

(iv) $A \backslash B$

5.4 Arrangement Problems

Sequences on a set

Suppose S is a set with s elements. We often need to know how many *ordered* sets or *sequences* of size k (often called *k-sequences*, or *permutations* of length k) can be chosen from S. This number is denoted $(s)_k$ or $P(s,k)$. In particular $(s)_s$ denotes the number of s-sequences that can be chosen from an s-set, which is the same as the number of different ways of ordering an s-set; it is given the special name "s factorial," and written $s!$. These s-sequences are called *permutations of* S.

Given an s-set $S = \{x_1, x_2, \ldots, x_s\}$, there are s different sequences of length 1 on S, namely (x_1), (x_2), ..., and (x_s). So $(s)_1 = s$. There are $s \cdot (s-1)$ sequences of length 2, because each sequence of length 1 can be extended to length 2 in $s-1$ different ways, and no two of these $s \cdot (s-1)$ extensions will ever be equal. So $(s)_2 = s(s-1)$. Similarly we find

$$\begin{aligned}
(s)_3 &= s \cdot (s-1) \cdot (s-2), \\
(s)_4 &= s \cdot (s-1) \cdot (s-2) \cdot (s-3), \\
\cdots & \\
(s)_k &= s \cdot (s-1) \cdot (s-2) \ldots (s-k+1).
\end{aligned}$$

So $(s)_k$ is calculated by multiplying, s, $s-1$, $s-2, \ldots$ until there are k factors.

It follows that $s! = s \cdot (s-1) \cdot (s-2) \ldots 3 \cdot 2 \cdot 1$ and that

$$(s)_k = s!/(s-k)!. \tag{5.4}$$

For convenience we define $0!$ to equal 1.

Sample Problem 5.14 *What are* $10!$ *and* $(10)_3$*?*

Solution. $10! = 10 \cdot 9 \cdot 8 \cdot 7 \cdot 6 \cdot 5 \cdot 4 \cdot 3 \cdot 2 \cdot 1 = 362880$.

There are two ways to calculate $(10)_3$. We could say $(10)_3 = 10 \cdot 9 \cdot 8 = 720$. Or we could use the formula: $(10)_3 = 10!/7! = 362880/5040 = 720$. The first way is easier.

Practice Exercise. Find $6!$ and $(6)_4$.

Sample Problem 5.15 *Find the number of different ways of arranging n women around a circular conference table.*

Solution. Suppose we arbitrarily label one seat at the table as "1," the one to its left as "2," and so on. There are $n!$ different ways of putting the n women into the n seats. However, we have counted two arrangements as different if one is obtained from the other by shifting every woman one place to the left, because these two arrangements put different women in "1"; but these are clearly the same arrangement for the purposes of the question. Each arrangement is one of a set of n, all obtained from the others by shifting in a circular fashion. So the number of really different arrangements is $n!/n$, which equals $(n-1)!$.

Practice Exercise. Find the number of different ways of arranging n men in one straight row of chairs.

Sample Problem 5.16 *A committee of three — Chair, Secretary and Treasurer — is to be elected by a club with* 14 *members. If every member is eligible to stand for each position, how many different committees are possible?*

Solution. We can treat the committee as an ordered set of three elements chosen from the 14-element set of members. So the answer is $(14)_3$, or 2184.

Practice Exercise. What is the number of possible committees if there are 12 members?

Combining arrangements with the multiplication principle

The methods used here are often combined with the multiplication principle. We illustrate with several examples.

Sample Problem 5.17 *Three boys and four girls are to sit along a bench. The boys must sit together, as must the girls. How many ways can this be done?*

Solution. First, decide whether the boys are to be on the left or on the right. This can be done in two ways. Then order the boys — there are $3! = 6$

ways. The girls can be ordered in $4! = 24$ ways. So there are $2 \cdot 6 \cdot 24 = 288$ arrangements.

Practice Exercise. How many ways could the three boys and four girls be arranged around a circular table if the boys must sit together and the girls as well?

Sample Problem 5.18 *A copying company has eight photocopying machines and seven employees who can operate them. There are four copying jobs to be done. How many ways are there to allocate these jobs to operators and machines?*

Solution. Call the jobs A, B, C, D. Choose two arrangements: first, which four operators should do the jobs; second, which machines should be used. The operator choice can be made in $P(7,4)$ ways, and the machines in $P(8,4)$ ways. In each case, the first member of the sequence is the one allocated to job A, the second to job B, and so on. There are $P(7,4) \times P(8,4)$ $= 7 \times 6 \times 5 \times 4 \times 8 \times 7 \times 6 \times 5 = 1411200$ ways.

Sample Problem 5.19 *A high-school class contains* 10 *boys and* 13 *girls. They have boys' and girls' charity drives, and wish to elect a Chair and Treasurer for each. How many combinations are possible?*

Solution. There are $(10)_2$ ways to select the boys' committee, and $(13)_2$ ways to select the girls' committee. So the total is $(10)_2 \cdot (13)_2 = 90 \cdot 156 =$ 14040.

Practice Exercise. What is the answer if each committee also has a Secretary?

Sample Problem 5.20 *The club in Sample Problem* 5.16 *wish to elect a by-laws committee with three members — Chair, Secretary and Legal Officer — and require that no members of the main club committee should be members. In how many different ways can the two committees be chosen?*

Solution. Suppose the main committee is chosen first. There are $(14)_3 =$ 2184 ways to do this. After the election, there are 11 members eligible for election to the by-laws committee, so it can be chosen in $(11)_3 = 990$ ways. So there are a grand total of $2184 \cdot 990 = 2162160$.

In the last Sample Problem we see that, even in small problems, the numbers get quite large. It is better to report the answer in its factored form, as $14 \cdot 13 \cdot 12 \cdot 11 \cdot 10 \cdot 9$. This form of answer also makes it clear that we could have solved the problem by treating the two committees as one six-member sequence, with $(14)_6$ possible solutions.

Even a small alteration to a problem can make it considerably more complicated. For example, suppose that one member of the club, Dr. Jones, decides she does not wish to stand for the by-laws committee, although she is happy to be on the main committee. In each case where Dr. Jones is a member of the main committee, there are $(11)_3$ possible by-laws committees, but in the other cases there are only $(10)_3$ possibilities. To proceed, it is necessary to know how many of the possible main committees contain Dr. Jones and how many do not. There are $(13)_3$ committees without Dr. Jones, so there are $((14)_3 - (13)_3)$ of which she is a member. So the total required is

$$\begin{aligned}
& [(13)_3 \cdot (10)_3] + [((14)_3 - (13)_3)] \cdot (11)_3 \\
= \; & [(13 \cdot 12 \cdot 11) \cdot (10 \cdot 9 \cdot 8)] + [(14 \cdot 13 \cdot 12) - (13 \cdot 12 \cdot 11)] \cdot (11 \cdot 10 \cdot 9) \\
= \; & [1716 \cdot 720] + [2184 - 1716] \cdot 990 \\
= \; & 1235520 + 463320 \; = \; 1698840.
\end{aligned}$$

Arrangements with repetitions

If repeated elements are allowed, the number of sequences that can be formed from a set is far larger. If there is no restriction, then s^k k-sequences can be formed from an s-set. To show just how great this difference is, even in small cases, Table 5.1 shows the relative values of $(6)_k$ and 6^k.

k	$(6)_k$	6^k	$(6)_k/6^k$
1	6	6	1.00000
2	30	36	0.83333
3	120	216	0.55556
4	360	1296	0.27778
5	720	7776	0.09259
6	720	46656	0.01543

Table 5.1: Numbers of sequences without and with repetitions

In some cases, repetitions are allowed but limited. Problems of this kind can be modeled by talking about sets with a certain number of copies of each element. For example, if you want to count the number of sequences of length five based on the set $\{A, B\}$ that contain no more than three A's and no more than four B's, you could instead talk about 5-sequences selected from the collection $\{A, A, A, B, B, B, B\}$.

To avoid the problems that can arise from having multiple copies of the same element, we often talk of *distinguishable* and *indistinguishable* elements. For example, many problems can be modeled in terms of selecting marbles from an urn; the usual convention is that two sequences are distinguishable only if they differ in color sequence: any two blue marbles are indistinguishable.

One way to tackle these problems is to assume the "indistinguishable" objects can be distinguished, and then take this into account. For example, consider the letters in the word *ASSESS*. In how many distinguishable ways can you order these letters?

Suppose the letters were written on tiles with numbers as subscripts, like scrabble tiles. Label them so that no two copies of the same letter get the same subscript, for example $A_1S_1S_2E_1S_3S_4$. Then all the six letters are different, and there are 6! orderings. Say you have each of these orderings written on slips of paper.

Now collect together into one pile all the slips that differ only in their subscripts. For example, $A_1E_1S_1S_2S_3S_4$ and $A_1E_1S_2S_1S_3S_4$ will be in the same pile, as will $A_1E_1S_2S_3S_1S_4$, $A_1E_1S_4S_2S_3S_1$, and several others. In fact, we can work out how many slips there are in a pile. There are four letters *S*, and one each of the others. Two slips will be in the same pile when they have the letters in the same order, but the subscripts on the *S*'s are in different order. There are $4! = 24$ ways to order the four subscripts, so there are 4! slips in each pile. Therefore there are $6!/4! = 30$ piles.

Two orderings can be distinguished if and only if their slips are in different piles, so there are $6!/4! = 30$ distinguishable orderings of *ASSESS*.

The same principle can be applied with several repeated letters. For example, if *SUCCESS* is written as $S_1U_1C_1C_2E_1S_2S_3$, we see that there are 2! ways of ordering the *C*'s and 3! ways of ordering the *S*'s, so each pile will contain $3!{\cdot}2!$ slips, and the number of distinguishable orderings is $7!/(3!{\cdot}2!) = 420$.

Sample Problem 5.21 *In how many distinguishable ways can you order the letters of the word MISSISSIPPI?*

Solution. There are one *M*, four *I*'s, four *S*'s and two *P*'s, for a total of 11 letters. So the number of orderings is $11!/(4!{\cdot}4!{\cdot}2!)$ or 34650.

Practice Exercise. In how many distinguishable ways can you order the letters of the word *BANANA*?

The problems are significantly harder if not all the available repetitions are used.

Sample Problem 5.22 *An experimenter has an urn containing 12 marbles: five red, two blue and five green. Assuming that marbles of the same color are indistinguishable, how many different sequences of length 4 can be chosen from the set?*

Solution. If unlimited numbers of each color were available, this would be the same as the problem of selecting sequences of length 4 from a 3-set with unlimited repetitions allowed, and there would be $3^4 = 81$ solutions. It is necessary to exclude those with three or four blue marbles. In the obvious notation, these are *BBBR*, *BBRB*, *BRBB*, *RBBB*, *BBBG*, *BBGB*, *BGBB*, *GBBB*, and *BBBB*, nine in total. So the answer is $81 - 9 = 72$.

Practice Exercise. What is the answer if there are four red, three blue and five green marbles?

Exercises 5.4

In Exercises 1 to 12, calculate the indicated quantity.

1. $5!$	**2.** $(8)_3$	**3.** $(4)_4$
4. $9!$	**5.** $(5)_4$	**6.** $(9)_2$
7. $8!$	**8.** $(8)_5$	**9.** $(4)_1$
10. $4!$	**11.** $(5)_2$	**12.** $(7)_2$

13. Ten midsize cars are available for rental. Three customers arrive, and each chooses a midsize. In how many different ways can the choice be made?

14. A *palindrome* is a "word" (any string of letters) that reads the same forwards or backwards, such as *CIVIC* or *AABAA*. How many 5-letter palindromes are there (using the ordinary, 26-letter alphabet)?

15. There are 13 contestants. In how many different ways can the judges award first, second and third prize?

16. How many ways are there of seating six people at a round table?

17. How many ways are there of seating six people at a round table so that two specific people sit together?

18. Five stereo systems are to be arranged in a line against the wall of the appliance department.
(i) How many ways can this be done?
(ii) How many ways can they be arranged if the most expensive model must be in the middle?

19. Three men and four women sit in a row. How many different ways can they do it if:
(i) the men must sit together?
(ii) the women must sit together?

20. The ACME company uses serial numbers consisting of three letters followed by five digits.
(i) How many possible serial numbers are there?
(ii) How many possible serial numbers are there, if no digit can be repeated?
(iii) How many if no repetitions of digits or letters are allowed?

21. John likes to arrange his books. He has four western, five mystery and six science fiction books, all different.

(i) In how many ways can he arrange them on a shelf?

(ii) In how many ways can he arrange them if all the books on the same subject must be grouped together?

In Exercises 22 *to* 29, *how many ways can you arrange the letters of the word?*

22. *BORROW*

23. *PROFESSOR*

24. *MOOSEWOOD*

25. *ACCESSORY*

26. *OFFERED*

27. *RUBBLE*

28. *BOOKKEEPER*

29. *DENDRONOID*

30. A bookseller has four copies of *Hearts in Atlantis*, five copies of *Bag of Bones* and three copies of *Needful Things* in her Stephen King section. Copies of the same book are identical. In how many different ways can she arrange them?

31. Prove that $(2n)!/n! = 2^n(1 \times 3 \times 5 \times \cdots \times (2n-1))$, and consequently that, for any positive integer n,

$$2 \times 6 \times 10 \times \cdots \times (4n-6) \times (4n-2) = (n+1) \times (n+2) \times \cdots \times (2n-1) \times 2n.$$

32. Write out a formal inductive proof of the value of $(s)_k$.

33. You have n different gemstones, and you want to string them together to make a necklace. Show that $(n-1)!/2$ different necklaces are possible.

34. You have a deck of nine cards: three (identical) aces of spades, three (identical) kings of hearts, and three (identical) queens of clubs. In how many different ways can you deal a sequence of four cards?

5.5 Selections

Selections

Given a set S, we are often interested in knowing how many different subsets of a given size are contained in S. This depends only on the size of S. We shall write $\binom{s}{k}$ or $C(s,k)$ for the number of k-subsets of an s-set; it is usual to read the symbol as "s choose k."

We can use the formula (5.4) to derive expressions for the numbers $C(s,k)$. Suppose S is a set with s elements. It is clear that every k-set that we choose from

S gives rise to exactly k! distinct k-sequences on S and that the same k-sequence never arises from different k-sets. So the number of k-sequences on S is $k!$ times the number of k-sets on S, or

$$\binom{s}{k} = \frac{(s)_k}{k!} = \frac{s!}{(s-k)!k!} \tag{5.5}$$

When calculating $\binom{s}{k}$ in practise, you would usually calculate $(s)_k$, then divide by $k!$. So

$$\binom{s}{k} = \frac{s \cdot (s-1) \cdot (s-2) \cdot \cdots \cdot (s-k+1)}{1 \cdot 2 \cdot 3 \cdot \cdots \cdot k}.$$

There are k factors in the denominator and in the numerator.

From the definition of 0! we get $\binom{s}{0} = 1$, which is quite consistent. It is possible to choose *no* elements from a set, but you can't imagine different ways of doing it. It will be convenient to define $\binom{s}{k} = 0$ if $k > s$ (again this makes sense; there is no way to choose more than s elements from an s-set).

Sample Problem 5.23 *Calculate* $C(8,4)$ *and* $\binom{6}{6}$.

Solution.

$$C(8,4) = \frac{8 \cdot 7 \cdot 6 \cdot 5}{4 \cdot 3 \cdot 2 \cdot 1} = 70.$$

$$\binom{6}{6} = \frac{6!}{0! \cdot 6!} = 1.$$

(There is no need for calculation: the terms 6! in the numerator and denominator cancel out.)

Practice Exercise. Calculate $C(9,5)$ and $\binom{6}{0}$.

Sample Problem 5.24 *A student must answer five of the eight questions on a test. How many different ways can she answer, assuming there is no restriction on her choice?*

Solution. $\binom{8}{5} = 56$ ways.

Practice Exercise. How many ways can she answer if she must choose five, one of which is Question 1?

Sample Problem 5.25 *Recall from Section* 1.5 *that computers read strings of* binary digits, *with every entry* 0 or 1. *How many 8-bit strings (strings of 8 binary digits) are there that contain exactly five* 1*'s?*

Solution. To specify a string, it is sufficient to say which positions have 1's. There are $C(8,5)$ choices, so the answer is $C(8,5) = 56$.

Practice Exercise. How many 8-bit strings contain exactly four 1's?

Sample Problem 5.26 *How many ways can a committee of three men and two women be chosen from six men and four women?*

Solution. The three men can be chosen in $\binom{6}{3}$ ways; the two women can be chosen in $\binom{4}{2}$ ways. Using the multiplication principle, the total number of committees possible with no restrictions is

$$\begin{aligned}\binom{6}{3} \times \binom{4}{2} &= \frac{6!}{3!3!} \times \frac{4!}{2!2!} \\ &= 120.\end{aligned}$$

Practice Exercise. You wish to borrow two mystery books and three westerns from your friend. He owns five mysteries and seven westerns. How many different selections can you make?

Sample Problem 5.27 *How many different "words" of five letters can you make from the letters of the word*

REPUBLICAN,

if every word must contain two different vowels and three different consonants?

Solution. The three consonants can be chosen in $\binom{6}{3} = 20$ ways, and the vowels in $\binom{4}{2} = 6$ ways. After the choice is made, the letters can be arranged in $5! = 120$ ways. So there are $20 \cdot 6 \cdot 120 = 14400$ "words."

Practice Exercise. What is the answer if you use the word

DEMOCRAT?

An illustrative example

The following example illustrates two important facts. First, the numbers that arise in selection problems can be very large — sometimes the intermediate steps are much greater than the answer. And second, there are often two (or more) ways to attack the same problem.

Suppose the sixteen members of the girls' track team are to be divided into four teams of four members each, to train for the relay. Assuming that you don't care who runs in which position, how many ways can they be divided?

Start by arbitrarily ordering the four teams I, II, III, IV. The members of Team I can be chosen in $C(16,4)$ ways. After they are chosen, there are twelve girls left, so there are $C(12,4)$ ways to choose team II. Team III can be chosen in $C(8,4)$ ways. Then Team IV is decided (there are only four girls left.) We have

$$C(16,4) \cdot C(12,4) \cdot C(8,4)$$

arrangements. But the order of the teams did not really matter, so every arrangement is one of 4! that all give the same solution. (4! is the number of ways of arbitrarily assigning the labels I, II, III, IV to the four teams.) So the answer is

$$\frac{C(16,4)\cdot C(12,4)\cdot C(8,4)}{4!} = \frac{16!}{12!\cdot 4!} \times \frac{12!}{8!\cdot 4!} \times \frac{8!}{4!\cdot 4!} \times \frac{1}{4!} = \frac{16!}{(4!)^5}$$

(the 12!'s and 8!'s cancel).

The equation (5.5) was useful, because of the way the factorials 12! and 8! canceled.

If you try to work out the exact answer to the problem, you'll find that 16! is greater than $2\cdot 10^{13}$. On many calculators this large number will only be represented approximately, and the divisions will give you at best an approximate answer. But if you write out

$$\frac{16\cdot 15\cdot 14\cdot 13\cdot 12\cdot 11\cdot 10\cdot 9\cdot 8\cdot 7\cdot 6\cdot 5\cdot 4\cdot 3\cdot 2\cdot 1}{(4\cdot 3\cdot 2\cdot 1)\cdot(4\cdot 3\cdot 2\cdot 1)\cdot(4\cdot 3\cdot 2\cdot 1)\cdot(4\cdot 3\cdot 2\cdot 1)\cdot(4\cdot 3\cdot 2\cdot 1)}$$

and cancel you'll get $15\cdot 7\cdot 13\cdot 11\cdot 5\cdot 7\cdot 5$, which comes to 2627625. This selection problem involved intermediate numbers nearly a million times as great as the solution. If, instead of cancelling the 12! and 8! we had calculated the three C-numbers, the multiplications would have been very large.

As we promised, there is another way to handle this problem. This time, select one girl at random. There are fifteen other girls from whom her three teammates can be chosen, so this can be done in $\binom{15}{3}$ ways. For each possible choice here, suppose you select one of the twelve remaining girls. Her team can be completed in $\binom{11}{3}$ ways. Again choose a girl from those remaining; her team can be completed in $\binom{7}{3}$ ways. And the remaining girls must make up the other team. The number of choices is

$$\binom{15}{3} \times \binom{11}{3} \times \binom{7}{3} = \frac{15\cdot 14\cdot 13}{3\cdot 2\cdot 1} \times \frac{11\cdot 10\cdot 9}{3\cdot 2\cdot 1} \times \frac{7\cdot 6\cdot 5}{3\cdot 2\cdot 1}$$

which comes to the same answer. This time the intermediate numbers were considerably smaller.

Problems combining different methods

Some problems can be solved by combining the methods we have been using with the rule of sum, which was given in Section 5.2 as

$$|S| = |S\cap T| + |S\backslash T|. \tag{5.2}$$

Sample Problem 5.28 *In Sample problem 5.26, how many ways are there to select the committee if one particular couple, say Mr. and Mrs. Smith, do not wish to be on the committee together?*

Solution. The total number of committees possible with no restrictions was 120. If Mr. and Mrs. Smith are both on the committee, then the rest of the committee is found by choosing two of the five remaining men and one of the three remaining women. This can be done in

$$\binom{5}{2} \times \binom{3}{1} = \frac{5!}{3!2!} \times \frac{3!}{2!1!} = 30$$

ways. So 30 of the 120 possible committees contain both Smiths; when these are excluded there remain 90 ways of choosing the committee.

(We have in fact applied the rule of sum: interpret S as the set of all possible committees, and T as the set of committees not containing both Smiths. Then we found $|S| = 120$ and $|S \cap T| = 30$, we required $|S \backslash T|$, which the rule tells us equals 90.)

Practice Exercise. A group contains three men and seven women. In how many ways can a committee of three people be chosen if it can contain no more than one man?

Sometimes both arrangement and selection techniques are applied to the same problem.

Sample Problem 5.29 *A copying company has eight photocopying machines and seven employees who can operate them. There are four identical copying jobs to be done. (There are 4000 booklets to be made; each person makes 1000 copies.) How many ways are there to allocate these jobs to operators and machines?*

Solution. We shall give two solutions to this problem.

(i) This is like Sample Problem 5.18, but the jobs are now identical. In that problem there were $P(7,4) \times P(8,4)$ arrangements. As the jobs are indistinguishable, the ordering of A,B,C,D is irrelevant, so we divide the number of arrangements by 4!. The answer is

$$\frac{(7)_4 \cdot (8)_4}{4!} = \frac{(7\cdot6\cdot5\cdot4) \times (8\cdot7\cdot6\cdot5)}{4\cdot3\cdot2\cdot1},$$

which works out to 58800.

(ii) There are $C(7,4)$ ways to choose four operators and $C(8,4)$ ways to choose four machines. When the choice is made, there are 4! ways to assign the four workers to four machines. So the answer is

$$P(7,4) \times P(8,4) \times 4! = \frac{7\cdot6\cdot5\cdot4}{4\cdot3\cdot2\cdot1} \times \frac{8\cdot7\cdot6\cdot5}{4\cdot3\cdot2\cdot1} \times (4\cdot3\cdot2\cdot1),$$

the same as before. (There is an extra factor 4! in the numerator and in the denominator).

Symmetry of the choice function

From (5.5) we see that

$$\binom{s}{k} = \binom{s}{s-k}. \tag{5.6}$$

This equality can also be derived as follows. If we wish to choose k things from a set of s, we could just as easily say which $k-s$ will *not* be included. In other words, to specify a subset of S, we could just as easily specify its complement (in the universal set S).

Sample Problem 5.30 *Calculate* $\binom{100}{98}$.

Solution. From (5.6),

$$\binom{100}{98} = \binom{100}{2} = \frac{100\cdot 99}{1\cdot 2} = 4950.$$

Practice Exercise. Calculate $\binom{81}{79}$.

Pascal's triangle

Theorem 5.6 *For all integers k and s such that* $1 \le k \le s$,

$$\binom{s}{k} = \binom{s-1}{k} + \binom{s-1}{k-1}.$$

Proof. Suppose S is an s-set, and suppose x is one particular member of S. Each subset of S falls into one of two categories: those that *do* contain x, and those that *do not* contain x. So we calculate the number of k-sets on S that contain x, and then the number that do not. Adding these numbers together, we get the total number of sets.

The k-sets on S that contain x each have $k-1$ other elements, and we get a different k-set whenever we choose a different $(k-1)$-set from those other objects. So their number equals the number of $(k-1)$-sets that can be chosen from $S\backslash\{x\}$, namely $\binom{s-1}{k-1}$.

The k-sets on S that do not contain x are precisely the k-sets of $S\backslash\{x\}$, and there are $\binom{s-1}{k}$ of them. Adding, we get the result. □

Theorem 5.6 can be used to generate a table of values of the numbers $\binom{s}{k}$. For convenience, we call the top row of the table row 0. In row s we write the values of

$$\binom{s}{0}\binom{s}{1}\binom{s}{2}\cdots\binom{s}{s}$$

in that order. We write the table in a triangular form: $\binom{s}{0}$ occurs a half space to the left of $\binom{s-1}{0}$. So the table is as shown. Observe that $\binom{s}{k}$ lies in the table with $\binom{s-1}{k}$ and $\binom{s-1}{k}$ above it, a half-step to the left and a half-step to the right respectively. So Theorem 5.6 tells us that every term can be constructed by adding together the two numbers above it.

$$
\begin{array}{ccccccc}
 & & & \binom{0}{0} & & & \\
 & & \binom{1}{0} & & \binom{1}{1} & & \\
 & \binom{2}{0} & & \binom{2}{1} & & \binom{2}{2} & \\
\binom{3}{0} & & \binom{3}{1} & & \binom{3}{2} & & \binom{3}{3} \\
\cdots & \cdots & \cdots & \cdots & \cdots & \cdots & \cdots
\end{array}
$$

This table is called *Pascal's triangle*. When the values are substituted in we get the following array.

$$
\begin{array}{ccccccccccccc}
 & & & & & & 1 & & & & & & \\
 & & & & & 1 & & 1 & & & & & \\
 & & & & 1 & & 2 & & 1 & & & & \\
 & & & 1 & & 3 & & 3 & & 1 & & & \\
 & & 1 & & 4 & & 6 & & 4 & & 1 & & \\
 & 1 & & 5 & & 10 & & 10 & & 5 & & 1 & \\
1 & & 6 & & 15 & & 20 & & 15 & & 6 & & 1 \\
 & \cdots & \cdots & \cdots & \cdots & \cdots & \cdots & \cdots & \cdots & \cdots & \cdots & \cdots &
\end{array}
$$

Exercises 5.5

In Exercises 1 *to* 12, *calculate the indicated quantity.*

1. $C(8,3)$ **2.** $C(7,7)$ **3.** $C(9,4)$

4. $\binom{6}{5}$ **5.** $\binom{6}{3}$ **6.** $\binom{6}{1}$

7. $C(7,3)$ **8.** $C(10,2)$ **9.** $C(9,7)$

10. $\binom{8}{7}$ **11.** $\binom{8}{6}$ **12.** $\binom{8}{0}$

13. The math department wishes to select four of its 16 members to teach the finite math course. In how many ways can this selection be made?

14. A test has ten questions, five in part A and five in part B.

(i) A student has to choose five questions, two from part A and three from part B. How many ways can she make her choice?

(ii) How many ways can she make her choice if she must choose five questions, at least two from each part?

15. The Student Council consists of six juniors and 12 seniors. A committee of two juniors and three seniors is to be formed. How many ways can this be done?

16. A regular deck of cards contains 52 cards: 13 spades, 13 hearts, 13 diamonds and 13 clubs. A five-card hand is dealt.

(i) How many hands contain only spades?

(ii) How many hands contain three spades and two clubs?

(iii) How many hands contain only spades and clubs, at least one of each?

17. How many 12-bit binary strings with five 1's are possible? How many of them start 101?

18. There are twelve appetizers on the menu, five cold and seven hot. How many ways can you choose four different appetizers for your table? In how many cases does this include two hot and two cold appetizers?

19. Ten points $A, B, \ldots, J$ in the plane are selected, no three of them collinear. How many triangles can be constructed with three of these points as the vertices? How many of the triangles have A as one vertex?

20. A club with 24 members wants to elect a committee consisting of President, Secretary and three Ordinary Members. How many committees are possible?

21. A state lottery requires you to choose five different numbers from $\{1, 2, \ldots, 49\}$.

(i) How many possible choices are there?

(ii) The state then draws six different numbers. You win if all five of your numbers are chosen. How many of your possible choices of five numbers will make you a winner?

(In both cases, don't worry about the order in which the numbers are chosen.)

22. The motor pool has eight drivers and nine vehicles. Four drivers must be assigned to four vehicles for today's duty. How many ways can it be done?

23. It is required to select, from the set of numbers $\{1, 2, 3, 4, 5, 6, 7, 8\}$, a subset of four that will contain either 1 or 2, but not both. How many selections are possible?

24. Suppose a class of r boys and s girls wishes to choose a committee with k members.

(i) Show that the total number of committees possible is
$$\binom{r+s}{k}.$$

(ii) Suppose there are to be t boys on the committee. Show that the number of possible committees is
$$\binom{r}{t}\times\binom{s}{k-t},$$
provided $0\le t\le k$.

(iii) Use these results to prove that
$$\binom{r+s}{k}=\binom{r}{k}\times\binom{s}{0}+\binom{r}{k-1}\times\binom{s}{1}+\cdots+\binom{r}{0}\times\binom{s}{k}.$$

25. Two lines are drawn on the plane, with intersection point A. From one line m points are selected, and from the other n are selected, but A is not in either set. How many triangles can be formed with vertices among the $m+n$ points? How many can be formed from the $m+n+1$ points if A is added to the set?

26. Prove the following identities algebraically.

(i) $k\binom{n}{k}=n\binom{n-1}{k-1}$ whenever $1\le k\le n$.

(ii) $k(k-1)\binom{n}{k}=n(n-1)\binom{n-2}{k-2}$ whenever $2\le k\le n$.

27. (i) It is required to select a committee with k members from a group of n candidates, and to select a chairman from the committee. Calculate the number of ways this can be done by two different methods:

(a) First find the number of ways that a committee can be chosen, and multiply by the number of ways the chairman can then be selected;

(b) First find the number of ways that the chairman can be selected, and multiply by the number of ways the rest of the committee can be chosen.

Use the fact that (a) and (b) must give the same answer to prove the identity in Exercise 26(i). (We call this a *combinatorial proof* of the identity.)

(ii) Find a combinatorial proof of the identity in Exercise 26(ii).

28. Prove Theorem 5.6 *by induction.*

29. Prove that
$$\binom{n}{r}+2\binom{n}{r-1}+\binom{n}{r-2}=\binom{n+2}{r}.$$

30. Construct the first ten rows of Pascal's triangle.

5.6 The Binomial Theorem and its Applications

The binomial theorem

Suppose s is any positive integer. We would like to find an expression for $(x+y)^s$, as a function of x and y.

Consider the following product of s factors:

$$(x_1+y_1)\cdot(x_2+y_2)\cdot\dots\cdot(x_s+y_s).$$

If you expand this you will get 2^s terms, and each will have s factors, with either x_i or y_i as the i-th factor. For example, if $s=2$ you get

$$x_1x_2+y_1x_2+x_1y_2+y_1y_2,$$

and if $s=3$ the product is

$$x_1x_2x_3+y_1x_2x_3+x_1y_2x_3+y_1y_2x_3+x_1x_2y_3+y_1x_2y_3+x_1y_2y_3+y_1y_2y_3.$$

In constructing the terms, you make s decisions:

- from the first factor, take either x_1 or y_1 ;
- from the second factor, take either x_2 or y_2 ;

and so on.

In the general case, how many terms are there with exactly k x's? The term with $x_1x_2\dots x_ky_{k+1}y_{k+2}\dots y_s$ can arise in one and only one way in the product: you must choose the x term at steps $1,2,\dots,k$ and the y term at every later step. Similarly, the term with $x_{i_1}x_{i_2}\dots x_{i_k}$ and all other terms y's occurs if and only if you choose x at steps $i_1,i_2,\dots,i_k$, and y at the other $s-k$ steps. So the number of terms that contain precisely k of the x's will equal the number of ways of selecting k objects from $\{x_1,x_2,\dots,x_s\}$, namely $\binom{s}{k}$.

Now put $x_1=x_2=\dots=x_s=x$ and $y_1=y_2=\dots=y_s=y$. There will be exactly $\binom{s}{k}$ terms equal to x^ky^{s-k}. So

$$(x+y)^s=\sum_{k=0}^{s}\binom{s}{k}x^ky^{s-k}.$$

This result is the *binomial theorem* for positive integer index s:

Theorem 5.7 *If x and y are any numbers and s is any positive integer, then*

$$(x+y)^s=\sum_{k=0}^{s}\binom{s}{k}x^ky^{s-k}.$$

Once you know the formula, it is easy to prove Theorem 5.7 by induction.

The first few cases of the theorem should be familiar to you:

$$(x+y)^2=x^2+2xy+y^2,$$

$$(x+y)^3 = x^3 + 3x^2y + 3xy^2 + y^3,$$
$$(x+y)^4 = x^4 + 4x^3y + 6x^2y^2 + 4xy^3 + y^4.$$

They should also be familiar when one of the two variables is replaced by 1, or when a negative sign or constant coefficient is included:

$$(1+x)^3 = 1 + 3x + 3x^2 + x^3,$$
$$(x-y)^3 = x^3 - 3x^2y + 3xy^2 - y^3,$$
$$(2x+y)^3 = 8x^3 + 12x^2y + 6xy^2 + y^3.$$

The later cases are easily worked out from the formula.

Sample Problem 5.31 *Write down an expression for* $(1+x)^6$.

Solution. The relevant coefficients are $\binom{6}{0} = \binom{6}{6} = 1$, $\binom{6}{1} = \binom{6}{5} = 6$, $\binom{6}{2} = \binom{6}{4} = 15$, $\binom{6}{3} = 21$. So

$$\begin{aligned} & (1+x)^6 \\ = \ & \tbinom{6}{0} + \tbinom{6}{1}x + \tbinom{6}{2}x^2 + \tbinom{6}{3}x^3 + \tbinom{6}{4}x^4 + \tbinom{6}{5}x^5 + \tbinom{6}{6}x^6 \\ = \ & 1 + 6x + 15x^2 + 21x^3 + 15x^4 + 6x^5 + x^6. \end{aligned}$$

Practice Exercise. Write down an expression for $(x-y)^6$.

The binomial theorem can be used to find the approximate values of powers of numbers close to 1.

Sample Problem 5.32 *Find the approximate value (to within* 10^{-3}*) of* 1.02^{10}.

Solution. We write $1.02^{10} = (1 + 2 \times 10^{-2})^{10}$. From the binomial theorem, this equals

$$1 + 10 \cdot 2 \cdot 10^{-2} + 45 \cdot 4 \cdot 10^{-4} + 120 \cdot 8 \cdot 10^{-6} + \cdots$$

Every subsequent term in at most one-tenth of the one before it. So the approximate value is

$$1 + 0.2 + 0.018 + 0.00096 + \ldots = 1.219 \text{ approx.}$$

Practice Exercise. Find the approximate value of 0.99^8.

You should observe the relationship between Pascal's triangle, in Section 5.5, and the binomial theorem. The coefficients in row s of the triangle are precisely the coefficients in the expression for $(x+y)^s$. For example, row 6 is

$$1, 6, 15, 21, 15, 6, 1.$$

(Remember, the first row is row 0.)

Consequences of the theorem

Two interesting summation formulae can be obtained by putting $x = y = 1$ and $x = -1, y = 1$ respectively in Theorem 5.7:

$$\sum_{k=0}^{s} \binom{s}{k} = 2^s, \tag{5.7}$$

$$\sum_{k=0}^{s} (-1)^k \binom{s}{k} = 0. \tag{5.8}$$

Equation (5.8) can be rewritten as

$$1 + \sum_{k=1}^{s} (-1)^k \binom{s}{k} = 0,$$

so

$$\begin{aligned} 1 &= -\sum_{k=1}^{s} (-1)^k \binom{s}{k} \\ &= \sum_{k=1}^{s} (-1)^{k-1} \binom{s}{k}; \end{aligned} \tag{5.9}$$

this form is sometimes more useful.

Equation (5.7) gives us an interesting way to work out the number of elements in the power set of a set.

Theorem 5.8 *Suppose S is any set with s elements. The number of subsets of S (that is, the number of elements of $\mathcal{P}(S)$) is 2^s.*

Proof. The number of k-element subsets of S is $\binom{s}{k}$. The total number of subsets equals the number of 0-element subsets (1, for the empty set), plus the number of 1-element subsets, plus the number of 2-element subsets, and so on up to S itself. So the number is $\sum_{k=0}^{s} \binom{s}{k}$, and the result follows from (5.7). □

Sample Problem 5.33 *How many ways are there to choose three or more people from a set of eleven people?*

Solution. There are 2^{11} possible ways to choose a subset of the eleven people. However, the subsets with 0, 1 or 2 elements are not allowed. So the number is

$$2^{11} - \binom{11}{0} - \binom{11}{1} - \binom{11}{2} = 2048 - 1 - 11 - 55 = 1981.$$

Practice Exercise. How many ways are there to choose at most seven books from a selection of ten books?

Sample Problem 5.34 *Evaluate* $\sum_{k=0}^{s} k\binom{s}{k}$.

Solution. We first evaluate $\sum_{k=0}^{s} k\binom{s}{k}x^k$ and then put $x = 1$. Since

$$\begin{aligned} k\binom{s}{k}x^k &= \frac{s!kx^k}{k!(s-k)!} \\ &= \frac{s!x^k}{(k-1)!(s-k)!} \\ &= s\frac{(s-1)!x^k}{(k-1)!(s-k)!} \\ &= sx\binom{s-1}{k-1}x^{k-1}, \end{aligned}$$

we have

$$\sum_{k=1}^{s} k\binom{s}{k}x^k = sx\sum_{k=1}^{s}\binom{s-1}{k-1}x^{k-1}.$$

Make the substitution $h = k-1$. The right-hand side becomes

$$sx\sum_{h=0}^{s-1}\binom{s-1}{h}x^h,$$

which equals $sx(1+x)^{s-1}$. Putting $x = 1$,

$$\sum_{k=0}^{s} k\binom{s}{k} = 0 + \sum_{k=1}^{s} k\binom{s}{k} = 2^{s-1}s.$$

(The fact used at the end of the sample problem, that the term (namely $sk\binom{s}{k}$) is zero when $k = 0$, is often useful in changing between an expression of the form $\sum_{k=0}^{s}$ and one of the form $\sum_{k=1}^{s}$.

Exercises 5.6

In Exercises 1 *to* 10, *use the binomial theorem to calculate the indicated quantity.*

1. $(x-1)^4$

2. $(x-3y)^3$

3. $(1-2z)^5$

4. $(1+x^2)^3$

5. $(2x+y)^4$

6. $(2x-5y)^3$

7. $(x+x^{-1})^4$

8. $(1+x^2)^2$

9. $(x+y+z)^3$

10. $(1+2x-y)^3$

11. How many 8-bit strings (strings of 8 binary digits) are there?

12. Use the binomial theorem with $x = -1$ and $y = 1$ to prove that

$$\binom{s}{0} + \binom{s}{2} + \binom{s}{4} + \cdots = \binom{s}{1} + \binom{s}{3} + \binom{s}{5} + \cdots$$

and consequently that

$$\binom{s}{0} + \binom{s}{2} + \binom{s}{4} + \cdots = 2^{s-1}.$$

13. Prove $\sum_{k=0}^{n} \binom{2n+1}{k} = 2^{2n}$.

14. Use the answer to the preceding exercise to find the number of distinguishable k-sets that can be chosen from a set S of $3k+1$ objects, if k members of S are indistinguishable and all other members can be distinguished.

15. Prove Theorem 5.7 by induction.

16. Show that

$$\sum_{k=1}^{n+r-1} k\binom{n-k}{r-1} = \binom{n+1}{r+1}.$$

Hence show that, if we consider all possible k-sets chosen from the set $\{1, 2, \ldots, n\}$, the average value of their smallest elements is

$$\frac{n+1}{r+1}.$$

5.7 Some Further Counting Results

Principle of inclusion and exclusion

We are now in a position to prove Theorem 5.1, the *principle of inclusion and exclusion*. We recall its statement:

The number of elements in $X_1 \cup X_2 \cup \cdots \cup X_n$ *is*

$$\begin{aligned}
&\sum_{i_1=1}^{n} |X_{i_1}| - \sum_{i_1=1}^{n} \sum_{i_2=i_1+1}^{n} |X_{i_1} \cap X_{i_2}| + \cdots \\
&+(-1)^{k-1} \sum_{i_1=1}^{n} \sum_{i_2=i_1+1}^{n} \cdots \sum_{i_k=i_{k-1}+1}^{n} |X_{i_1} \cap X_{i_2} \cap \cdots \cap X_{i_k}| \\
&+\cdots+(-1)^{n-1}|X_{i_1} \cap X_{i_2} \cap \cdots \cap X_{i_n}|.
\end{aligned} \tag{5.10}$$

Proof. Suppose the object x belongs to r of the sets, namely $X_{j_1}, X_{j_2}, \ldots, X_{j_r}$. Then x is counted r times in calculating $\sum |X_{i_1}|$, and contributes r to that part of the sum

(5.10). It appears $\binom{r}{2}$ times in subsets of the form $X_{i_1} \cap X_{i_2}$, since it appears in precisely those for which $\{i_l, i_2\}$ is a 2-set of $\{j_l, j_2, \dots, j_r\}$; so it contributes $-\binom{r}{2}$ to that part of (5.10). Continuing, we find its total contribution is

$$\sum_{k=1}^{r} (-1)^{k-1} \binom{r}{k}, \tag{5.11}$$

which equals 1 by (5.9). (The sum in (5.11) stops at $k = r$ because x cannot occur in any intersection of r+l or more sets.) So the sum (5.10) contains a contribution 1 for every member of $X_1 \cup X_2 \cup \cdots \cup X_n$, and its total is $|X_1 \cup X_2 \cup \cdots \cup X_n|$. □

It is easiest to think of (5.10) as

sum of the (sizes of) the sets, minus sum of the (sizes of) intersections of two sets, plus sum of the (sizes of) intersections of three sets, minus ....

There are n terms in the first summation, $\binom{n}{2}$ in the second, $\binom{n}{3}$ in the third, and so on. This is particularly useful when all the sets X_i are the same size, as are all the sets $X_{i_1} \cap X_{i_2}$, and so on: if the intersection of any k of the sets X_i always has s_k elements, then (5.10) becomes

$$ns_1 - \binom{n}{2}s_2 + \binom{n}{3}s_3 - \cdots + (-1)^{n-1}s_n. \tag{5.12}$$

Sample Problem 5.35 *A social security number is a string of nine digits. If any digit can be used in any position, how many such numbers are possible? How many are there that contain all the odd digits?*

Solution. There are 10^9 possibilities. To solve the second part, it is easiest to find out how many *do not* contain all the odd digits, then subtract the answer from 10^9. Write S_x for the set of all possibilities that do not contain x. Then we want $|S_1 \cup S_3 \cup S_5 \cup S_7 \cup S_9|$. For any x, $|S_x| = 9^9$, if $x \neq y$, $|S_x \cap S_y| = 8^9$. Similarly $|S_x \cap S_y \cap S_z| = 7^9$, and so on. So, from (5.12) the sum is

$$5 \cdot 9^9 - \binom{5}{2} \cdot 8^9 + \binom{5}{3} \cdot 7^9 - \binom{5}{4} \cdot 6^9 + 5^9$$

and the answer we required is

$$10^9 - 5 \cdot 9^9 + 10 \cdot 8^9 - 10 \cdot 7^9 + 5 \cdot 6^9 - 5^9.$$

Practice Exercise. How many possible social security numbers contain all the digits 6, 7, 8 and 9?

Counting powers

There are some interesting applications of the principle of inclusion and exclusion in elementary number theory. For example, suppose we want to know how many of the numbers from 1 to n are perfect squares, perfect cubes, or any perfect higher power. We illustrate the technique for finding this in a relatively small case so that the result can be checked by hand, but the method can be applied in general.

We shall find out how many of the numbers $\{1,2,\ldots,100\}$ are perfect squares, perfect cubes, or perfect higher powers. The number 1 is a perfect n-th power for every n. For convenience, we shall find the number of perfect powers other than 1, and add 1 to the answer. In this case we need only consider up to sixth powers: $2^7 = 128 > 100$, so every seventh or higher power (other than 1) is greater than 1000. Write X_i for the set of all i-th powers in the range$\{2,3,\ldots,100\}$, for $i = 2,3,\ldots,6$. Then, from the theorem, the number of integers that belong to at least one of these sets is

$$\begin{array}{rl} & |X_2| + |X_3| + |X_4| + |X_5| + |X_6| \\ - & (|X_2 \cap X_3| + |X_2 \cap X_4| + \ldots + |X_4 \cap X_6| + |X_5 \cap X_6|) \\ + & (|X_2 \cap X_3 \cap X_4| + |X_2 \cap X_3 \cap X_5| + \ldots + |X_4 \cap X_5 \cap X_6|) \\ - & (|X_2 \cap X_3 \cap X_4 \cap X_5| + \ldots + |X_3 \cap X_4 \cap X_5 \cap X_6|) \\ + & |X_2 \cap X_3 \cap X_4 \cap X_5 \cap X_6|. \end{array}$$

All of these sets and their sizes are easy to calculate. The largest square up to 100 is 10^2, so $X_2 = \{2,3,4,\ldots,10\}$ and $|X_2| = 9$, $X_3 = \{2,3,4\}$ so $|X_3| = 3$, $|X_4| = 2$, $|X_5| = |X_6| = 1$. In general $X_i \cap X_j = X_l$, where l is the least common multiple of i and j, and similarly for intersections of three or more, so $X_2 \cap X_3 = X_2 \cap X_6 = X_3 \cap X_6 = X_2 \cap X_3 \cap X_4 = X_2 \cap X_3 \cap X_6 = X_6$, $X_2 \cap X_4 = X_4$, and all the others are empty. Therefore the total is

$$\begin{array}{rl} & |X_2| + |X_3| + |X_4| + |X_5| + |X_6| \\ - & |X_2 \cap X_3| - |X_2 \cap X_4| - |X_2 \cap X_6| - |X_3 \cap X_6| \\ + & |X_2 \cap X_3 \cap X_4| + |X_2 \cap X_3 \cap X_6| \end{array}$$

which equals $9+3+2+1+1-1-2-1-1+1+1 = 13$. So the answer to the original question is 14 (adding 1 for the integer 1).

The same technique can be applied to problems about factors; see the exercises.

Derangements

A *derangement* of an ordered set of objects is a way of rearranging the objects so that none appears in its original position. A derangement of $(1,2,\ldots,n)$, for example, is an arrangement $(\alpha_1,\alpha_2,\ldots,\alpha_n)$ of the first n integers in which $\alpha_i = i$ never occurs. We write D_n for the number of possible derangements of n objects; clearly D_n does not depend on the type of objects in the set.

Theorem 5.9 *For every positive integer n,*

$$D_n = n![1 - \frac{1}{1!} + \frac{1}{2!} - \frac{1}{3!} + \cdots + (-1)^n \frac{1}{n!}].$$

Proof. Write S_i for the set of all arrangements $(\alpha_1, \alpha_2, \ldots, \alpha_n)$ of $(1, 2, \ldots, n)$ in which $\alpha_i = i$. We want to know how many arrangements are *not* in S_i for any i, so we want

$$n! - |S_1 \cup S_2 \cup \cdots \cup S_n|. \tag{5.13}$$

Each arrangement with $\alpha_1 = 1$ can be derived from an arrangement of $(2, 3, \ldots, n)$ by putting 1 in front of it, and each arrangement of $(2, 3, \ldots, n)$ gives precisely one member of S_1; so

$$|S_1| = (n-1)!.$$

Each of the sets S_i has the same number of elements as S_1, so

$$\sum_{i=1}^{n} |S_i| = n!.$$

Similarly, $S_1 \cap S_2 \cap \cdots \cap S_j$ has $(n-j)!$ elements. The intersection of any j sets has the same size, and there are $\binom{n}{j}$ ways of choosing j sets. So

$$\sum\sum\cdots\sum |S_{i_1} \cap S_{i_2} \cap \cdots \cap S_{i_j}| = \binom{n}{j}(n-j)! = \frac{n!}{j!}.$$

The theorem follows upon substituting this formula into (5.10) and applying the result to (5.13). □

For consistency we define $D_0 = 1$. The other small derangement numbers are easily calculated.

Sample Problem 5.36 *Calculate D_1, D_2, D_3, and D_4.*

Solution. Obviously $D_1 = 0$ and $D_2 = 1$. From the formula (or from listing cases) we get $D_3 = 2$ and $D_4 = 9$.

If the i-th element of the ordered set S occurs in position i of a permutation of S, we say i is a *fixed point* of the permutation. For example, the set S_i in the proof of Theorem 5.9 is the set of all permutations of S with fixed point i.

Suppose the number of permutations of an n set with exactly k fixed permutations is $f(n,k)$. To calculate $f(n,k)$ we first ask in how many ways could one choose the k positions to remain fixed? The answer is obviously $\binom{n}{k}$. The other $n-k$ entries must form a derangement, so they can be filled in D_{n-k} ways. So the number is $\binom{n}{k}D_{n-k}$.

Sample Problem 5.37 *How many permutations of $\{1,2,3,4\}$ have fewer than three fixed points?*

Solution. Let P_i be the set of permutations with i fixed points. Then the answer is

$$\begin{aligned} |P_0| + |P_1| + |P_2| &= f(4,0) + f(4,1) + f(4,2) \\ &= \binom{4}{0}D_4 + \binom{4}{1}D_3 + \binom{4}{2}D_2 \\ &= 9 + 4\cdot 2 + 6\cdot 1 = 23. \end{aligned}$$

Practice Exercise. How many permutations of $\{1,2,3,4\}$ have two or more fixed points?

A number of theorems can be proved about the derangement numbers. we present one example.

Theorem 5.10 $D_n = (n-1)(D_{n-1} + D_{n-2})$.

Proof. The D_n derangements of $\{1,2,\dots,n\}$ can be partitioned into $n-1$ classes $S_1, S-2, \dots, S_{n-1}$, where S_k consists of all the derangements with last element k. (No derangement will have last element n.) We shall prove that each set S_k has $D_{n-1} + D_{n-2}$ elements.

The elements of S_k can be classified further. Let A_k be the subset of S_k consisting of all the derangements in S_k with k-th element n and B_k be the set of those whose k-th element is not n.

Of the members of A_k, if entries n and k are deleted, the remaining elements must be a derangement of the $(n-2)$-set $T = \{1,2,\dots,k-1,k+1,\dots,n-1\}$, and different arrangements give rise to different members of A_k. So the number of elements of A_k equals the number of derangements of T, or D_{n-2}. The members of B_k are just the derangements of $\{1,2,\dots,k-1,n,k+1,\dots,n-1\}$, so $|B_k| = D_{n-1}$.

So $|S_k| = |A_k| + |B_k| = D_{n-2} + D_{n-1}$, and $D_n = \sum_{i=1}^{n-1} |S_k|$, giving the result. □

Exercises 5.7

1. Calculate D_5 and D_6.

2. Calculate D_7 and D_8.

3. Find the number of perfect powers (other than first powers)
   (i) between 1 and 200 (inclusive);
   (ii) between 1 and 500 (inclusive).

4. How many integers in the range from 1 to 250 are divisible by at least one of 2, 3 and 5? How many are divisible by at least one of 2, 3, 5 and 7?

5. How many positive integers less than 60 are relatively prime to 60?

6. Find the number of (positive) prime numbers less than 100.

7. Seven job applicants are to be interviewed by seven members of a hiring committee. Each will be interviewed twice. To avoid bias, each candidate is assigned their first interviewer at random; the second interviewer is also chosen at random, but no candidate is to get the same interviewer twice.
   (i) How many different ways are there to schedule the first round of interviews?

(ii) How many different ways are there to schedule the whole process?

8. Prove that, if $n \geq 2, D_n - nD_{n-1} = (-1)^n$. Show that Theorem 5.9 can be proved from this result.

9. Prove that D_{n+1} is even if and only if D_n is odd.

10. Our mailman is not very good at his job. When he delivers letters to the six apartments in our building, he delivers them at random. Today there is exactly one letter addressed to each apartment, and he delivers exactly one to each apartment. How many possible ways are there for him to deliver the letters? In how many of these does he get:

(i) none right?

(ii) at least one right?

(iii) exactly five right?

(iv) all of them right?

6

Probability

We use the word "chance" frequently in our everyday conversation. For example, we meet somebody "by chance"; we "chance upon" the solution to a problem; one team has a "better chance" of reaching the Super Bowl than another; the Weather Channel announces a "30% chance" of precipitation.

There are two ideas here. In every case there is the feature of unpredictability — a chance occurrence is one for which we cannot be certain of the outcome. The other feature is that sometimes chance is *quantitative* — either it can be measured exactly (the weather forecast) or else we can at least say one "chance" is greater than another (the football teams). We shall refer to these two aspects of chance as *randomness* and *probability* respectively.

For example, say you flip a coin. There is no way to tell whether it will fall heads or tails, so we would say this is a *random* occurrence. If the coin is made uniformly, so that it is equally likely to show a head or a tail, we normally call it a *fair* coin, and we say the probability of a head is $\frac{1}{2}$, and so is the probability of a tail. (Sometimes we say "50%" instead of "$\frac{1}{2}$.")

Suppose the coin in our example was not uniform; for example, say it was made as a sandwich of disks of metal, like a quarter, but the different disks were of different densities. It might be that, if we flipped it enough times, heads would come up 70% of the time and tails only 30%. Then we would say the probability of a head is $\frac{7}{10}$. However, this still qualifies as random, because no one knows beforehand whether a particular flip will be one of the (more common) ones that results in a head or one that gives a tail. The probabilities do not have to be equal in order for an event to be random.

In order to discuss randomness formally, we shall use the idea of an "event" just as we did in Section 5.1. The ideas and notation of set theory — and in particular the number of elements — will be very useful.

6.1 Probability Measures

Probability distributions

Consider a random event: for example, the fall of a head when a fair coin is flipped. We do not know whether a head will show on any particular flip, but we expect to see a head half the time when a large number of attempts are made. We use the word "probability" for this numerical measure of the likelihood of an event. The probabilities 1 and 0 indicate absolute certainty and impossibility, respectively.

We shall write $P(E)$ for the probability that the event E occurs. For example, if H means "a head shows" and T means "a tail shows" when a fair coin is flipped, then

$$P(H) = P(T) = \tfrac{1}{2}.$$

A list of the probabilities of all outcomes of an experiment is called the *probability distribution* of the experiment.

One important case is where each of the possible outcomes is equally likely, as in the case of the fair coin. In this case we say the experiment has *equally likely outcomes*, and set their probabilities equal. In this case the probability distribution is called *uniform*, and we also refer to the experiment as a *uniform experiment*. If a uniform experiment has a sample space with n elements, then we shall assign probability $\frac{1}{n}$ to each outcome.

If a uniform experiment has sample space S, then an event E will occur if the outcome is one of the $|E|$ outcomes in E. We expect that E will occur in fraction $|E|/|S|$ of cases, if the experiment is repeated. So we define

$$P(E) = \frac{|E|}{|S|}.$$

Sample Problem 6.1 *A fair die is rolled. S_i is the event that the number i is rolled for $i = 1,2,3,4,5,6$. E means the roll of an odd number, and F the roll of a number less than* 3. *Find the probabilities of these events.*

Solution. Since the die is fair, the outcomes are equally likely:

$$P(S_1) = P(S_2) = P(S_3) = P(S_4) = P(S_5) = P(S_6) = \tfrac{1}{6}.$$

E and F contain three and two outcomes respectively, so $P(E) = \frac{3}{6} = \frac{1}{2}$, and $P(F) = \frac{2}{6} = \frac{1}{3}$.

Practice Exercise. A single die is rolled. What are the probabilities of the following events?

E : a 4 or a 5 is rolled.

F : an even number is rolled.

G : an odd number greater than 2 is rolled.

If E is any event in a uniform experiment, then $P(E)$ will be a proper fraction. If E and F are two disjoint events in such an experiment, then the event $E \cup F$ ("either E or F occurs") has $|E| + |F|$ elements, so

$$P(E \cup F) = \frac{|E| + |F|}{|S|} = \frac{|E|}{|S|} + \frac{|F|}{|S|} = P(E) + P(F).$$

We shall now apply these ideas to the general situation to make a general definition of probability.

Consider an experiment with sample space $S = \{s_1, s_2, \ldots, s_m\}$. A *probability distribution* for the experiment is a function P with the following properties:

1. For each $s_i, 1 \le i \le n, P(s_i)$ is a real number and $0 \le P(s_i) \le 1$;
2. $P(s_1) + P(s_2) + \cdots + P(s_m) = 1$.

Sample Problem 6.2 *A black die and a white die are thrown simultaneously. What is the probability that the sum of the numbers shown is 8, given that the dice are fair?*

Solution. Write (x, y) to mean that x shows on the black die and y shows on the white die. Then there are 36 possible outcomes, namely $(1,1), (1,2), \ldots,$ $(1,6), (2,1), \ldots, (6,6)$; they are equally likely. Five of them — $(6,2), (5,3),$ $(4,4), (3,5)$, and $(2,6)$ — have sum 8. So

$$P(E) = \frac{|E|}{|S|} = \frac{5}{36}.$$

Practice Exercise. A quarter and a nickel are flipped simultaneously. What is the probability that exactly one head shows, assuming that both coins are fair?

Sample Problem 6.3 *A deck of cards is shuffled and one card is dealt. What is the probability that it is a spade?*

Solution. There are 52 cards in a deck, of which 13 are spades. So

$$P(E) = \frac{|E|}{|S|} = \frac{13}{52} = \frac{1}{4}.$$

Practice Exercise. A deck is shuffled and one card dealt face up. What is the probability that it is a picture card (king, queen or jack)?

Non-uniform experiments

Not all experiments are uniform. But in many cases, given an experiment A, we can find a uniform experiment B such that the outcomes of A are events (not necessarily simple) of B.

For example, consider an experiment in which two fair dice are rolled, and the total of the points on them is recorded. This experiment has eleven possible outcomes $s_2, s_3, \ldots, s_{12}; s_i$ means the total is i. To calculate the probability of s_i, we shall look at a slightly different experiment. In this one, two fair dice are rolled, and the result is recorded as an ordered pair of digits — $(1,3)$ means "1 on die 1, 3 on die 2." There are $6^2 = 36$ outcomes, and each has probability $\frac{1}{36}$. In this experiment we write E_i for the event that the two numbers showing add to i. Then

$$P(E) = \frac{|E_i|}{36}$$

and we can calculate all the probabilities; see Table 6.1. But in each case the probability of the outcome s_i in the first experiment is obviously equal to the probability of E_i in the second experiment. So $P(s_2) = \frac{1}{36}, P(s_3) = \frac{2}{36}$, and so on.

$$\begin{array}{lll lll lllll}
E_2 &=& \{11\} & |E_2| &=& 1 & P(E_2) &=& \frac{1}{36} & & \\
E_3 &=& \{12,21\} & |E_3| &=& 2 & P(E_3) &=& \frac{2}{36} &=& \frac{1}{18} \\
E_4 &=& \{14,22,31\} & |E_4| &=& 3 & P(E_4) &=& \frac{3}{36} &=& \frac{1}{12} \\
E_5 &=& \{14,23,32,41\} & |E_5| &=& 4 & P(E_5) &=& \frac{4}{36} &=& \frac{1}{9} \\
E_6 &=& \{15,24,33,42,51\} & |E_6| &=& 5 & P(E_6) &=& \frac{5}{36} & & \\
E_7 &=& \{16,25,34,43,52,61\} & |E_7| &=& 6 & P(E_7) &=& \frac{6}{36} &=& \frac{1}{6} \\
E_8 &=& \{26,35,44,53,62\} & |E_8| &=& 5 & P(E_8) &=& \frac{5}{36} & & \\
E_9 &=& \{36,45,54,63\} & |E_9| &=& 4 & P(E_9) &=& \frac{4}{36} &=& \frac{1}{9} \\
E_{10} &=& \{46,55,64\} & |E_{10}| &=& 3 & P(E_{10}) &=& \frac{3}{36} &=& \frac{1}{12} \\
E_{11} &=& \{56,65\} & |E_{11}| &=& 2 & P(E_{11}) &=& \frac{2}{36} &=& \frac{1}{18} \\
E_{12} &=& \{66\} & |E_{12}| &=& 1 & P(E_{12}) &=& \frac{1}{36} & &
\end{array}$$

Table 6.1: Probabilities when rolling two fair dice

Sample Problem 6.4 *Three coins are flipped, and the number of heads is recorded. What are the possible outcomes and what is the probability distribution?*

Solution. The outcomes are the four numbers 0, 1, 2 and 3. To calculate probabilities, suppose three coins were flipped and all results recorded. E_i is the event "there are i heads showing" Then $|S| = 8$ and

$$\begin{array}{lll lll}
E_0 &=& \{TTT\}, & P(E_0) &=& \frac{1}{8}, \\
E_1 &=& \{HTT, THT, TTH\}, & P(E_1) &=& \frac{3}{8}, \\
E_2 &=& \{HHT, HTH, THH\}, & P(E_2) &=& \frac{3}{8}, \\
E_3 &=& \{HHH\}, & P(E_3) &=& \frac{1}{8}.
\end{array}$$

Practice Exercise. A game is played using two dice each of which has the numbers 1, 2, 3 on its faces (so each number appears twice per die). An experiment consists of rolling the dice and adding the numbers showing. What is the sample space? What is the probability distribution?

Sample Problem 6.5 *There are five marbles — two blue, three red — in a box. One is selected at random. What is the probability that it is blue?*

Solution. If the marbles were marked v, w, x, y, z, where v and w are blue and the others are red, then the event "blue is chosen" is $\{v, w\}$ and its probability is

$$P(E) = \frac{|E|}{|S|} = \frac{2}{5}.$$

Practice Exercise. There are two red, three white and four blue marbles in a box. One is drawn at random. What is the probability that it is *not* blue?

Non-uniform probabilities

Even when the outcomes are not equally likely, we calculate the probability of an event from the formula

$$P(E) = \sum_{s \in E} P(s).$$

Several important facts can be deduced from this; in particular, if E is any event, $0 \leq P(E) \leq 1; P(E) = 0$ if and only if E is impossible, and $P(E) = 1$ if and only if E is certain. If E and F are mutually exclusive events, then

$$P(E \cup F) = P(E) + P(F),$$

and for general E and F,

$$P(E \cup F) = P(E) + P(F) - P(E \cap F).$$

The probability of the complement $\overline{E}$ of E is

$$P(\overline{E}) = 1 - P(E).$$

Sample Problem 6.6 *E and F are events in a sample space with $P(E) = .6, P(F) = .4$, and $P(E \cap F) = .2$. What is $P(E \cup F)$? What is $P(\overline{E})$?*

Solution.

$$\begin{aligned} P(E \cup F) &= P(E) + P(F) - P(E \cap F) \\ &= .6 + .4 + .2 \\ &= .8 \\ P(\overline{E}) &= 1 - P(E) \\ &= 1 - .6 \\ &= .4. \end{aligned}$$

Practice Exercise. E and F are events in a sample space with $P(E) = .7, P(F) = .3$, and $P(E \cap F) = .1$. What are $P(E \cup F), P(\overline{E})$?

These probabilities can be represented in a Venn diagram — in each area of the diagram, write the probability of the corresponding event. Then the answer can be obtained by addition. The Venn diagram for the preceding Sample Problem is

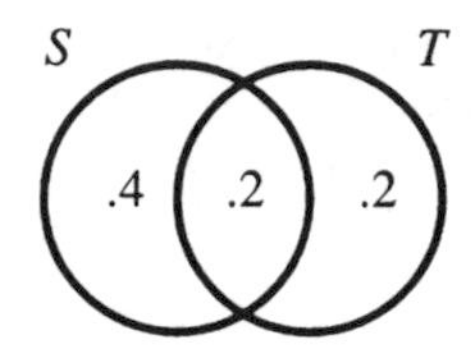

Sample Problem 6.7 *A discrete mathematics class is restricted to business and math majors, all freshmen or sophomores. The percentage makeup of the class is*

Freshman business majors	30%,
Freshman math majors	35%,
Sophomore business majors	21%,
Sophomore math majors	14%.

A name is chosen at random from the class list. What are the probabilities of the following events?

(i) *The student is a freshman.*

(ii) *The student is not a freshman.*

(iii) *The student is either a freshman or a math major.*

Solution. We write F for the event that the student is a freshman, S for sophomore, B for business major and M for math major. Then the data mean

$$\begin{aligned} P(F \cap B) &= .3, & P(F \cap M) &= .35, \\ P(S \cap B) &= .21, & P(S \cap M) &= .14. \end{aligned}$$

(i) Since $M = \overline{B}$ in this problem (all students are math or business majors, and none can be both or else the total would add to more than 100%), the usual equation

$$F = (F \cap B) \cup (F \cap \overline{B})$$

becomes

$$F = (F \cap B) \cup (F \cap M)$$

and, since this is a disjoint union,

$$\begin{aligned} P(F) &= P(F \cap B) + P(F \cap M) \\ &= .3 + .35 \\ &= .65. \end{aligned}$$

(ii) $P(\overline{F}) = 1 - P(F) = 1 - .65 = .35.$

(iii) $F \cup M = F \cup (S \cap M)$, and this is a disjoint union, so

$$\begin{aligned} P(F \cup M) &= P(F) + P(S \cap M) \\ &= .65 + .14 \\ &= .79. \end{aligned}$$

The data could also be represented as

	B	M
F	.30	.35
S	.21	.14

The questions can then be answered by adding the probability in the appropriate cells of the diagram. For example,

$$P(F \cup M) = .30 + .15 + .14 = .79.$$

Practice Exercise. An entymologist has found that the butterflies in a certain area can be classified as follows.

striped males	19%
striped females	22%
unstriped males	28%
unstriped females	31%

Represent the data in a diagram. Assuming the butterflies appear at random, what is the probability that the next butterfly to be sighted will be: (i) striped; (ii) female; (iii) either striped or female?

Exercises 6.1

1. A fair coin is flipped four times. Find the probabilities that

(i) at least three heads appear;

(ii) an even number of heads appear;

(iii) the first result is a head.

In Exercises 2 *to* 7 *two fair dice are rolled. What is the probability of the indicated event?*

2. The total is 9.

3. The total is 6.

4. The total is even.

5. The total is odd.

6. Both scores are even.

7. One odd and one even number are shown.

8. A box contains 12 cards, one for each month of the year. A card is drawn at random.

(i) What is the probability that the selected card is March?

(ii) What is the probability that the selected card is from a month with r in its name?

9. A box contains six red and four white balls. One ball is drawn at random. What is the probability that it is white?

10. A box contains seven red and five white balls. One ball is drawn at random. What is the probability that it is white?

11. In the game of roulette, a wheel is divided into 38 equal parts, labelled with the numbers from 1 to 36, 0 and 00. The spin of the wheel causes the ball to be randomly placed in one of the parts; the chances of the ball landing on any part are equal. Half of the parts numbered from 1 to 36 are red and half are black; the 0 and 00 are green. If the chances of the ball landing in any one part are equal, what are the probabilities of the following events?

(i) The ball lands on a red number.

(ii) The ball lands on a black number.

(iii) The ball lands on a green number.

(iv) The ball lands on 17.

(v) The ball lands on a number from 25 to 36 inclusive.

12. In a lottery there are 90 losing tickets and 10 winning tickets. You draw one ticket at random. What is the probability that it is a winner?

13. The moose population in a Canadian park are 45% plain brown, 35% mottled and 20% spotted. One moose is captured at random for the Bronx Zoo. What are the probabilities that

(i) It is spotted?

(ii) It is not spotted?

(iii) It is not mottled?

14. A random number from 0 to 99 is chosen by a computer. What are the probabilities of the following events?

(i) The number is even.

(ii) The number ends in 5.

(iii) The number is divisible by 11.

15. Box A contains tickets numbered 1, 2, 3, 4. Box B contains tickets numbered 2, 3, 4, 5. One ticket is selected from each box.

(i) List all the elements of the sample space of this experiment.

(ii) Find the probability that the tickets have the same number.

(iii) Find the probability that the sum of the two selected numbers is even.

16. A box contains one red, two blue and two white marbles. One marble is selected at random. What are the probabilities that it is

(i) red?

(ii) blue?

(iii) white?

17. A box contains four red, two blue and three white marbles. One is selected at random. What is the probability that it is

(i) blue?

(ii) not blue?

18. There are 20 students in a class. There are twelve men (eight physics and four chemistry majors) and eight women (five physics and three chemistry majors). One student's name is selected at random. What is the probability that

(i) The student is a chemistry major?

(ii) The student is male?

19. A sample space contains two events, E and F, and

$$P(E) = .70, P(F) = .25, P(E \cap F) = .15.$$

Determine

$$P(\overline{E}), P(E \cup F), P(\overline{E \cup F}).$$

20. A sample space contains two events, E and F, and

$$P(E) = .5, P(F) = .3, P(E \cap F) = .2.$$

Determine

$$P(\overline{F}), P(E \cup F), P(\overline{E} \cap F).$$

21. The events E, F and G satisfy

$$P(E) = .6, P(F) = .6, P(G) = .8,$$
$$P(E \cup F) = .8, P(E \cap G) = .5, P(F \cap G) = .5.$$

Determine

$$P(E \cap F), P(E \cup G), P(F \cup G).$$

22. The events E, F and G satisfy

$$P(E) = .6, P(F) = .5, P(G) = .5,$$
$$P(E \cup F) = .8, P(E \cup G) = .8, P(F \cap G) = .2.$$

Determine

$$P(E \cap F), P(E \cap G), P(F \cup G).$$

23. An examination has two questions. Of 100 students, 75 do Question 1 correctly and 72 do Question 2 correctly. Sixty-four do both questions correctly.

(i) Represent the data in a Venn diagram.

(ii) A student' s answer book is chosen at random. What is the probability that

(a) Question 1 contains an error?
(b) Exactly one question contains an error?
(c) At least one question contains an error?

24. Of 1000 researchers at Microsoft, 375 have a degree in mathematics and 450 have a degree in computer science. 150 of the researchers have degrees in both fields. One researcher's name is selected at random.

(i) What is the probability that the researcher has a degree in mathematics, but not in computer science?

(ii) What is the probability that the researcher has no degree in either mathematics or computer science?

25. In a market survey, 50% of those polled said that they usually buy medications at a pharmacy, and the others that they buy them at a supermarket. 80% buy meat at the supermarket, 20% at the butcher's. 40% buy both meat and medications at the supermarket. One shopper is selected at random from the survey group. What is the probability that he

(i) buys meat at the supermarket but buys medications at the pharmacy?

(ii) does not buy either meat or medications at the supermarket?

26. During the 1992 presidential elections, two hundred voters were surveyed in the St. Louis area. Voters were classified according to whether they lived in Missouri (MO), Illinois (IL) or other states (OS), and whether they intended to vote for Bush (REP), Clinton (DEM) or Perot (IND). The results were as follows:

	REP	DEM	IND
MO	44	33	21
IL	20	22	14
OS	14	20	12

One of the voter's responses is selected at random. What are the probabilities of the following events?

(i) The voter is an Illinois democrat.

(ii) The voter is from Missouri.

(iii) The voter intends to vote for Perot.

(iv) The voter is from outside Illinois.

(v) The voter is either a Democrat or is from Missouri.

27. Two hundred new automobile buyers were surveyed. Their purchases were as follows:

	Sedan	Pickup	Van
Ford	50	10	10
G M	35	10	15
Other	40	10	20

One survey form is chosen at random. What is the probability that the vehicle was

(i) a sedan;

(ii) a Ford;

(iii) a General Motors van or pickup?

28. Five hundred researchers at the National Security Agency were surveyed. It was found that 300 have at least one degree in computer science, 173 have a Bachelor's degree in computer science, 123 have a Master's degree in computer science, and 99 have a Ph.D. in computer science. Additionally, it was found that 63 have both a Bachelor's and Master's degree in computer science, 22 have both a Master's and Ph.D. degree in computer science, and 19 have both Bachelor's and Ph.D. degrees in computer science. One employee is selected at random. What are the probabilities of the following events?

(i) She has all three degrees in computer science.

(ii) She has the Bachelor's and Ph.D. degrees but not the Master's.

29. Last year, there were 200 students enrolled in Calculus II. 140 of these students also enrolled in Linear Algebra. Their majors were distributed as follows:

Enrolment	Major		
	Math	Science	Business
Calculus only	20	35	5
Calculus and Linear Algebra	90	45	5

Suppose the same trend is followed this year. When the next student comes to enroll in Calculus II, what is the probability that this student

(i) is a science major?

(ii) is not enrolling in Linear Algebra this year?

(iii) is not a mathematics major but will take Linear Algebra?

30. 150 Business students tried to sign up at registration for required Marketing, Accounting, and Finance courses. 70 registered for Marketing, 60 for Accounting, and 50 for Finance. 25 registered for both Marketing and Accounting, 15 for Finance and Marketing, 15 for Finance and Accounting, and fivesigned up for all three courses. One student's registration form is chosen at random. What is the probability that the student

(i) was unable to sign up for any of the courses?

(ii) signed up for exactly one of the courses?

(iii) signed up for Marketing and Accounting but not for Finance?

31. Jerry's Gas sells three grades of gasoline at both their full and self-service pumps. One day they kept track of their first 100 customers and their purchases in the following table:

	Regular Leaded	Regular Unleaded	Premium Unleaded
Full service	15	15	20
Self service	20	25	5

If this trend continues, what is the probability that the next customer who comes in:

(i) goes to the self-service pumps?

(ii) buys regular unleaded at the self-service pumps?

(iii) buys premium unleaded at the full-service pumps?

6.2 Repeated Experiments

Stochastic processes

Sometimes an experiment can be viewed as a sequence of smaller experiments. The outcome consists of a sequence: "outcome of subexperiment 1" followed by "outcome of subexperiment 2" .... An experiment of this kind is called a *stochastic process.*

For example, suppose a jar contains three red and two blue marbles. An experiment consists of drawing a marble from the jar, noting its color, drawing another marble, and noting the second marble's color. (The first marble is not replaced.) The possible outcomes are the four ordered pairs of colors: *RR*, *RB*, *BR* and *BB*.

To analyze the experiment, we first observe that the first marble drawn is either red (three-fifths of the cases) or blue (two-fifths). If the first marble is red, then the remaining marbles are two red and two blue, so in half of these cases ($\frac{1}{2} \times \frac{3}{5} = \frac{3}{10}$ of the original cases) the second marble drawn is red and in the other half ($\frac{3}{10}$ of the orginal) it is blue. So, in the obvious notation,

$$\begin{aligned} P(RR) &= \frac{3}{10}, \\ P(RB) &= \frac{3}{10}. \end{aligned}$$

In the same way, if the first marble is blue, then the second marble drawn is red in $\frac{3}{4}$ and blue in $\frac{1}{4}$ of the cases. So

$$\begin{aligned} P(BR) &= \frac{2}{5} \times \frac{3}{4} = \frac{3}{10}, \\ P(BB) &= \frac{2}{5} \times \frac{1}{4} = \frac{1}{10}. \end{aligned}$$

So the four outcomes are *RR*, *RB*, *BR* and *BB*, and

$$P(RR) = P(RB) = P(BR) = \frac{3}{10}, P(BB) = \frac{1}{10}.$$

Sample Problem 6.8 *There are two bags A and B. Bag A contains two red balls and one green ball; bag B contains two red and three green balls. First a bag is chosen at random, then a ball is chosen at random from it. What are the possible outcomes, and what are their probabilities? What is the probability that a red ball is selected?*

Solution. The outcomes are *AR*, *AG*, *BR* and *BG* where *A* and *B* are the bags and *R* and *G* denote the color of the ball selected. The possible outcomes of the first subexperiment — the selection of bag — are *A* and *B*, with probabilities each $\frac{1}{2}$. If *A* is selected, the probability of *R* is $\frac{2}{3}$ and the probability of *G* is $\frac{1}{3}$. If *B* is selected, *R* has probability $\frac{2}{5}$ and *G* has probability $\frac{3}{5}$. So

$$P(AR) = \frac{1}{2} \times \frac{2}{3} = \frac{1}{3},$$

$$P(AG) = \frac{1}{2} \times \frac{1}{3} = \frac{1}{6},$$
$$P(BR) = \frac{1}{2} \times \frac{2}{5} = \frac{1}{5},$$
$$P(BG) = \frac{1}{2} \times \frac{3}{5} = \frac{3}{10}.$$

So the probability of red is

$$P(AR) + P(BR) = \frac{1}{3} + \frac{1}{5} = \frac{8}{15}.$$

Notice that there are four red and four green balls. If the two bags were emptied and one ball chosen, the probability of red would be $\frac{1}{2}$. The process of dividing the balls into two bags, whose contents are not identical, changes the probabilities.

Practice Exercise. A die is chosen from a pair of dice, and rolled. One die (die A) is standard; the other (die B) has three faces marked 1 and three marked 6. What are the outcomes? Assuming that the die is chosen at random, what is the probability of a 6? Of a 3?

Tree diagrams

The use of tree diagrams to represent experiments (see Section 5.1) is particularly useful for stochastic processes. Let us go back to the example of three red and two blue marbles. The first stage of the experiment can be represented as a branching into two parts, labeled "red" and "blue," and the branches can be also be labeled with the probabilities $\frac{3}{5}$ (on red) and $\frac{2}{5}$ (on blue):

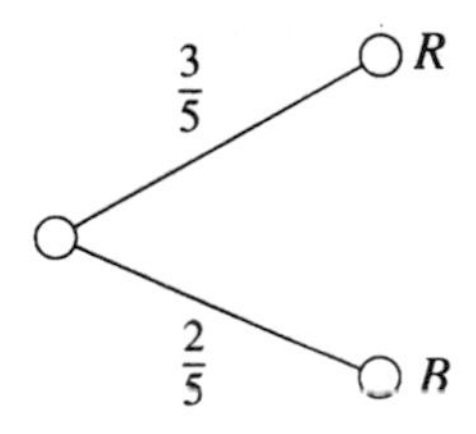

In each case the second stage can be represented similarly. The tree diagram for the whole experiment is shown in Figure 6.1.

Finally, the outcomes are sequences of the results of the subexperiments. Each endpoint on the right of the tree represents an outcome, and the sequence can be read by tracing back through the branch to the root. The probability of an outcome can be calculated by multiplying the probabilities along the branch. So the experiment under discussion yields the diagram Figure 6.2:

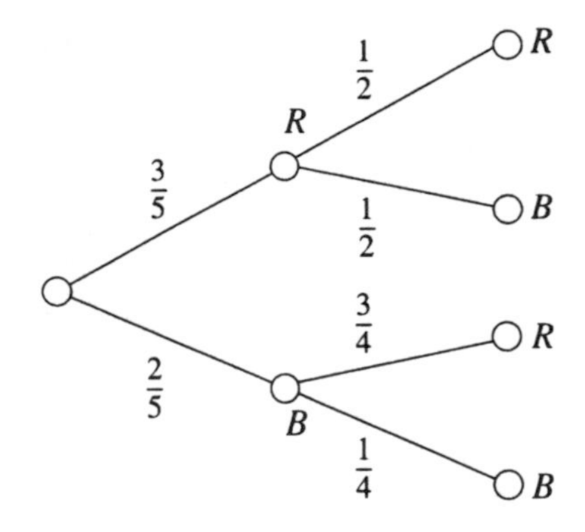

Figure 6.1: Tree diagram for the whole experiment

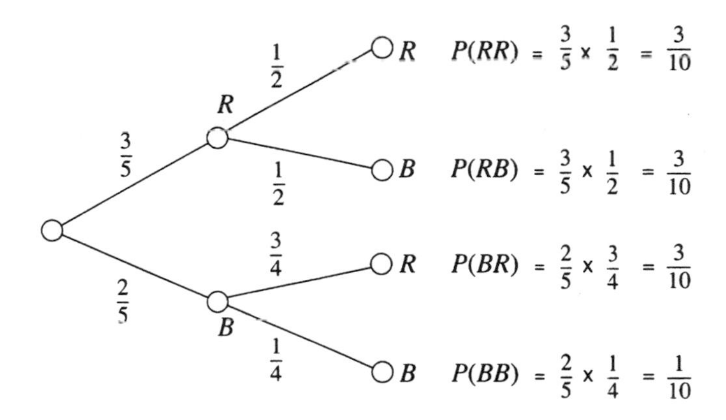

Figure 6.2: Tree diagram with branch probabilities

Bernoulli trials and binomial experiments

Consider the experiment: "flip a coin three times and count how many heads occur." This could be viewed as repeating one basic experiment that has exactly two outcomes ("flip a coin") three times and then counting the results.

We shall define a *Bernoulli trial* to be an experiment in which there are exactly two possible outcomes. In many applications it makes sense to think of one of these events as "success" and the other as "failure" (abbreviated to S and F). We shall often denote the probability of success by p. A *binomial experiment* is one in which a Bernoulli trial is repeated a certain number of times, and the outcome is the number of successes in total. So our initial example was a binomial experiment in which the trial is repeated three times. Since the number of heads is to be counted, we would probably refer to a head as a "success." If the coin was fair, p will equal $\frac{1}{2}$.

In a binomial experiment, if the trial is repeated n times, there are $n+1$ possible outcomes, from "0 successes" to "n successes."

Sample Problem 6.9 *A trial consists of throwing a die; a result of 1 or 2 is a success, other throws are failures. What is the probability of two successes in three trials?*

Solution. There are three ways in which two successes can occur: the sequences SSF, SFS and FSS. Now

$$\begin{aligned} P(SSF) &= \tfrac{1}{3} \times \tfrac{1}{3} \times \tfrac{2}{3} &= \tfrac{2}{27}, \\ P(SFS) &= \tfrac{1}{3} \times \tfrac{2}{3} \times \tfrac{1}{3} &= \tfrac{2}{27}, \\ P(FSS) &= \tfrac{2}{3} \times \tfrac{1}{3} \times \tfrac{1}{3} &= \tfrac{2}{27}. \end{aligned}$$

So the probability of two successes is

$$\frac{2}{27} + \frac{2}{27} + \frac{2}{27} = \frac{6}{27}.$$

Practice Exercise. In the experiment described in this sample problem, what is the probability of exactly one success in three trials?

More generally, the number of sequences of n trials that contain k successes and $n-k$ failures is $\binom{n}{k}$. The probability of any such sequence is $p^k(1-p)^{n-k}$. So the probability of k successes is

$$\binom{n}{k} p^k (1-p)^{n-k}.$$

Sample Problem 6.10 *A sales representative estimates that a sale results from one in four of his calls to companies. If he makes five calls today, what is the probability of at least two sales?*

Solution. Each call is a Bernoulli trial with $p = \frac{1}{4}$, so the five calls in a day can be thought of as a binomial experiment with $p = \frac{1}{4}, n = 5$. So

$$\begin{aligned} P(2 \text{ successes}) &= \tbinom{5}{2}(\tfrac{1}{4})^2(\tfrac{3}{4})^3 &= 10 \times \tfrac{3^3}{4^5} &= \tfrac{270}{1024}, \\ P(3 \text{ successes}) &= \tbinom{5}{3}(\tfrac{1}{4})^3(\tfrac{3}{4})^2 &= 10 \times \tfrac{3^2}{4^5} &= \tfrac{90}{1024}, \\ P(4 \text{ successes}) &= \tbinom{5}{4}(\tfrac{1}{4})^4(\tfrac{3}{4})^1 &= 5 \times \tfrac{3}{4^5} &= \tfrac{15}{1024}, \\ P(5 \text{ successes}) &= \tbinom{5}{5}(\tfrac{1}{4})^5(\tfrac{3}{4})^0 &= 1 \times \tfrac{1}{4^5} &= \tfrac{1}{1024}, \end{aligned}$$

and the probability of at least two successes is the sum of these:

$$\frac{270 + 90 + 15 + 1}{1024} = \frac{376}{1024} = \frac{47}{128}$$

or about 37%. Alternatively, we could have noticed that "he makes *at least* two sales per day" is the complement of "he makes 0 or 1 sales per day."

The probability of 0 or 1 sales is

$$\begin{aligned} P(0 \text{ sales}) + P(1 \text{ sale}) &= \binom{5}{0}(\frac{1}{4})^0(\frac{3}{4})^5 + \binom{5}{1}(\frac{1}{4})^1(\frac{3}{4})^4 \\ &= 1 \times \frac{3^5}{4^5} + 5 \times \frac{3^4}{4^5} \\ &= \frac{243 + 5 \times 81}{1024} = \frac{243 + 405}{1024} \\ &= \frac{648}{1024} = \frac{81}{128} \end{aligned}$$

so the probability of at least two sales is

$$1 - \frac{81}{128} = \frac{47}{128}.$$

Practice Exercise. If the salesman could improve his record to give him a 50% chance of a sale from each call, what is his probability of at least three successes in the day?

Note that, for practical purposes, the exact value $\frac{47}{128}$ in the preceding sample problem is not important. It would normally be enough to say "his chance of two or more sales is a little better than one in three." In probability problems, it is important to have the correct formula, and to have an approximate idea of the numerical value. If, for any reason, a precise value is needed, it can be calculated from the formula.

Sample Problem 6.11 *The salesman in the preceding sample problem would like to have at least a 60% chance of two successes per day. How many calls should he make per day?*

Solution. Suppose he makes n calls. The probability of 0 or 1 success is

$$\binom{n}{0}(\frac{1}{4})^0(\frac{3}{4})^n + \binom{n}{1}(\frac{1}{4})^1(\frac{3}{4})^{n-1} = \frac{(1 \times 3^n)}{4^n} + \frac{n(1 \times 3^{n-1})}{4^n}.$$

So what he wants is

$$\frac{3^{n-1}(3+n)}{4^n} < \frac{4}{10}.$$

We do some experimental arithmetic:

$$\begin{aligned} n &= 5: & \frac{3^{n-1}(3+n)}{4^n} &= \frac{81 \times 8}{1024} &= .63\ldots, \\ n &= 6: & \frac{3^{n-1}(3+n)}{4^n} &= \frac{243 \times 9}{4096} &= .53\ldots, \\ n &= 7: & \frac{3^{n-1}(3+n)}{4^n} &= \frac{729 \times 10}{16384} &= .44\ldots, \\ n &= 8: & \frac{3^{n-1}(3+n)}{4^n} &= \frac{2187 \times 11}{65536} &= .36\ldots. \end{aligned}$$

So eight calls are needed.

Practice Exercise. How many calls are needed to give the salesman at least a 70% chance of two successes per day?

The birthday coincidence

Sometimes probabilities can be very surprising. One well-known example is the "birthday coincidence" problem. Suppose there are 30 people in a room. What is the probability that two of them share the same birthday (day and month)?

For simplicity, let us ignore leap years. (The answer will still be approximately correct.) What is the probability of the complementary event — that no two have the same birthday? (The probability of a birthday coincidence will be found by subtracting this probability from 1.) If we assume the people are ordered in some way (alphabetical order, for example), then there are 365 choices for the first person's birthday, 364 for the second, 363 for the third, and so on. So the total number of possible lists of birthdays (with no repeats) is

$$P(365,30) = 365 \times 364 \times \cdots \times 336.$$

If there is no restriction, each person has 365 possible birthdays. So there are 365^{30} possibilities. Therefore the probability of the event "no two have the same birthday" is

$$P(E) = \frac{|E|}{|S|} = \frac{P(365,30)}{365^{30}}$$

which is approximately .294. So the probability of the "birthday coincidence" is about .706.

Similar calculations can of course be carried out for any number of people. Table 6.2 shows the probability of a "birthday coincidence" for various numbers of people. It is about a 50% chance when there are 23 people in the room.

n	P_n
5	.027
10	.117
15	.253
20	.411
23	.507
25	.569
30	.706
40	.891
50	.970

Table 6.2: P_n is the probability of a "birthday coincidence" among n people

Most students find this very surprising, and would guess that many more people would be needed to bring the probability up to 50%.

Exercises 6.2

1. A jar contains six marbles, four red and two white. In stage 1 of an experiment, one marble is selected at random and its color noted; it is not replaced. In stage 2, another marble is drawn and its color noted.

(i) Draw a tree diagram for this experiment.

(ii) What are the probabilities of the four outcomes (RR, RW, WR, WW)?

(iii) What is the probability that the two marbles selected are different colors?

2. A cage contains six mice: three white females, one white male and two gray males. In an experiment, two mice are selected one after the other, without replacement, and sex and color are noted.

(i) Draw a tree diagram for this experiment.

(ii) Find the probabilities of all the outcomes.

(iii) What is the probability that two males are selected?

(iv) What is the probability that both a white and a gray mouse are selected?

(v) What is the probability that the second mouse is gray?

3. Three cards are dealt in order from a standard deck. In each case it is only recorded whether the card is a face card (ace, king, queen, jack) or a minor card (10 through 2).

(i) Draw a tree diagram for this experiment. Find the probabilities of the different outcomes.

(ii) What is the probability that at least two face cards are dealt?

4. Three cards are dealt in order from a standard deck. Only the color (red or black) of the card is recorded. The cards are not replaced.

(i) Draw a tree diagram for this experiment.

(ii) What is the probability that the three cards are red?

5. Eight evenly-matched horses run a race. Three are bay colts, one is a bay filly, two are brown colts and two are brown fillies. The color and sex of the first and second finishers are recorded.

(i) Draw a tree diagram for this experiment.

(ii) What is the probability that both the horses recorded are colts?

(iii) What is the probability that the two horses are of different colors?

(iv) What is the probability that the second horse is a filly?

6. A jar contains four red, three white and two blue marbles. One marble is drawn and placed aside; then another marble is drawn. Only the colors are recorded. Draw a tree diagram for this experiment. What is the probability that the two marbles are of different colors?

7. A coin is weighted so that a head is twice as likely to occur as a tail. It is tossed four times, and the result is noted.

(i) What is the probability that exactly one head occurs?

(ii) What is the probability that at least three heads occur?

8. A die is rolled six times. What is the probability of

(i) exactly one six?

(ii) at most one six?

9. A fair coin is tossed six times. Which is more probable — that heads and tails occur three times each, or that they divide 4 and 2 (either 4 heads or 4 tails)?

10. A multiple choice test contains five questions, and each question has three possible answers. A passing grade is three or more correct answers. What is the probability of passing if you guess the answers at random?

11. In a twenty-question, true-false test, what is the probability of getting exactly sixteen answers correct by random guessing?

12. A pair of fair dice are rolled and the total is recorded. If this is done ten times, what is the probability that 7 is recorded at least three times?

13. On average, one light bulb in twenty is defective. What is the probability that there is more than one defective bulb in a box of ten?

14. A drug is effective in nine cases out of ten. If it is tested on twelve patients, what is the probability that it will be ineffective in more than two cases?

15. A baseball pitcher estimates that he can throw his fast ball for a strike seven times in every eight.

(i) What is the probability that he throws exactly seven strikes in eight pitches?

(ii) What is his probability of throwing at least seven strikes in eight pitches?

16. A children's game uses a die that has 1 on three of its faces, 2 on two of its faces and a 3 on the last face. If it is rolled six times, what is the probability of rolling 2 exactly three times?

17. The probability that a new tire will last 35,000 miles is .9. If you replace all four tires now, what is the probability that you will need to replace at least one within the next 35,000 miles?

18. There are five people in a room. What is the probability that two were born in the same month (but not necessarily the same year)?

6.3 Counting and Probability

All of the elementary counting principles that we studied in Chapter 5 are useful in calculating probabilities. We shall show this using several examples.

Choosing marbles

Many elementary problems involve random drawing of marbles from a jar (or bag or box). These obviously arise in lotteries and other games of chance, but this is not the only reason for studying such problems. The mathematics is the same if we consider many other types of random selection. As an example, suppose there are 500 Democrat supporters and 400 Republicans in your area. Working out the probabilities of various combinations being contacted in an opinion poll involves exactly the same calculations as the ones made in studying random drawings from a jar of 500 black and 400 white marbles. There are also examples in physical science — for example, molecules escaping from a container of heated gas behave in the same way.

Sample Problem 6.12 *There are twelve red marbles and three blue marbles in a jar. Three are selected at random. What is the probability that all are red?*

Solution. There are 15 marbles, so the number of ways of selecting three is $\binom{15}{3}$. So $|S| = \binom{15}{3}$. The number of ways of selecting three red balls is $|E| = \binom{12}{3}$. So

$$\begin{aligned} P(E) &= \frac{\binom{12}{3}}{\binom{15}{3}} = \frac{12!}{9!3!} \times \frac{12!3!}{15!} \\ &= \frac{12 \cdot 11 \cdot 10}{15 \cdot 14 \cdot 13} = \frac{44}{91}. \end{aligned}$$

Practice Exercise. An urn contains five red, four blue and two green marbles. Three marbles are chosen at random. What is the probability that the three are all different colors?

Sample Problem 6.13 *A box contains four red, three white and two blue balls. Two balls are chosen at random. What is the probability that they are the same color?*

Solution. There are $\binom{9}{2}$ selections available. There are $\binom{4}{2} = 6$ ways to choose two red balls, $\binom{3}{2} = 3$ ways to choose two white balls, and $\binom{2}{2} = 1$ ways to choose two blue balls. So

$$P(\text{two are the same}) = \frac{12+6+1}{\binom{9}{2}} = \frac{19}{36}.$$

Practice Exercise. A jar contains six white and four blue balls. Two balls are chosen at random. What is the probability of selecting at least one white ball?

Card problems

Remember that in a standard deck there are 52 cards, 13 in each of the four suits. So there are four cards in each of the 13 denominations: Ace, King, $\ldots, 4, 3, 2$. In most card games, the hands of cards are dealt face down to the players, so the order in which cards are received does not matter.

Sample Problem 6.14 *A poker hand of five cards is dealt from a standard deck. What are the probabilities of the following hands?*

(i) *The four kings and the ace of spades.*

(ii) *A full house (three cards of one denomination, two of another).*

(iii) *A hand with no pair (all different denominations).*

Solution. In each case, the hand dealt is one of the $\binom{52}{5}$ possible selections of five cards from the full deck of fifty two. So, in each case, $|S| = \binom{52}{5}$.

(i) The event E: "the hand consists of four kings and the ace of spades" contains only one outcome. So

$$\begin{aligned} |E| &= 1, \\ P(E) &= \frac{|E|}{|S|} = \frac{1}{\binom{52}{5}} \\ &= \frac{5 \cdot 4 \cdot 3 \cdot 2 \cdot 1}{52 \cdot 51 \cdot 50 \cdot 49 \cdot 48} = \frac{1}{2598960}. \end{aligned}$$

(ii) Let's say a full house is of "type (x, y)" if it contains three x's and two y's—one could have "type (7,5)," "type (king, 4)" and so on. There are thirteen ways to choose the denomination of the three, and for each such choice there are twelve ways to select the two. So there are $13 \times 12 = 156$ types. Once the type is known, we can calculate the number of hands of that type. For example, if there are three kings and two fours, then there are $\binom{4}{3}$ ways of choosing which kings are to be used (we must choose three out of the four possible kings), and $\binom{4}{2}$ ways of choosing which fours. So there are $\binom{4}{3} \times \binom{4}{2} = 4 \times 6 = 24$ hands of each type. So the number of full houses is

$$\begin{aligned} & 13 \times 12 \times \binom{4}{3} \times \binom{4}{2} \\ = \; & 13 \times 12 \times 24 \end{aligned}$$

and the probability is

$$\frac{13\cdot 12\cdot 24}{\binom{52}{5}} = \frac{6}{4165}.$$

(iii) If there is no pair, the hands have one card of each of five denominations. This collection of denominations can be chosen in $\binom{13}{5}$ ways. The card in each denomination can be selected in four ways. So the number of hands is

$$\binom{13}{5} \times 4^5,$$

and the probability is

$$\frac{\binom{13}{5}\cdot 4^5}{\binom{52}{5}} = \frac{2112}{4165},$$

which is a little greater than 50%.

Practice Exercise. A poker hand of five cards is dealt from a standard deck. What is the probability that it contains a flush (all five cards of the same suit)?

Choosing a committee

Sample Problem 6.15 *A committee of four people is to be chosen at random from a club with twelve members, four men and eight women.*

(i) *What is the probability that at least one man is chosen?*

(ii) *What is the probability that one particular member, Jack Smith, is chosen?*

Solution. The number of possible outcomes — all possible committees — is $\binom{12}{4}$. This is $|S|$. They are equally likely.

(i) Let E be the event that at least one man is chosen. Then $\overline{E}$ is the event that no man is chosen. $|\overline{E}| = \binom{8}{4}$. So $|E| = \binom{12}{4} - \binom{8}{4}$ and

$$P(E) = 1 - \frac{\binom{8}{4}}{\binom{12}{4}} = 1 - \frac{14}{99} = 1 - \frac{85}{99}.$$

(ii) The number of committees containing a given individual is $\binom{11}{3}$. (Jack Smith must be a member; the remaining three are chose from the other eleven members and that can be done in $\binom{11}{3}$ ways.) Therefore

$$P(\text{Jack Smith is chosen}) = \frac{\binom{11}{3}}{\binom{12}{4}} = \frac{1}{3}.$$

Practice Exercise. A class has eighteen members — six math majors, five in economics and seven in computer science. A class committee of three is chosen at random. What is the probability that all three have the same major?

Derangement problems

The number of derangements of n objects is

$$D_n = n!\left[1 - \frac{1}{1!} + \frac{1}{2!} - \frac{1}{3!} + \cdots + (-1)^n \frac{1}{n!}\right].$$

The number of ways to arrange n objects is $n!$, so the probability that a randomly selected arrangement is a derangement is

$$\frac{D_n}{n!} = 1 - \frac{1}{1!} + \frac{1}{2!} - \frac{1}{3!} + \cdots + (-1)^n \frac{1}{n!}.$$

Sample Problem 6.16 *Four students have identical backpacks. As they leave the library, each chooses a backpack at random. What is the chance that none gets the correct backpack?*

Solution. The probability is

$$\frac{D_4}{4!} = 1 - \frac{1}{1} + \frac{1}{2} - \frac{1}{6} + \frac{1}{24} = \frac{3}{8}.$$

Practice Exercise. What is the probability that some, but not all, get their correct backpacks?

Exercises 6.3

1. There are three red, three blue and four white marbles in a jar. Two marbles are chosen at random. What are the probabilities that
   (i) Both are blue?
   (ii) The two are different colors?

2. There are ten red, three white and four blue marbles in a jar. Two are chosen at random. What are the probabilities that
   (i) Neither is red?
   (ii) Exactly one is red?
   (iii) Both are red?

3. From the jar in the preceding exercise, three marbles are chosen. What are the probabilities that
   (i) The three are red?
   (ii) the three are of different colors?

4. A jar contains four red, five blue and three white marbles. Three marbles are chosen at random. What are the probabilities that
   (i) All three are red?
   (ii) None is red?

5. A student must select two courses from a list of two humanities and three science couses. He selects at random. What is the probability that both are science courses?

6. A box of 50 matches contains three broken matches. Two matches are drawn at random. What is the probability that neither is broken?

7. Your midterm test contains ten questions, with true-false answers. If you select your answers at random, what are the probabilities of getting

(i) exactly four correct answers?

(ii) exactly two correct answers?

(iii) at most two correct answers?

8. A committee of six people (four men and two women) need to select a chairman and secretary. They do so by drawing names from a hat. What are the probabilities that

(i) Both are men?

(ii) Exactly one is a man?

9. In the game Yahtzee, five fair dice are rolled at once. What are the probabities of getting

(i) a Yahtzee, that is, all five dice the same?

(ii) four of a kind (that is four of one denomination, with the other different)?

(iii) a full house (that is three of one denomination, and two of another)?

10. An electrician knows that two switches are faulty out of a batch of five. He tests them one at a time. What is the probability that he finds the two faulty switches in the first two trials?

11. A five-card poker hand is dealt at random from a standard deck. What is the probability that it contains a straight (A2345, 23456, 34567, ..., 10JKQA)? (In a straight, suits do not matter.)

12. Suppose licence plates consist of two letters followed by four numbers. Suppose all possible combinations have been used. If you select a car at random, what is the probability that the plate has

(i) Both letters the same?

(ii) No repeated symbol (letter or number)?

(iii) An even number?

13. A congressional committee contains five Democrats and four Republicans. To choose a subcommittee they list all groups of four that contain at least one Republican and one Democrat, and then select one at random. What is the probability that the selected subcommittee consists of two Democrats and two Republicans?

14. A secretary types three letters and the three corresponding envelopes. She is in a hurry to quit work for the day, and forgets to check that she puts the correct letters in the correct envelopes, so it is as though she puts the letters in the envelopes at random. What are the probabilities that

(i) Every envelope gets the correct letter?

(ii) No envelope gets the correct letter?

15. A committee of six men and four women select a steering committee of three people. If all combinations are equally likely, what is the probability that all three will be of the same gender?

16. Four people apply for two jobs. Among them are a husband and wife. Assume that the four are equally well qualified and choices are made at random.

(i) What is the probability that both husband and wife are appointed?

(ii) What is the probability that at least one is appointed?

17. In a lottery, you must select three different numbers from 1, 2, 3, 4, 5, 6, 7, 8, 9. If your three are the numbers drawn, you win first prize; if you have two right out of three, you win second prize. (The order of the numbers does not count.)

(i) What is the probability of winning first prize?

(ii) What is the probability of winning second prize?

18. A club has twelve members, including Mr. and Mrs. Smith. A committee of three is to be chosen at random.

(i) What is the probability that both are chosen?

(ii) What is the probability that neither are chosen?

19. Your discussion group chooses its chairman at each monthly meeting by drawing names from a hat. If there are twelve members, what is the probability that you will not be chosen at any meeting this year?

20. A pinball machine selects a number from 0 to 9 at random; you are awarded a free game if that number equals the last digit in your score.

(i) What is the probability that you win a game in this way on your first attempt?

(ii) What is the probability that you do not win a game on your first attempt but you win a game on your second attempt?

(iii) What is the probability that you will get no "match" in five games?

6.4 Conditional Probabilities

Conditional probability defined

Suppose you draw a marble at random from a jar containing two red and two blue marbles. You look at this marble, then discard it, and draw another marble. What is the probability of drawing a blue marble the second time? There are two answers. If the first marble is red, then the probability is $\frac{2}{3}$; if it is blue, the probability is $\frac{1}{3}$. We write

$$\begin{aligned} P(\text{second blue} \mid \text{first red}) &= \tfrac{2}{3}, \\ P(\text{second blue} \mid \text{first blue}) &= \tfrac{1}{3}. \end{aligned}$$

The sign "|" is read as the word "given," so the first probability above is "probability that the second is blue given that the first is red." These probabilities are called *conditional*, because they express the probability of a result in the second draw if a certain condition (the result of the first draw) is satisfied.

In general, suppose the fact that event A occurs gives extra information about the probability that B occurs. The *probability that B occurs, given that A occurs*, which is the *conditional probability* $P(B \mid A)$, is the best measure to use in deciding whether or not B will happen.

Sample Problem 6.17 *Two cards are dealt from a standard deck. The first card is not replaced before the second is dealt. What is the probability that the second card is a king, given that the first card is an ace? What is the probability that the second card is a king, given that the first card is a king?*

Solution. Given that the first card is an ace, the second card is a random selection from 51 equally likely possibilities. Four of these outcomes are kings. So

$$P(K \mid A) = \frac{n(E)}{n(S)} = \frac{4}{51}.$$

If the first card is a king, there are only three kings among the cards for the second trial, so

$$P(K \mid K) = \frac{3}{51}.$$

Practice Exercise. Two cards are dealt, without replacement, from a standard deck. If the first is a spade, what is the probability that the second is (i) a spade? (ii) a heart?

The probabilities written on the later branches of a tree diagram are all conditional probabilities. If an experiment consists of two stages in which the first stage has possible outcomes A and B and the second stage has possible outcomes C and D, the tree diagram is shown in Figure 6.3. Of course, this all works in the same way when there are more than two stages, or when there are more than two outcomes at a stage.

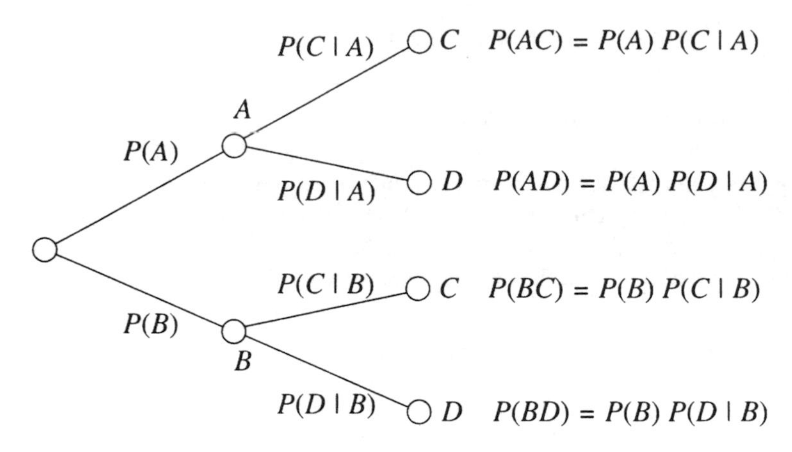

Figure 6.3: Tree diagram with branch probabilities

Sample Problem 6.18 *Two cards are dealt from a standard deck without replacement, and it is noted whether or not the cards are kings. What are the outcomes of this experiment, and their probabilities? Represent the information in a tree diagram.*

Solution. We shall write K and N for "a king" and "a card other than a king." By a similar argument to Sample Problem 6.17 we see that the probability of a king on the second draw, given a king on the first draw, is

$$P(K \mid K) = \tfrac{3}{51}$$

and similarly

$$P(K \mid N) = \tfrac{4}{51}.$$

Given that the first card is a king, the second card is either a king or not, so

$$P(K \mid K) + P(N \mid K) = 1;$$

therefore

$$P(N \mid K) = \frac{48}{51}$$

and similarly

$$P(N \mid N) = \frac{47}{51}.$$

So the diagram is the one shown in Figure 6.4 (where, for example, "K, N" means "king first, non-king second").

Practice Exercise. Repeat this sample Problem, in the case where the information noted is whether or not the card is a spade.

The conditional probability formula

The probability of the sequence "A followed by B" is

$$P(AB) = P(A)P(B|A). \tag{6.1}$$

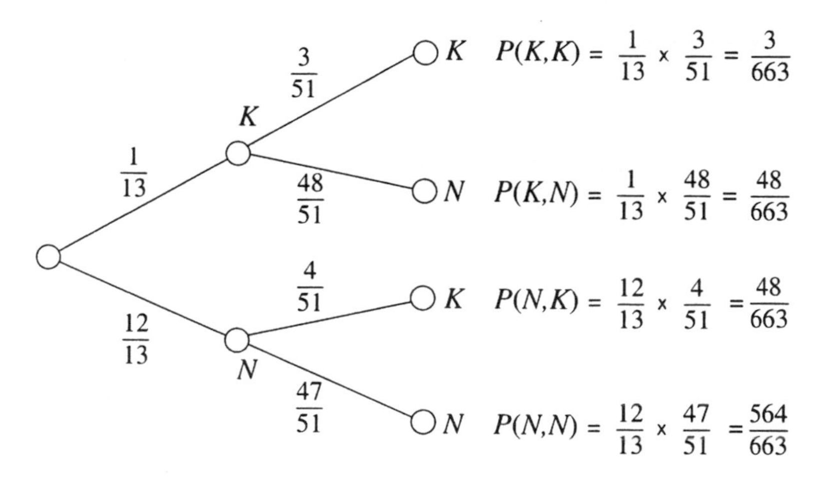

Figure 6.4: Tree diagram for Sample Problem 6.18

This formula can be used more generally, even when A does not happen before B. Whenever we get more information about whether an event B occurs by knowing whether or not event A occurs, we define the conditional probability of B given A by

$$P(B|A) = \frac{P(A \cap B)}{P(A)}. \tag{6.2}$$

Sample Problem 6.19 *A car pool contains both ten red and ten white Fords and fifteen red and five white Buicks. A car is chosen at random, by picking up the keys.*

1. *What is the probability that a red car is chosen?*
2. *You notice that the keys belong to a Buick. What is the probability that a red car is chosen?*

Solution.

1. There are 40 cars, of which 25 are red. So

$$P(R) = \frac{|R|}{|S|} = \frac{25}{40} = \frac{5}{8}.$$

2. There are 20 Buicks of which 15 are red. So

$$\begin{aligned} P(B) &= \frac{20}{40} = \frac{1}{2}, \\ P(R \cap B) &= \frac{15}{40} = \frac{3}{8}, \\ P(R \mid B) &= \frac{P(R \cap B)}{P(B)} = \frac{3}{8} \times \frac{2}{1} = \frac{3}{4}. \end{aligned}$$

Practice Exercise. There are five Republicans and three Democrats on a committee. A subcommittee of two is chosen by a random drawing.

1. What is the probability that both are Democrats?
2. You are told that the committee contains at least one Democrat. What is the probability that both are Democrats?

Sample Problem 6.20 *A carpenter uses screws made by two companies, X and Y; 40% of his stock comes from X and the rest from Y. About 2% of the screws from Y are faulty. If he chooses a screw at random, what is the probability that it is a faulty screw from company Y?*

Solution. We have

$$P(Y) = .6, \; P(F \mid Y) = .02.$$

So

$$\begin{aligned} P(Y \cap F) &= P(Y)P(F \mid Y) \\ &= .6 \times .02 = .012 \end{aligned}$$

and the probability is .012 or 1.2%.

Practice Exercise. Records show that two March days out of five are dull, and it rains on one out of every three dull days. What is the probability that March 12th next year will be dull and rainy?

Independence

We say two events are independent in everyday English if they have no effect on each other. In probability theory, this idea is formalized as follows: Events A and B are *independent* when

$$P(A \mid B) = P(A).$$

Provided neither A nor B is impossible, the relation

$$P(B \mid A) = \frac{P(A \cap B)}{P(A)}$$

leads us to the equivalent definition: A and B are independent when

$$P(A \cap B) = P(A) \times P(B).$$

Sample Problem 6.21 *Two dice are rolled and their sum is recorded. Consider the events:*

A: sum is 2, 8 *or* 11*;*

B: sum is even;

C: sum is 4, 7 *or* 10.

Which of these events are independent?

Solution. We write E_i for the outcome: "the sum is i." These outcomes are the events shown with their probabilities in Table 6.1. Then

$$\begin{aligned} A &= \{E_2, E_8, E_{11}\}, \\ B &= \{E_2, E_4, E_6, E_8, E_{10}, E_{12}\}, \\ C &= \{E_4, E_7, E_{10}\}, \\ A \cap B &= \{E_2, E_8\}, \\ A \cap C &= \emptyset, \\ B \cap C &= \{E_4, E_{10}\}, \end{aligned}$$

and from the table,

$$\begin{aligned} P(A) &= (1+5+2)/36 = 8/36, \\ P(B) &= (1+3+5+5+3+1)/36 = 18/36, \\ P(C) &= (3+6+3)/36 = 12/36, \\ P(A \cap B) &= \frac{(1+5)}{36} = 6/36, \\ P(A \cap C) &= 0, \\ P(B \cap C) &= (3+3)/36 = 6/36. \end{aligned}$$

Now

$$\begin{aligned} P(A) \times P(B) &= 8/36 \times 18/36 = 4/36 \neq P(A \cap B), \\ P(A) \times P(C) &= 8/36 \times 12/36 = 8/108 \neq P(A \cap C), \\ P(B) \times P(C) &= 18/36 \times 12/36 = 6/36 = P(B \cap C), \end{aligned}$$

so B and C are independent, but A and B are dependent, and so are A and C.

Practice Exercise. An urn contains eight balls numbered 1 through 8. Balls 1, 2 and 3 are red; 4, 5, 6, and 7 are white; 8 is blue. One ball is drawn. Consider the events:

A: the ball is red;

B: the ball is blue;

C: the number is odd.

Which of these events are independent?

A summation formula

Suppose two events, A and B are considered. Since

$$A = (A \cap B) \cup (A \cap \overline{B})$$

and the latter events are mutually exclusive, we have

$$P(A) = P(A \cap B) + P(A \cap \overline{B})$$

so

$$P(A) = P(A \mid B)P(B) + P(A \mid \overline{B})P(\overline{B}).$$

More generally, if the possible outcomes of an experiment are $B_1, B_2, \ldots, B_k$, then

$$P(A) = \sum_{i=1}^{k} P(A \mid B_i)P(B_i). \tag{6.3}$$

In terms of tree diagrams, this formula is a way of stating in symbols the following rule: "to find the probability of A, find the probabilities associated with all branches that contain A, and sum them."

Sample Problem 6.22 *There are six boxes, two of them round and four square. Each round box contains two green marbles and three blue marbles. Each square box contains one green marble and three blue marbles. A box is chosen at random and a marble is chosen at random from it. What is the probability that the marble is blue? What is the probability that the box was round, given that the marble was blue? Represent the probabilities in a tree diagram.*

Solution. We write R, S, G, B for round, square, green, blue. Then the probabilities of choosing round and square boxes are

$$P(R) = 1/3, P(S) = 2/3.$$

So

$$\begin{aligned} P(B) &= P(B \mid R)P(R) + P(B \mid S)P(S) \\ &= (3/5)(1/3) + (3/4)(2/3) \\ &= 1/5 + 1/2 = 7/10. \end{aligned}$$

We can calculate $P(R \cap B)$:

$$P(R \cap B) = P(B \mid R)P(R) = (3/5)(1/3) = 1/5.$$

So

$$P(R \mid B) = \frac{P(R \cap B)}{P(B)} = \frac{1/5}{7/10} = \frac{2}{7}.$$

The diagram is

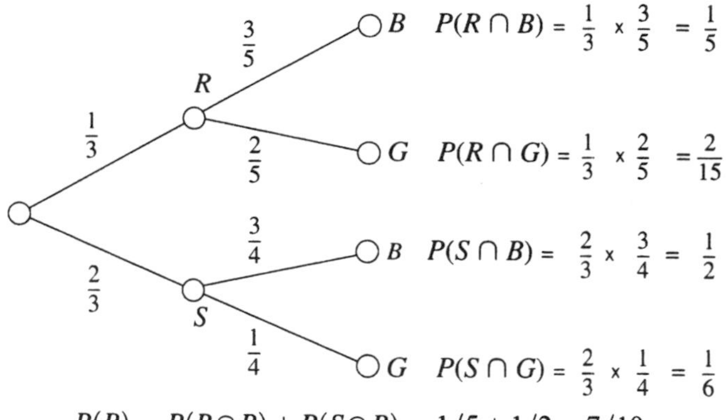

$$P(B) = P(R \cap B) + P(S \cap B) = 1/5 + 1/2 = 7/10.$$

(This last equation is the sum of the probabilities of all branches that contain B.)

Practice Exercise. Consider the builder in Sample Problem 6.20. Suppose further that about 1.5% of the screws from company X are faulty. If he selects a screw at random, what is the probability that it is faulty?

Exercises 6.4

1. Box A contains three red pens and four blue pens. Box B contains two red, one green and one blue pen. A pen is selected from Box A at random, and placed in Box B. Then a pen is selected at random from Box B. What is the probability that this pen is

   (i) red? (ii) blue? (iii) green?

2. A city worker can either go to work by car or by bus. If she goes by car, she uses the tunnel 40% of the time and the bridge 60% of the time. If she takes the bus, it is equally likely that her bus will use the tunnel or the bridge.
   (i) Draw a tree diagram to represent her possible routes to work.
   (ii) Suppose she drives on three days out of five and takes the bus twice out of every five times. What are the probabilities of the various outcomes? What is the probability, on a given day, that she crosses the bridge on her way to work?

3. Box 1 contains five red balls and three white balls. Box 2 contains two red and two white balls. An experiment consists of selecting two balls at random from Box 1 and placing them in Box 2, then selecting one ball from Box 2 at random.
   (i) Draw a tree diagram for this experiment.
   (ii) What is the probability that the ball chosen from Box 2 is red?

4. A card is dealt at random from a regular deck.

(i) What is the probability that it is a jack given that it is a picture card (king, queen or jack)?

(ii) What is the probability that it is a 5, given that it is not a picture card?

5. Two fair dice are rolled. Consider the following events.

A: The sum is 6.

B: Both dice show even numbers.

C: At least one die shows a 4.

What are

$$P(A \mid B), P(A \mid C), P(B \mid A), P(B \mid C), P(C \mid A), P(C \mid B)?$$

6. A pair of fair dice are rolled. Consider the following events.

D: The sum is 7.

E: The sum is odd.

F: At least one die shows a 4.

What are

$$P(D \mid E), P(E \mid D), P(D \mid F), P(F \mid D)?$$

7. Two cards are selected from a regular deck, without replacement. Consider the following events.

A: Both cards are red.

B: The second card is red.

C: At least one card is red.

D: At least one card is a heart.

What are

$$P(A), P(D), P(A \mid B), P(A \mid C), P(D \mid B), P(D \mid C)?$$

8. Two fair dice are rolled. Consider the following events.

A: The sum is 8.

B: Both dice show even numbers.

C: At least one die shows a 5.

What are

$$P(A \mid B), P(A \mid C), P(B \mid A), P(B \mid C), P(C \mid A), P(C \mid B)?$$

9. A jar contains three white and three red marbles. Two are drawn in succession, without replacement, and the colors are noted.

(i) Draw a tree diagram for this experiment.

(ii) Let A, B and C denote the following events.

A: Both marbles are red.

B: The second marble is red.
C: At least one marble is red.

What are $P(A), P(A \mid B), P(A \mid C), P(B \mid C)$?

10. In Exercise 6.2.2, what are the conditional probabilities of the following events?
  (i) Two males are selected, given that at least one male is selected.
  (ii) Two females are selected, given that at least one white mouse is selected.

11. In Exercise 1, what are the conditional probabilities of the following events?
  (i) The pen selected from Box B is blue, given that the pen selected from Box A was blue.
  (ii) The pen selected from Box B is blue, given that the pen selected from Box A was red.

12. In Exercise 6.4.2, what are the conditional probabilities of the following events?
  (i) The worker traveled over the bridge on her way to work, given that she traveled by bus.
  (ii) She used the tunnel, given that she came by car.

13. In Exercise 6.4.3, what are the probabilities of the following events?
  (i) The ball chosen from Box 2 is red, given that both balls chosen from Box 1 were red.
  (ii) The ball chosen from Box 2 is red, given that at least one ball chosen from Box 1 was red.
  (iii) The ball chosen from Box 2 is red, given that at most one ball chosen from Box 1 was red.

In Exercises 14 *to* 21, *illustrate the data with a Venn diagram and calculate* $P(A \mid B)$ *and* $P(B \mid A)$.

14. $P(A) = .7, P(B) = .6, P(A \cup B) = .9.$

15. $P(A) = .7, P(B) = .4, P(A \backslash B) = .4.$

16. $P(A) = .6, P(B) = .4, P(A \cap B) = .1.$

17. $P(A \cup B) = .7, P(A \cap B) = .3, P(B) = .3.$

18. $P(A) = .6, P(B) = .3, P(A \cap B) = .2.$

19. $P(A \cup B) = .8, P(A \cap B) = .4, P(A) = .6.$

20. $P(A) = .4, P(B) = .6, P(A \cup B) = .9.$

21. $P(A) = .6, P(B) = .6$, and $P(A \backslash B) = .4.$

In Exercises 22 to 27, A and B are independent events. Find $P(A \cup B)$ and $P(A \cap B)$ when the probabilities are as given.

22. $P(A) = .6, P(B) = .4.$

23. $P(A) = .5, P(B) = .8.$

24. $P(A) = .8, P(B) = .8.$

25. $P(A) = .7, P(B) = .5.$

26. $P(A) = .4, P(B) = .4.$

27. $P(A) = .9, P(B) = .7.$

28. Two cards are selected without replacement from a regular deck. It is recorded whether or not each card is a spade. Define the events A and B as

A: The first card is a spade;

B: The second card is a spade.

(i) Draw a tree diagram for this experiment.

(ii) Find $P(A), P(B), P(A \mid B), P(B \mid A)$, and $P(A \cap B)$.

(iii) Are A and B independent?

29. Repeat Problem 28 in the case where the first card is replaced and the deck is shuffled before the second selection.

30. A box contains six red and four blue marbles. Two marbles are drawn from the box (without replacement) and their colors are noted. Define the events

E: The first marble is red;

F: The second marble is red.

(i) Draw a tree diagram for this experiment.

(ii) Find $P(E), P(F), P(E \mid F), P(F \mid E)$, and $P(E \cap F)$.

(iii) Are E and F independent?

31. Repeat problem 30 in the case where the first marble is replaced before the second drawing.

32. A weather forecasting station predicts that on December 18 it will snow with a probability of .6, there will be a change in the wind direction with a probability of .8, and there will be both snow and a change in the wind direction with a probability of .4. If the prediction is correct:

(i) What is the probability that there will be neither snow nor a change in the wind direction?

(ii) What is the probability of snow given that the wind direction has changed?

33. A random sample of 100 people were asked their income level and whether they regularly invest in the stock market. The following table gives the results of the survey.

Income level	Regularly invest in stock market	
	Yes	No
High	18	6
Medium	21	16
Low	15	24

A person in the survey is chosen at random. Let H be the event that the person has a high income. Let Y be the event that the person regularly invests in the stock market, and let N be the event that the person does not regularly invest in the stock market. Find

(i) $P(H)$;
(ii) $P(Y)$;
(iii) $P(H \cup Y)$;
(iv) $P(N \mid H)$.

34. Professor Jones has taught the same course for the last 20 years. Each time he teaches it he gives a pretest on the first day of the class. At the end of each semester he compares course grades with whether or not students have passed the pretest. 70% of the course grades he gives are C or better. He finds that 80% of those students who got C or better in the course have passed the pretest while 30% of the students who got less than a C in the course also passed the pretest.

(i) Draw and label a tree diagram illustrating the process.
(ii) Find the probability that a student who takes the pretest will *not* pass it.
(iii) If a student has not passed the pretest, what is his probability of getting less than a C in the course?

35. Two dice are rolled and the sum of the top faces is recorded. Let E be the event that the sum is an odd number, let F be the event that the sum is 4, 7 or 10, and let G be the event that the sum is 2, 7 or 11.

(i) Find $P(G \mid E)$.
(ii) Determine whether E and F are independent.

36. Suppose an urn contains two red, four green and five blue marbles. A marble is selected from the urn and its color noted. If it is red or green it is withdrawn, otherwise it is replaced. Then a second marble is selected from the urn. Determine the probability the first marble was red, given the second was red.

37. Two fair dice, one red and one green, are rolled. Events A, B and C are defined as follows.

A: The sum of the two dice is 7.
B: The green die shows a 3.
C: The green die shows a 1.

(i) Are A and B independent?

(ii) Are A and C independent?

38. Three jars, labelled A, B and C, contain red and green jelly beans as follows: A has two red and two green, B has four red and three green, and C has two red and five green. A jar is selected at random, and one jelly bean is selected at random from it. Represent this experiment in a tree diagram, and find the probability that the jelly bean is red.

39. Suppose E, F and G are any three events. Prove that

$$P(E \cap F \cap G) = P(E) \times P(F \mid G) \times P(G \mid E \cap F).$$

40. Forty percent of those who take drugs also have an alcohol problem and five percent of those who do not take drugs have an alcohol problem. If 32% of the population take drugs, what is the probability that a person who has an alcohol problem also takes drugs?

41. Two fair dice, one red and one blue, are rolled. Events A, B and C are defined as follows.

A: The sum of the two numbers shown is 6.

B: The number showing on the blue die is 2.

C: The blue die shows a 3.

(i) Are A and B independent?

(ii) Are A and C independent?

42. Two jars, labelled A and B contain marbles as follows: A has two red, one white and four blue, and B contains six red, two white, and two blue. A jar is selected at random, and a marble is selected at random from it. Represent this experiment in a tree diagram. What is the probability that the marble chosen is blue?

In Exercises 43 *to* 50, E *and* F *are events in the sample space* S. *You may assume that* $0 < P(E) < 1$ *and* $0 < P(F) < 1$. *Is the statement true, or false, or is it impossible to decide?*

43. $P(E \mid E) - 1$.

44. $P(E \mid \overline{E}) = 1$.

45. $P(F \mid S) = 1$.

46. $P(S \mid F) = 1$.

47. $P(F \mid S) = P(F)$.

48. $P(E \mid E \cap F) = 0$.

49. $P(E \mid F) = P(F \mid E)$.

50. $P(E \cup F) \geq P(E \cap F)$.

6.5 Bayes' Formula and Applications

Bayes' formula

If A and B are any two events, then $A \cap B$ and $B \cap A$ are the same event. However, the conditional probability formula gives two different looking expressions for their probabilities:

$$P(A \cap B) = P(B \mid A)P(A),$$
$$P(B \cap A) = P(A \mid B)P(B).$$

It follows that the two expressions must be equal:

$$P(B \mid A)P(A) = P(A \mid B)P(B).$$

This formula is usually written in the form

$$P(B \mid A) = \frac{P(A \mid B)P(B)}{P(A)}. \tag{6.4}$$

Sample Problem 6.23 *A store gets* 40% *of its stock of light bulbs from factory* X, 35% *from* Y, *and* 25% *from* Z. *Some bulbs are faulty:* 1% *of the output from* X, 2% *from* Y *and* 3% *from* Z. *A light bulb is chosen at random from stock and found to be faulty. What is the chance that it comes from factory* Z?

Solution. Write Z for the event "the bulb comes from factory Z" and A for "the bulb is faulty." Then $P(Z \mid A)$ is required. From the data we know $P(Z) = .25$ and $P(A \mid Z) = .03$. To calculate $P(A)$ we use the formula (6.3):

$$\begin{aligned} P(A) &= P(A \mid X)P(X) + P(A \mid Y)P(Y) \\ &\quad + P(A \mid Z)P(Z) \\ &= (.01)(.40) + (.02)(.35) + (.03)(.25) \\ &= .004 + .007 + .0075 \\ &= .0185. \end{aligned}$$

So

$$P(Z \mid A) = \frac{(.25)(.03)}{.0185} = \frac{75}{185} = \frac{15}{37},$$

which is between 40 and 41%.

Practice Exercise. You are given three coins; one is biased so that it shows heads two-thirds of the time, and the other two are fair (heads and tails are equally likely). You select a coin at random and flip; it shows heads. What is the probability that it is the biased coin?

The formula (6.4) is called *Bayes' Formula*. It is often combined with (6.3) as follows. Suppose the possible outcomes of an experiment are $B_1, B_2, \ldots, B_k$. Then

$$P(B_i \mid A) = \frac{P(A \mid B_i)P(B_i)}{P(A \mid B_1)P(B_1) + \cdots + P(A \mid B_k)P(B_k)}. \tag{6.5}$$

(The denominator of (6.5) comes directly from (6.3).)

When Bayes' formula is used, very often the outcomes $B_1, B_2, \ldots, B_k$ are the outcomes of an experiment that occurred earlier, and A is an outcome, or set of outcomes, of a later experiment. Bayes' formula answers the question, "given the outcome of the second experiment, what is the probability that the outcome of the first experiment was so-and-so?" This sometimes confuses students, because it seems to suggest that the outcome of the later experiment can somehow affect the earlier experiment, and that is impossible — causes must come before effects. What in fact happens is that knowledge of the later experiment can increase our (incomplete) knowledge of the earlier experiment.

Sample Problem 6.24 *The car pool contains ten Fords (five red and five white) and fifteen Pontiacs (five red and ten white). You are allocated a car at random. You see from a distance that it is red. What is the probability that you have been given a Ford?*

Solution. We require $P(F \mid R)$. There are 25 cars of which ten are Fords, so $P(F) = \frac{10}{25} = .4$, and ten are red, so $P(R) = .4$ also. The probability of a red car, given that it is a Ford, is .5, since half the Fords are red. So

$$\begin{aligned} P(R)P(F \mid R) &= P(F)P(R \mid F), \\ .4 \times P(F \mid R) &= .4 \times .5, \\ P(F \mid R) &= .5. \end{aligned}$$

Practice Exercise. Box X contains two blue pens and three red pens. Box Y contains two red pens and three blue pens. A box is chosen at random and a pen is chosen at random from it. If the pen is blue, what is the probability that the box was box X?

Tree diagrams

It is often convenient to use tree diagrams in Bayes' formula problems. When the diagram is completed, the terms to be added for the denominator of (6.5) are on the right-hand side.

The computation of Bayes' formula can be carried out as follows:

1. First construct a tree diagram with the possible outcomes $B_1, B_2, \ldots, B_k$ as branches.

2. To each of these branches add the branches A ("the event A occurs") and $\overline{A}$ ("A does not occur"). The paths ending in A together represent all the circumstances in which A can occur.

3. Now calculate the probability — the product of the conditional probabilities — for each branch that ends in A. The sum of these probabilities will be $P(A)$, the denominator of (6.5).

4. The numerator $P(B_i \mid A)$ will be written next to one of the branches.

Sample Problem 6.25 *Construct a tree diagram for Sample Problem* 6.23.

Solution. In the terminology of experiments, Sample Problem 6.23 could be described as follows. First, a supplier is chosen; then a light bulb is chosen from that supplier. So the outcomes of the first experiment (the $B_1, B_2, \ldots$) are factories X, Y or Z. The first part of the tree diagram has three branches labelled X, Y and Z, with probabilities .4, .35 and .25 respectively. Event A is "the bulb is faulty," and the three probabilities are .01, .02 and .03. So we obtain the diagram of Figure 6.5.

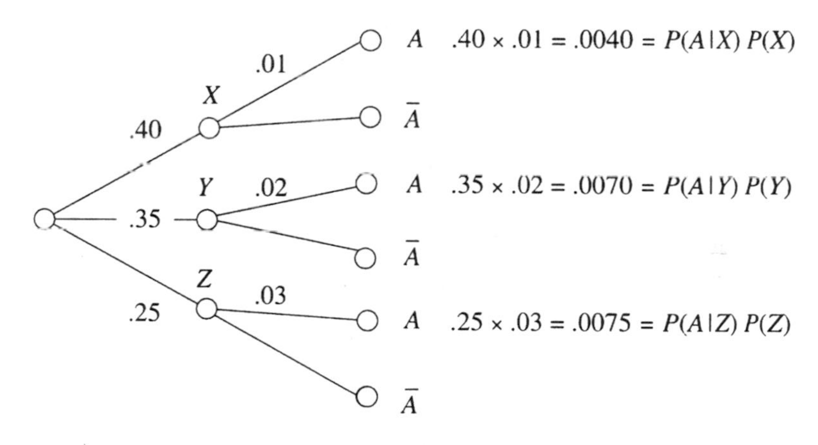

Figure 6.5: Tree diagram for Sample Problem 6.25

From the Figure, the denominator is .0185. So

$$P(Z \mid A) = \frac{P(A \mid Z)P(Z)}{.0185} = \frac{.0075}{.0185} = .676.$$

Practice Exercise. Repeat the above for Sample Problem 6.24.

Sample Problem 6.26 *The following table shows the proportion of people over* 18 *who are in various age categories, together with the probability that a person in a given category will vote in a given election. A vote is selected at random. What is the probability that the voter was from the* 18–24 *age group?*

Age	Proportion	Probability
A: 18–24	18.2%	.49
B: 25–44	38.6%	.71
C: 45–64	28.2%	.63
D: 65+	15.2%	.74

Solution. We use the diagram

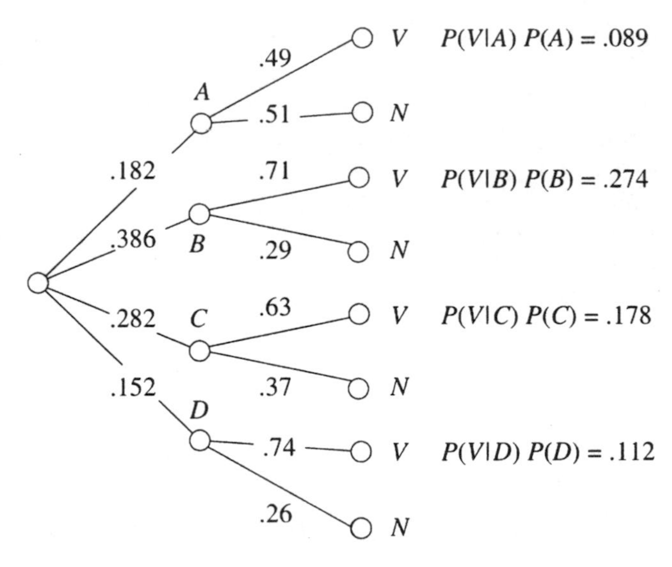

(V and N mean "voter" and "non-voter" respectively).

$$P(A \mid V) = \frac{.089}{.653} = .136.$$

Practice Exercise. What is the probability that the voter was 65 or older?

Box diagrams

In many examples it is easiest to represent a two-stage experiment using a box diagram. The outcomes of the first experiment are used as labels for rows, and the outcomes of the second experiment are used as labels for columns. Where row A meets column B, we put a box containing $P(A \cap B)$. The sum of entries in row A, which equals $P(A)$, is written at the end of row A, and so on.

As an example, consider Sample Problem 6.23. The first experiment — determination of factory — has outcomes X, Y and Z; the second experiment — "check: is it faulty" — has outcomes A (faulty) and B (okay). Then

$$\begin{aligned} P(A \cap X) &= P(A \mid X)P(X) = .004, \\ P(A \cap Y) &= P(A \mid Y)P(Y) = .007, \\ P(A \cap Z) &= P(A \mid Z)P(Z) = .0075. \end{aligned}$$

(See Figure 6.5 for these calculations.)

We can calculate $P(B \cap X)$ in various ways. For example, A and B are complements, so

$$\begin{aligned} P(A \cap X) + P(B \cap X) &= P(X), \\ P(B \cap X) &= P(X) - P(A \cap X) \\ &= .4 - .004 = .396; \end{aligned}$$

similarly $P(B \cap Y) = .343, P(B \cap Z) = .2425$. So the box diagram is as shown in Figure 6.6.

	A	B	
X	.004	.396	.4
Y	.007	.343	.35
Z	.0075	.2425	.25
	.0185	.9815	

Figure 6.6: Box diagram of data for Sample Problem 6.23

Sample Problem 6.27 *Use the box diagram in Figure* 6.6 *to calculate* $P(Z \mid A)$ *in Sample Problem* 6.23.

Solution. To find $P(Z \mid A)$, look at column A, which has a total of .0185. Then look at the (Z,A) entry .0075, and take the ratio

$$P(Z \mid A) = \frac{P(A \cap Z)}{P(Z)} = \frac{0075}{.0185} = .406.$$

Practice Exercise. Produce a box diagram for the Practice Exercise following Sample Problem 6.23 and *use it* to find the probability that you have tossed the biased coin, given that it shows heads.

Medical testing

Probability theory is often misused in the press. One particular area where probabilistic ideas are mentioned but not properly analyzed is in the discussion of tests for disease and drugs. This is a very important topic nowadays, when compulsory drug testing is becoming more common in everyday life and testing for diseases such as AIDS is also very important.

You may see in a newspaper that a certain test is "90% accurate." And you would most likely think this means that, if you test positive, there is a 90% probability that you have the disease in question. But this is not so, as the next example shows.

Sample Problem 6.28 *A test for a venereal disease is* 90% *accurate: if you have the disease, the probability is* .9 *that you will test positive; and if you do not have the disease, the probability is* .9 *that you will test negative. In the whole population, one person in a hundred has the disease. If a given person tests positive, what is the chance that he has the disease?*

Solution. We shall use the abbreviations T (tests positive), N (tests negative), D (has the disease) and H (is healthy). We want to find $P(D \mid T)$.

From the data, we know

$$\begin{aligned} P(T \mid D) &= .9, & P(N \mid D) &= .1, \\ P(T \mid H) &= .1, & P(N \mid H) &= .9, \\ P(D) &= .01, & P(H) &= .99. \end{aligned}$$

So

$$\begin{aligned} P(D \mid T) &= \frac{P(T \mid D)P(D)}{P(T \mid H)P(H) + P(T \mid D)P(D)} \\ &= \frac{(.9)(.01)}{(.1)(.99) + (.9)(.01)} \\ &= \frac{.009}{.099 + .009} = \frac{.009}{.108} = .083. \end{aligned}$$

So, even if you test positive, it is still most unlikely that you have the disease. This example can be represented by the following box diagram:

	P	N	
D	.009	.001	.01
H	.099	.891	.99
	.108	.892	

Practice Exercise. A disease has infected 5% of the population. The test is 95% effective—95% of those with the disease test positive, and 95% of those without the disease test negative. If your test shows a positive result, what is the probability that you have the disease?

The Monty Hall problem

Marilyn vos Savant's *Ask Marilyn* column in *Parade* Magazine for September 9th, 1990 contained a puzzle that generated a lot of interest. It is in fact a version of an older problem called the *Monty Hall Problem*, named for the game show host.

> The climax of a TV game show is run as follows. The contestant is given a choice of three numbered doors. Behind one closed door is a valuable new car; behind each of the others is a nearly worthless goat. The contestant is to choose a door, and wins the prize behind it. However, after the choice is announced but before the door is opened, the host opens one of the other two doors (not the one she chose) and reveals a goat. (Of course the host knows where the car is.) He then asks, "Do you want to stay with your original choice? Or would you rather switch to the third door?"

Well, should the contestant stay or switch? Or doesn't it matter?

Before analyzing the problem, we need to agree on three points. First, the car is placed behind the doors at random, so that on any given night the chance that it is behind any particular door is $\frac{1}{3}$. Second, the game always proceeds in the same way: the host *always* opens a door to show a goat, then offers the switch. Third, on those nights when the contestant's first choice is the door with the car, there is an equal chance that the host will open either of the other two doors.

Without loss of generality, let us suppose the contestant chooses door 1 and the host opens door 2. We write $C1$, $C2$ and $C3$ as abbreviations for "the car is behind door 1," "the car is behind door 2," and "the car is behind door 3," and $H2, H3$ for "the host opens door 2," "the host opens door 3." (He cannot open door 1.) Then what we want to know is:

$$\text{is } P(C3 \mid H2) > P(C1 \mid H2)?$$

If so, the contestant should switch, otherwise not.

We know the probabilities of the host's actions, given the position of the car. If the car is behind door 1, she is equally likely to open either door, so

$$P(H2 \mid C1) = P(H3 \mid C1) = \tfrac{1}{2}.$$

In the other cases she must open the remaining "goat" door, so

$$P(H2 \mid C2) = P(H3 \mid C3) = 0,$$
$$P(H2 \mid C3) = P(H3 \mid C2) = 1.$$

Moreover we know $P(C1) = P(C2) = P(C3) = \frac{1}{3}$.

Now calculation of the probabilities is a simple application of Bayes' formula. First

$$\begin{aligned} P(H2) &= P(H2|C1)P(C1) + P(H2|C2)P(C2) + P(H2|C3)P(C3) \\ &= (\frac{1}{2})(\frac{1}{3}) + (0)(\frac{1}{3}) + (1)(\frac{1}{3}) = \frac{1}{2}, \end{aligned}$$

and similarly $P(H3) = \frac{1}{2}$. So

$$P(C1 \mid H2) = \frac{P(H2 \mid C1)P(C1)}{P(H2)} = \frac{(\frac{1}{2})(\frac{1}{3})}{\frac{1}{2}} = \frac{1}{3},$$

$$P(C3 \mid H2) = \frac{P(H2 \mid C3)P(C3)}{P(H2)} = \frac{(1)(\frac{1}{3})}{(\frac{1}{2})} = \frac{2}{3}.$$

So the odds are 2 to 1 in favor of switching. You may find this very surprising — intuitively, you might argue that "the car was equally likely to be behind any of the doors, so the probabilities are still equal," but the host's choice of doors has actually given you some information.

The most important assumption in this discussion is that when the contestant has chosen correctly, the host is equally likely to open either door. Suppose this were not true — for example, suppose he always chooses the lowest numbered available door. Then if the contestant chooses door 1 and the host opens door 3, she should always switch; but if he opens door 2, there is no advantage (or disadvantage) in switching.

This problem is particularly well suited to analysis by box diagrams. Assuming that the contestant chooses door 1, then the fact that the car is equally likely to be behind any door means that the diagram looks like

	$C1$	$C2$	$C3$	
$H2$	$*$	$*$	$*$	$*$
$H3$	$*$	$*$	$*$	$*$
	$\frac{1}{3}$	$\frac{1}{3}$	$\frac{1}{3}$	

(where asterisks represent the numbers that have not yet been determined). Since the host never opens the door that hides the car, the $(H2, C2)$ and $(H3, C3)$ entries must be zero, and to make the column sums come out right we have

	$C1$	$C2$	$C3$	
$H2$	$*$	0	$\frac{1}{3}$	$*$
$H3$	$*$	$\frac{1}{3}$	0	$*$
	$\frac{1}{3}$	$\frac{1}{3}$	$\frac{1}{3}$	

Assuming the host chooses at random when the car is behind door 1, we can make the two missing numbers in column $C1$ equal, and we have

	$C1$	$C2$	$C3$	
$H2$	$\frac{1}{6}$	0	$\frac{1}{3}$	$\frac{1}{2}$
$H3$	$\frac{1}{6}$	$\frac{1}{3}$	0	$\frac{1}{2}$
	$\frac{1}{3}$	$\frac{1}{3}$	$\frac{1}{3}$	

When the host opens door 2, we need only look at the $H2$ row to see that the odds are 2 to 1 ($\frac{1}{3}$ to $\frac{1}{6}$) in favor of switching, and similarly if he opens door 3.

Exercises 6.5

1. The events E and F satisfy $P(E) = .6, P(F \mid E) = .5$, and $P(F \mid \overline{E}) = .75$. Find $P(E \mid F)$ and $P(\overline{E} \mid F)$.

2. The events E and F satisfy $P(E) = .85, P(F \mid E) = .8$, and $P(F \mid \overline{E}) = .8$. Find $P(E \mid F)$ and $P(\overline{E} \mid F)$.

3. One experiment has the possible outcomes A, B and C; a second experiment has the possible outcomes E and F. Find $P(A \mid E)$, $P(B \mid E)$, $P(C \mid E)$, $P(A \mid F)$, $P(B \mid F)$ and $P(C \mid F)$ in each of the following cases:

(i) $P(A) = .4, P(B) = .4, P(E \mid A) = .25, P(E \mid B) = .75$ and $P(E) = .6$.
(ii) $P(A) = .25$, $P(B) = .25$, $P(E \mid A) = .4$, $P(E \mid B) = .4$ and $P(E) = .5$.
(iii) $P(A) = \frac{3}{5}, P(B) = \frac{1}{10}, P(E \mid A) = \frac{1}{3}, P(E \mid B) = 0$ and $P(E) = \frac{1}{2}$.

4. One experiment has the possible outcomes A, B and C; a second experiment has the possible outcomes E and F. Find $P(A \mid E), P(B \mid E), P(C \mid E)$, $P(A \mid F), P(B \mid F)$ and $P(C \mid F)$ in each of the following cases:
(i) $P(A) = \frac{3}{10}, P(B) = \frac{1}{2}, P(E \mid A) = \frac{2}{3}, P(F \mid B) = \frac{2}{5}$ and $P(F \mid C) = \frac{1}{2}$.
(ii) $P(A) = .25$, $P(B) = .25$, $P(E \mid A) = .6$, $P(F \mid B) = .6$, and $P(F \mid C) = .5$.
(iii) $P(A) = .5$, $P(B) = .3$, $P(E \mid A) = .6$, $P(F \mid B) = 0$, and $P(F \mid C) = .5$.

5. An automobile dealership finds that 4% of their customers default on payments, so that the car must be repossessed. On analyzing the records, it is found that among those who did not default, 40% made a large down-payment ($2,000 or more), while only 10% of those who later defaulted made a large down-payment.
(i) Suppose a customer makes a large down-payment. What is the probability that he will default on payments?
(ii) If a customer makes a small down-payment, what is the probability that he will default on payments?

6. A builder buys tiles from two companies, A and B; he gets 80% from A and 20% from B. He finds that 98% of the tiles he gets from A are undamaged, while 96% of those from B are undamaged. If he finds a damaged tile, what is the probability that it came
(i) from A?
(ii) from B?

7. In manufacturing metal office equipment, your company uses nuts supplied by three companies, A, B and C. Company A supplies 30%, company B supplies 45% and company C supplies 25%. It is known that on average the following percentages of the nuts are defective: 1% of those from A, 1.5% of those from B, and 0.5% of those from C.
(i) If a nut is selected at random, what is the probability that it is defective?
(ii) If a nut is selected at random and is found to be defective, what is the probability that it was made by company B?

8. Suppose the shipping department of a company has three workers who prepare shipping labels. They prepare 60%, 30% and 10% of the labels respectively. The respective percentages of errors are 3%, 5% and 10%. find the probability that an incorrect label is due to the first person. Also find the probability that an incorrect label is due to each of the other persons.

9. In a certain city, it is found that equally many people have fair and dark hair. A survey shows that 20% of people with dark hair and 40% of people with fair hair have blue eyes. A person is chosen at random from the population and is found to have blue eyes. What is the probability that this person has fair hair?

10. Box 1 contains three red pens and four blue pens. Box 2 contains four red pens, one green pen, and three black pens. A pen is chosen from Box 1 and then it is placed in Box 2. Then a box is chosen at random and a pen is selected.

(i) What is the probability that the pen is blue?

(ii) If the pen is blue, what is the probability it originally came from Box 1?

(*Hint:* Work this problem as though there were two types of blue pen: light blue initially in Box 1, dark blue initially in Box 2.)

11. In Exercise 10, suppose Box 2 is the one from which a pen in finally selected.

(i) If the pen drawn is blue, what is the probability that the pen drawn earlier from Box 1 was red?

(ii) If the pen drawn is red, what is the probability that the pen drawn earlier from Box 1 was red?

12. In a manufacturing plant, three machines, A, B and C, produce 50%, 35% and 15% respectively of the total production. The quality control department of the company has determined that 1% of the items produced by Machine A and 2% of the items produced by each of Machines B and C are defective. If an item is selected at random and found to be defective, what is the probability that it was produced by Machine B?

13. In a factory, 30% of the workers smoke. It is found that smokers have three times the absentee rate of other workers. If a worker is absent, what is the probability that he is a smoker?

14. An auto insurance company classifies 20% of its drivers as good risks, 60% as medium risks and 20% as bad risks (called classes A, B and C respectively). The probability of at least one accident in a given year is 1% for class C, .5% for class B and .1% for class A. If one of their insureds has an accident this year, what is the probability that he is a class B driver?

15. A class contains 60% women. It is found that 12% of the men students and 7% of the women students are left-handed. A student is chosen at random. If the student is left-handed, what is the probability that the student is male?

16. A school tax proposition is submitted to voters. The voters' registered party affiliation as a percentage of all voters, and the percentage of each group who voted in favor of the proposition, are as follows:

Party	Registration	In Favor
Democrat	40	70
Republican	40	20
Independent	10	80
Other	10	50

A voter is selected at random.

(i) What is the probability that she voted in favor of the proposition?

(ii) If she voted in favor of the proposition, what is the probability that she is a Democrat?

(iii) If she voted against the proposition, what is the probability that she is not a Republican?

17. Among one strain of cattle in Texas, it is estimated that 30% of the bulls suffer from Ruckel's disease. A blood test has been developed to detect the disease. If a bull has the disease, then the test gives a positive response 80% of the time. If a bull does not have the disease, the test is positive in 10% of cases. If a bull is tested for the disease and the test is positive, what is the probability that he really *does not* have the disease?

18. A test is positive for 94% of subjects who have a certain disease and negative for 96% of those who do not have the disease. If 4% of the population suffers from the disease, then:

(i) If the test is positive, what is the probability that the patient actually has the disease?

(ii) If the test is negative, what is the probability that the patient does not have the disease?

19. A smallpox vaccine produces total immunity in 95% of cases. If the vaccination does not produce immunity, then it has no effect — the person's probability of catching smallpox is the same as that of an unvaccinated person. Suppose 30% of the population have been vaccinated. If a person contracts smallpox, what is the probability that she had been vaccinated?

20. A laboratory test for a particular disease tests positive 95% of the time when a person has the disease, and tests negative 98% of the time when the person does not have the disease. In general 0.1% of the population has the disease.

(i) What is the probability that a randomly selected person who tested positive did not have the disease?

(ii) What is the probability that a randomly selected person who tested negative actually had the disease?

21. Suppose the "Monty Hall" game involved four doors, not three. What is the probability of winning if you switch? What is the probability of winning if you do not switch?

22. Suppose there are three cards in a hat. One is red on both sides, one black on both sides, and the third has one side red and one side black. One card is withdrawn, and one side is observed to be red. What is the chance that the other side is red?

7

Graph Theory

The word *graph* is used in discrete mathematics for a mathematical object consisting of a set of objects together with a way in which some pairs of objects are related. So we could say it represents the most general type of symmetric binary relation.

This simple structure has very widespread applications. For example, graphs can be used to represent road systems, airline routes and water pipelines. They have a number of applications in routing and scheduling. They are important in computer science as models of structures (such as VLSI chips) and of processes (in the design of operating systems, for example) and also as data structures.

7.1 Introduction to Graphs

Graphs

Any binary relation can be represented by a diagram. Suppose $\sim$ is a binary relation on the set S. The elements of S are shown as points (called *vertices*), and if $x \sim y$ is true, then a line (called an *edge*) is shown from x to y, with its direction indicated by an arrow. Provided the set S is finite, all information about any binary relation on S can be shown in the diagram.

The diagram of a binary relation on a finite set is called a *directed graph* or *digraph*. If $x \sim x$ is true for at least one vertex x, the diagram is a *looped* digraph; if $x \sim x$ never occurs ($\sim$ is antireflexive), then the digraph is called *simple*.

In the case where the relation is symmetric, the arrows can be omitted from its diagram. The diagram of a symmetric, antireflexive binary relation on a finite set is called a *graph*. The phrase "looped graph" can be used to describe the diagram of a symmetric relation in which $x \sim x$ is sometimes true, but be careful: a "looped graph" is not really a graph! Similarly, "infinite graph" means an object defined like a graph in which the vertex set can be infinite.

Two vertices are either related or not, so the diagram of a graph can never contain two lines joining the same pair of vertices. However, there are some applications where two edges between the same vertices might make sense. (For example, see the applications to road systems in the next section.) For this reason we sometimes talk about *multigraphs* in which there can be several edges joining the same pair of vertices; those edges are called *multiple edges*. (Similarly one can have multiple edges in the same direction in a directed structure; the somewhat clumsy word "multidigraph" has been used.) When discussing multigraphs, the word "simple" means that neither loops nor multiple edges occur.

Sample Problem 7.1 *The binary relations $<$, $\equiv$, $\sim$ and $\approx$ are defined on the set $S = \{1,2,3,4\}$ as follows:*

(i) $x < y$ *means* x *is less than* y;

(ii) $x \equiv y$ *means* x *is congruent to* y $(\text{mod } 2)$ *and* $x \neq y$;

(iii) $x \sim y$ *means* $x = y \pm 1$;

(iv) $x \approx y$ *means* $y = x^2$.

Draw the diagrams associated with $<$, $\equiv$, $\sim$ and $\approx$. Which are graphs or simple digraphs?

Solution. The diagrams are shown below. Relations $\equiv$ and $\sim$ yield graphs, $<$ gives a simple digraph, and $\approx$ a looped digraph.

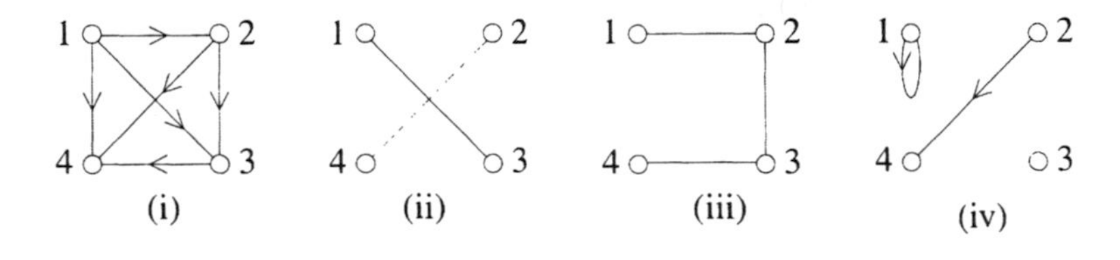

Practice Exercise. The binary relations $\equiv$, $\sim$ and $\approx$ are defined on S by

(i) $x \equiv y$ means $x + y = 4$;

(ii) $x \sim y$ means $x = y + 1$;

(iii) $x \approx y$ means $y \neq x$.

Draw the diagrams associated with $\equiv$, $\sim$ and $\approx$. Which are graphs or simple digraphs?

We make a formal definition and set up some notation as follows: a *graph* G consists of a finite set V of objects called *vertices*, together with a symmetric, antireflexive binary relation called *adjacency*, written $\sim$. Associated with G is a set E of unordered pairs of elements of V, called *edges*. $\{x,y\}$ is an edge if and only if $x \sim y$. Vertices x and y are called the *endpoints* of the edge $\{x,y\}$, and we say this edge *joins* x to y. If a graph G is under discussion, we usually write $V(G)$ to mean the set of vertices of G and $E(G)$ for its set of edges.

In the diagram representing a graph, an edge $\{x,y\}$ is shown as a line from x to y. This is usually drawn as a straight line segment, but not always; the only important thing is that it joins the correct points. Moreover, the position of the points is not fixed. For these two reasons, one graph can give rise to several drawings that look quite dissimilar. For example, the three diagrams in Figure 7.1 all represent the same graph. Although the two diagonal lines cross in the first picture, their point of intersection does not represent a vertex of the graph.

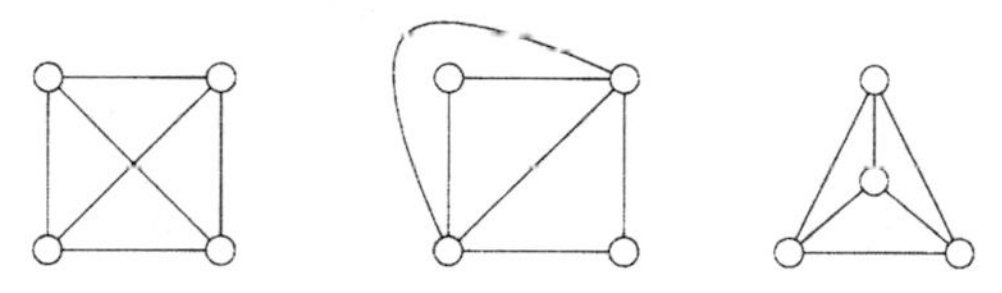

Figure 7.1: Three representations of the same graph

Some special graphs

If G is a graph, it is possible to choose some of the vertices and some of the edges of G in such a way that these vertices and edges again form a graph, say H. H is then called a *subgraph* of G; one writes $H \leq G$. Clearly every graph G has itself as a subgraph; we say a subgraph H is a *proper* subgraph of G, and write $H < G$, if it does not equal G.

Given a set S of v vertices, the graph formed by joining each pair of vertices in S is called the *complete* graph on S and denoted K_S. We also write K_v to mean any complete graph with v vertices. The three drawings in Figure 7.1 are all representations of K_4.

Given any graph G, the set of all edges of $K_{V(G)}$ that are *not* edges of G will form a graph with $V(G)$ as its vertex set; this new graph is called the *complement* of G, and written $\overline{G}$. More generally, if G is a subgraph of H, then the graph formed by deleting all edges of G from H is called the *complement of G in H*, denoted $H - G$. The complement $\overline{K_S}$ of the complete graph K_S on vertex set S is called a *null graph*; we also write $\overline{K}_v$ for a null graph with v vertices.

The *complete bipartite graph* on V_1 and V_2 has two disjoint sets of vertices, V_1 and V_2; two vertices are adjacent if and only if they lie in different sets. We write $K_{m,n}$ to mean a complete bipartite graph with m vertices in one set and n in the other. Figure 7.2 shows $K_{3,4}$; $K_{1,n}$ in particular is called an *n-star*. Any subgraph of a complete bipartite graph is called *bipartite*.

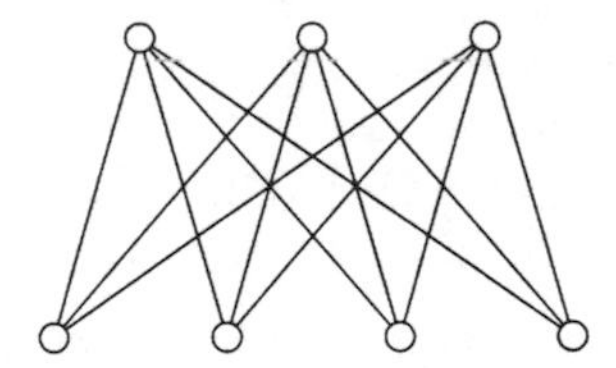

Figure 7.2: $K_{3,4}$

A graph is called *disconnected* if its vertex set can be partitioned into two subsets, V_1 and V_2, that have no common element, in such a way that there is no edge with one endpoint in V_1 and the other in V_2; if a graph is not disconnected, then it is *connected*. A disconnected graph consists of a number of disjoint subgraphs; a maximal connected subgraph is called a *component*.

The complement of the complete bipartite graph $K_{m,n}$ is a disconnected graph. It consists of two components, a K_m and a K_n. This graph is denoted $K_m \cup K_n$. More generally, if g and H are any two graphs, the graph formed by putting the two together is called the *union* of G and H, and is written $G \cup H$. When G and H have no common vertex, we say it is a *disjoint* union. So we could say, "the complement of $K_{m,n}$ is the disjoint union of a K_m and a K_n."

Degree

We define the *degree* or *valency* $d(x)$ of the vertex x to be the number of edges that have x as an endpoint. If $d(x) = 0$, then x is called an *isolated* vertex, while a vertex of degree 1 is called a *pendant* vertex. The edge incident with a pendant vertex is called a *pendant edge*. A graph is called *regular* if all its vertices have the same degree. If the common degree is r, it is called *r-regular*. In particular, a 3-regular graph is called *cubic*. We write $\delta(G)$ for the smallest of all degrees of vertices of G, and $\Delta(G)$ for the largest. (One also writes either $\Delta(G)$ or $\delta(G)$ for the common degree of a regular graph G.)

Sample Problem 7.2 *What are the degrees of the vertices in graphs* (ii) *and* (iii) *of Sample Problem 7.1? What are the degrees of the vertices of* $K_{3,4}$*?*

Solution. In Figure 7.1(ii), every vertex has degree 2. In 7.1(iii), vertices 1 and 4 have degree 1, while 2 and 3 have degree 2. $K_{3,4}$ has three vertices of degree 4 (the upper vertices in Figure 7.2), while the other four have degree 3.

Practice Exercise. What are the degrees of the vertices on the graphs in Figure 7.3?

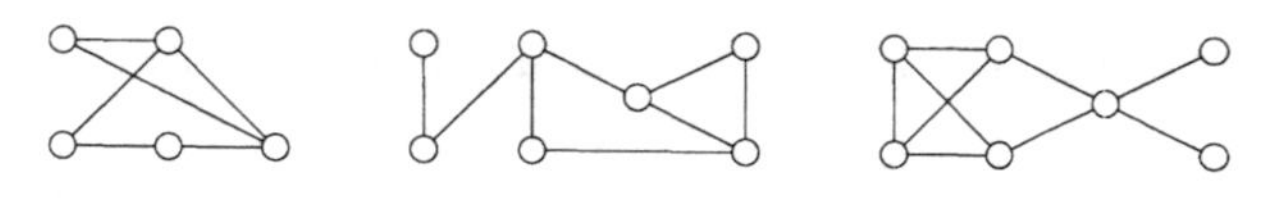

Figure 7.3: What are the degrees?

Theorem 7.1 *In any graph or multigraph, the sum of the degrees of the vertices equals twice the number of edges (and consequently the sum of the degrees is an even integer).*

Proof. Suppose the graph or multigraph has e edges; label the edges, say $y_1, y_2, \ldots, y_e$. Consider a list in which each edge appears twice, once for each endpoint. For example, if y_1 has endpoints x_4 and x_7, you might make entries $y_1 : x_4$ and $y_1 : x_7$. Vertex x will appear in precisely $d(x)$ entries, so the total number of entries equals the sum of the degrees of the vertices. On the other hand, each edge appears twice, so the total number of entries equals twice the number of edges. □

Corollary 7.1.1 *In any graph or multigraph, the number of vertices of odd degree is even. In particular, a regular graph of odd degree has an even number of vertices.*

A collection of v non-negative integers is called *graphical* if and only if there is a graph on v vertices whose degrees are the members of the collection. A graphical collection is called *valid* if and only if there is a *connected* graph with those degrees.

Sample Problem 7.3 *Which of the following collections are graphical?*

$3,3,2,2,0 \quad 4,3,1,0 \quad 3,2,2,2,2 \quad 4,3,3,3,2,2,1 \quad 4,4,3,2,1$

Solution. The first and fourth are graphical; see (i) and (ii).

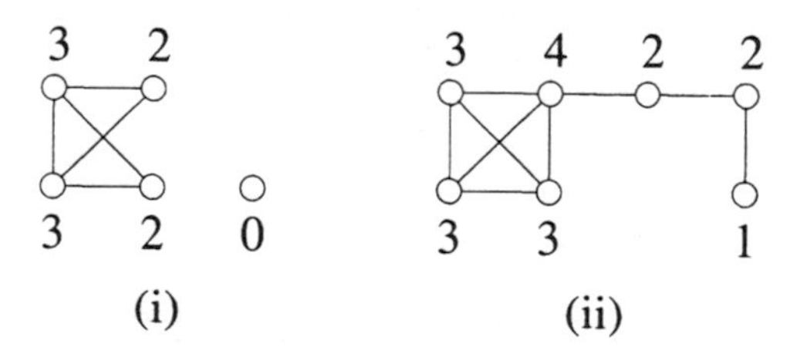

(Notice the isolated vertex, corresponding to degree 0.) $4,3,1,0$ is impossible because there are only four vertices, so no vertex can have degree greater than 3. The collection $3,2,2,2,2$ is impossible because the sum of the degrees is odd. And $4,4,3,2,1$ is impossible; if there are five vertices and two of them have degree 4, the other vertices must all be adjacent to those two; certainly no vertex can have degree smaller than 2.

Practice Exercise. Which of the following collections are graphical?

$$2,2,2,1,1 \quad 5,3,3,1,1,1 \quad 4,3,2,2,1,1,1$$

This is one situation where the distinction between graphs and multigraphs is important. For example, even though $4,3,1,0$ is not graphical, there is a *multi*graph with those degrees:

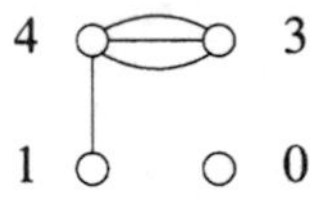

Exercises 7.1

In Exercises 1 *to* 8, *draw the diagram of the binary relation* $\sim$, *defined on* $\{-2,-1,0,1,2\}$. *Is it simple? Is it directed?*

1. $x \sim y$ means $x+y \leq 2$.

2. $x \sim y$ means $x+y \leq 6$.

3. $x \sim y$ means $x = y+1$.

4. $x \sim y$ means $x = \pm y$.

5. $x \sim y$ means $x \leq y^2$.

6. $x \sim y$ means $x+y \geq 0$.

7. $x \sim y$ means $x+y$ is odd.

8. $x \sim y$ means xy is odd.

In Exercises 9 *to* 15, $\sim$ *is defined on the positive integers. Is the corresponding diagram simple? Is it directed?*

9. $x \sim y$ means $x+y$ is odd.

10. $x \sim y$ means x divides y.

11. $x \sim y$ means x and y have greatest common divisor 1.

12. $x \sim y$ means $x+y \leq 4$.

13. $x \sim y$ means x and y are both prime numbers.

14. $x \sim y$ means $x = \pm y$.

15. $x \sim y$ means xy is odd.

16. What are the degrees of the vertices in these graphs?

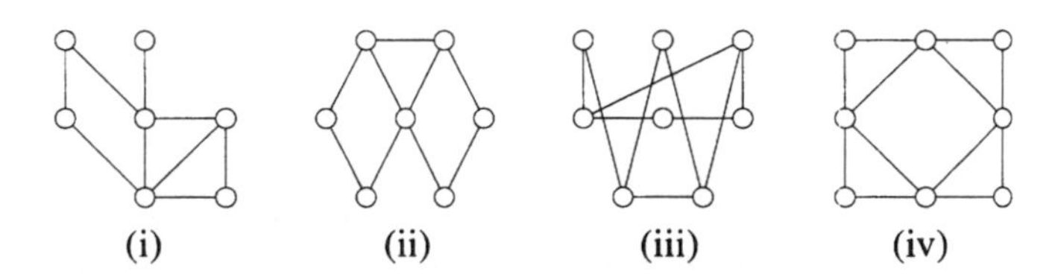

17. What are the degrees of the vertices in these graphs?

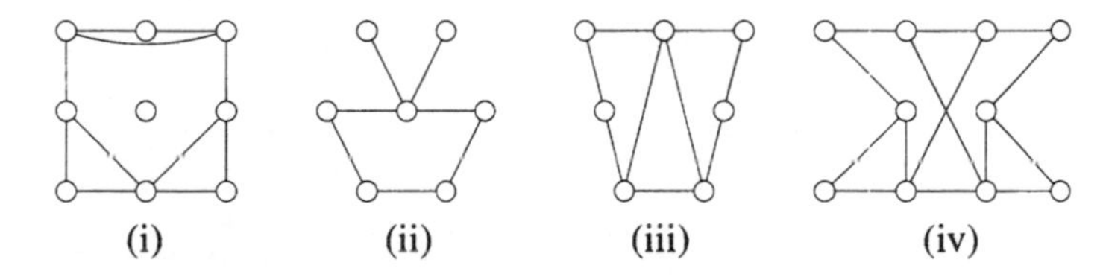

18. Suppose G has v vertices and $\delta(G) \geq \frac{v-1}{2}$. Prove that G is connected.

19. How many edges does the star $K_{1,n}$ have? What are the degrees of its vertices?

20. The *n-wheel* W_n has $n+1$ vertices $\{x_0, x_1, \ldots, x_n\}$; x_0 is joined to every other vertex and the other edges are

$$x_1x_2, x_2x_3, \ldots, x_{n-1}x_n, x_nx_1.$$

How many edges does W_n have? What are the degrees of its vertices?

21. Prove that the collection $\{3,3,2,1,1\}$ is graphical and valid. Find a connected graph with this collection of degrees.

22. Prove that the collection $\{3,2,2,2,1\}$ is graphical and valid. Find a connected graph with this collection of degrees.

In Exercises 23 *to* 30, *is the collection graphical? If it is graphical, is it valid?*

23. $\{5,5,4,4,2,2\}$

24. $\{3,3,2,2,2,1\}$

25. $\{5,4,4,3,3,3\}$

26. $\{4,3,2,2,2,1\}$

27. $\{5,4,4,4,3,3\}$

28. $\{5,4,4,4,2,1\}$

29. $\{2,2,1,1,1,1\}$

30. $\{3,3,2,2,1,1\}$

31. Prove that if G is not a connected graph, then $\overline{G}$ is connected.

32. Prove that a regular graph of odd degree can have no component with an odd number of vertices.

33. Prove that no graph has all its vertices of different degrees.

7.2 The Königsberg Bridges; Traversability

Modeling road systems

Suppose we want to describe the roads joining various towns in order to plan an itinerary. If our only desire is to list the towns that will be visited, in order, we do not need to know the physical properties of the region, such as hills, whether different roads cross, or whether there are overpasses. The important information is whether or not there is a road joining a given pair of towns. In these cases we could use a complete road map, with the exact shapes of the roads and various other details shown, but it would be less confusing to make a diagram as shown in Figure 7.4(a), that indicates roads joining B to C, A to each of B and C, and C to D, with no direct roads joining A to D or B to D. The diagram is obviously a graph, whose vertices are the towns and whose edges are the road links.

In some cases there is a choice of routes between two towns. This information may be important: you might prefer the freeway, which is faster, or you might prefer the more scenic coast road. In this case the graph does not contain enough information. For example, if there had been two different ways to travel from B to C, this information could be represented as in the diagram of Figure 7.4(b). This new diagram is a *multigraph*; that is, there is a *multiple edge* (of *multiplicity* 2) joining B to C.

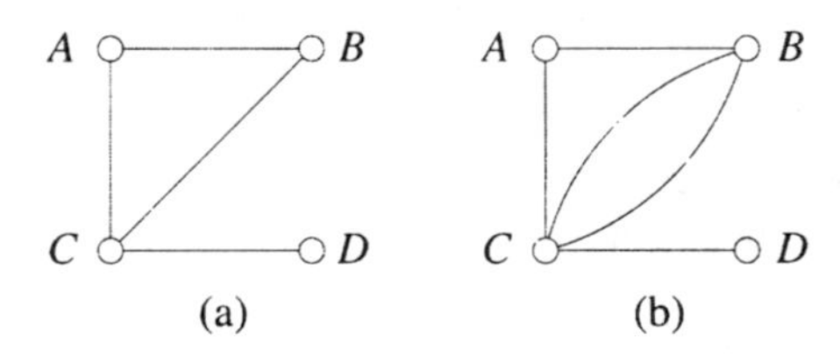

Figure 7.4: Representing a road system

Sample Problem 7.4 *Suppose the road system connecting towns A, B and C consists of two roads from A to B, one road from B to C, and a bypass road directly from A to C. Represent this road system graphically.*

Solution.

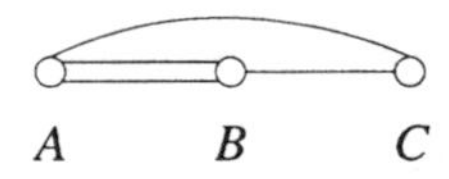

Practice Exercise. Represent graphically the road system connecting towns A, B, C, D, with one road from A to B, one from A to C, two from A to D, and one from B to C.

The Königsberg bridges

The first mathematical paper on graph theory was published by the great Swiss mathematician Leonhard Euler in 1736, and had been delivered by him to the St. Petersburg Academy one year earlier.

Euler's paper grew out of a famous old problem. The town of Königsberg in Prussia is built on the river Pregel. The river divides the town into four parts, including an island called The Kneiphof, and in the eighteenth century the town had seven bridges; the layout is shown in Figure 7.5. The question under discussion was whether it is possible from any point on Königsberg to take a walk in such a way as to cross each bridge exactly once.

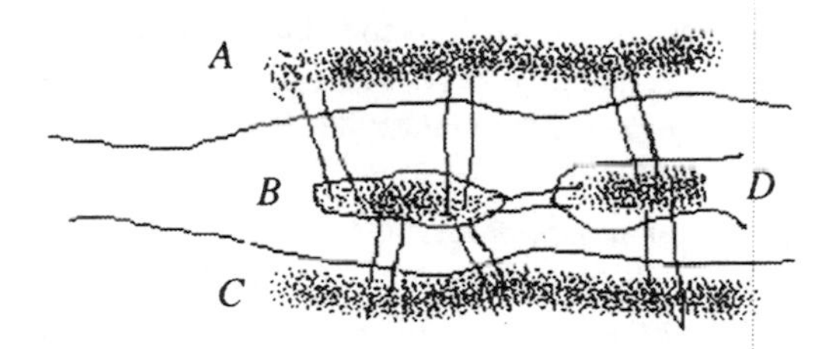

Figure 7.5: The Königsberg bridges

Euler set for himself the more general problem: given any configuration of river, islands and bridges, find a general rule for deciding whether there is a walk that covers each bridge precisely once.

We first show that it is impossible to walk over the bridges of Königsberg. Suppose there was such a walk. There are three bridges leading to the area C: you can traverse two of these, one leading into C and the other leading out, at one time in your tour. There is only one bridge left: if you cross it going into C, then you cannot leave C again, unless you use one of the bridges twice, so C must be the finish of the walk; if you cross it in the other direction, C must have been the start of the walk. In either event, C is either the place where you started or the place where you finished.

A similar analysis can be applied to A, B and D, since each has an odd number of bridges — five for B and three for the others. But any walk starts at one place and finishes at one place; either there can be two endpoints, or the walk can start and finish at the same place. Therefore it is impossible for A, B, C and D all to be either the start or the finish.

Euler started by finding a multigraph that models the Königsberg bridge problem (considered as a road network, with the islands and river banks as separate "towns"). Vertices A, B, C and D represent the parts A, B, C and D of the town, and each bridge is represented by an edge. The multigraph is shown in Figure 7.6. In terms of this model, the original problem becomes: can a simple walk be found that contains every edge of the multigraph?

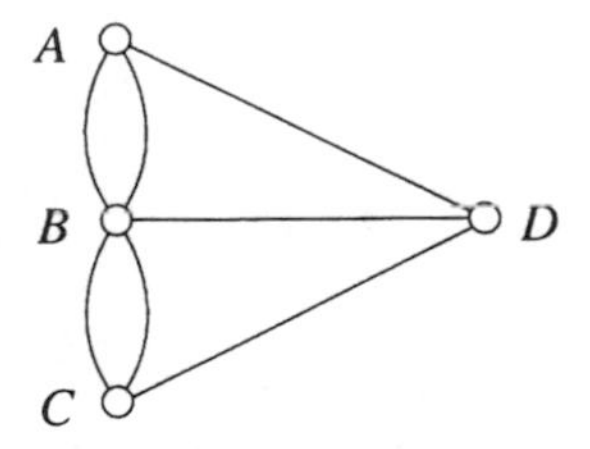

Figure 7.6: The multigraph representing the Königsberg bridges

These ideas can be applied to more general configurations of bridges and islands, and to other problems. A simple walk that contains every edge of a given multigraph will be called an *Euler walk* in the multigraph. Finding whether a given multigraph (or a given road network) has an Euler walk is called the *traversability problem.*

Sample Problem 7.5 *The islands A, B, C, D are joined by seven bridges — two joining A to B, two from C to D, and one each joining the pairs AC, AD and BD. Represent the system graphically. Could one walk through this system, crossing each bridge exactly once?*

Solution. Islands B and D each have an odd number of bridges, so one must be the start of the walk and the other the finish. With a little experimentation you will find a solution — one example is $BACDABDC$, and there are others.

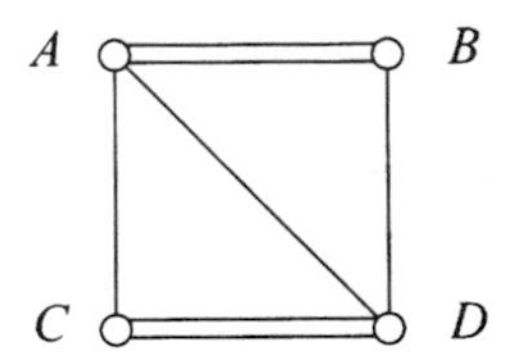

Practice Exercise. The islands X, Y, Z, T are joined by seven bridges — three joining X to Y, and one each joining the pairs XZ, XT and YZ. Represent the system graphically. Could one walk through this system, crossing each bridge exactly once?

In proving that a solution to the Königsberg bridge problem is impossible, we used the fact that certain vertices had an odd number of edges incident with them. (The precise number was not important; oddness was the significant feature.) Let us call a vertex *even* if its degree is even, and *odd* otherwise. It was observed that an odd vertex must be either the first or the last point in the walk. In fact, if a multigraph is traversable, then either the multigraph has two odd vertices, the start and finish of the Euler walk, or the multigraph has no odd vertices, and the Euler walk starts and finishes at the same point. Another obvious necessary condition is that the multigraph must be connected. These two conditions are together sufficient.

Theorem 7.2 *If a connected multigraph has no odd vertices, then it has an Euler walk starting from any given point and finishing at that point. If a connected*

multigraph has two odd vertices, then it has an Euler walk whose start and finish are the odd vertices.

(A closed Euler walk — one that starts and ends at the same point — is called an *Euler circuit.*)

Proof. Consider any simple walk in a multigraph that starts and finishes at the same vertex. If one erases every edge in that walk, one deletes two edges touching any vertex that was crossed once in the walk, four edges touching any vertex that was crossed twice, and so on. (For this purpose, count "start" and "finish" combined as one crossing.) In every case an *even* number of edges is deleted.

First, consider a multigraph with no odd vertex. Select any vertex x, and select any edge incident with x. Go along this edge to its other endpoint, say y. Then choose any other edge incident with y. In general, on arriving at a vertex, select any edge incident with it that has not yet been used, and go along the edge to its other endpoint. At the moment when this walk has led into the vertex z, where z is not x, an odd number of edges touching z has been used up (the last edge to be followed, and an even number previously). Since z is even, there is at least one edge incident with z that is still available. Therefore, if the walk is continued until a further edge is impossible, the last vertex must be x — that is, the walk is closed. It will necessarily be a simple walk and it must contain every edge incident with x.

Now assume that a connected multigraph with every vertex even is given, and a simple closed walk has been found in it by the method just described. Delete all the edges in the walk, forming a new multigraph. From the first paragraph of the proof it follows that every vertex of the new multigraph is even. It may be that we have erased every edge in the original multigraph; in that case we have already found an Euler walk. If there are edges still left, there must be at least one vertex, say c, that was in the original walk and that is still on an edge in the new multigraph — if there were no such vertex, then there could be no connection between the edges of the walk and the edges left in the new multigraph, and the original multigraph must have been disconnected. Select such a vertex c, and find a closed simple walk starting from c. Then unite the two walks as follows: at one place where the original walk contained c, insert the new walk. For example, if the two walks are

$$x, y, \dots, z, c, u, \dots, x$$

and

$$c, s, \dots, t, c,$$

then the resulting walk will be

$$x, y, \dots, z, c, s, \dots, t, c, u, \dots, x.$$

(There may be more than one possible answer if c occurred more than once in the first walk. Any of the possibilities may be chosen.) The new walk is a closed simple walk in the original multigraph. Repeat the process of deletion, this time deleting the newly formed walk. Continue in this way. Each walk contains more edges than the preceding one, so the process cannot go on indefinitely. It must stop. But this will only happen when one of the walks contains all edges of the original multigraph, and that walk is an Euler walk.

Finally, consider the case where there are two odd vertices p and q and every other vertex is even. Form a new multigraph by adding an edge pq to the original. This new multigraph has every vertex even. Find a closed Euler walk in it, choosing p as the first vertex and the new edge pq as the first edge. Then delete this first edge; the result is an Euler walk from q to p. □

It is clear that loops make no difference as to whether or not a graph has an Euler walk. If there is a loop at vertex x, it can be added to a walk at some traversing of x.

A good application of Euler walks is planning the route of a highway inspector or mail contractor, who must travel over all the roads in a highway system. Suppose the system is represented as a multigraph G, as was done earlier. Then the most efficient route will correspond to an Euler walk in G.

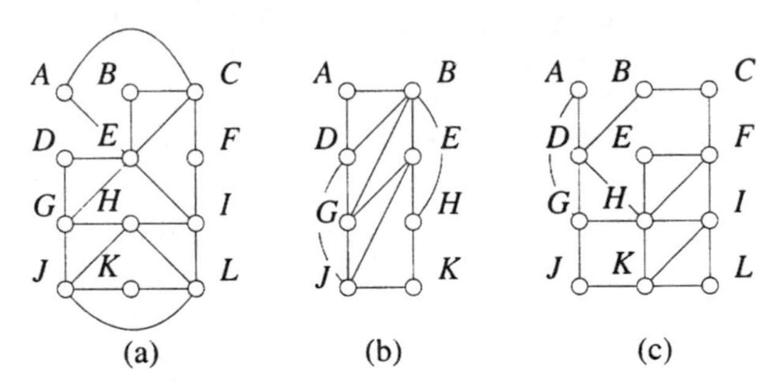

Figure 7.7: Find Euler walks

Sample Problem 7.6 *Find an Euler walk in the road network represented by Figure* 7.7(a,b).

Solution. In the first example, starting from A, we find the walk $ACBEA$. When these edges are deleted (see Figure 7.8(a)) there are no edges remaining through A. We choose C, a vertex from the first walk that still has edges adjacent to it, and trace the walk $CFIHGDEC$, after which there are no edges available at C (see Figure 7.8(a)). E is available, yielding walk $EGJLIE$. As is clear from Figure 7.8(c), the remaining edges form a walk $HJKLH$.

We start with $ACBEA$. We replace C by $CFIHGDEC$, yielding

$$ACFIHGDECBEA,$$

then replace the first E by $EGJLIE$, with result

$$ACFIHGDEGJLIECBEA$$

(we could equally well replace the second E). Finally H is replaced by $HJKLH$, and the Euler walk is

$$ACFIHJKLHGDEGJLIECBEA.$$

In the second example, there are two odd vertices, namely B and F, so we add another edge BF and make it the first edge used. The first walk found was

$$BFHGCABFDBCEB,$$

and the second is $DEGD$, exhausting all the vertices and producing the walk

$$BFHGCABFDEGDBCEB.$$

The first edge (the new one we added) is now deleted, giving Euler walk

$$FHGCABFDEGDBCEB$$

in the original graph.

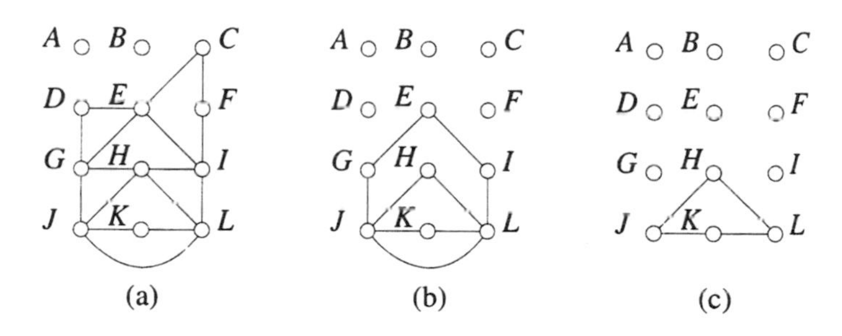

Figure 7.8: Constructing an Euler walk

Practice Exercise. Find an Euler walk in the road network represented by Figure 7.7(c).

Eulerization

If G contains no Euler walk, the highway inspector must repeat some edges of the graph in order to return to his starting point. Let us define an *Eulerization* of G to be a multigraph, with a closed Euler walk, that is formed from G by duplicating some edges. A *good* Eulerization is one that contains the minimum number of new edges, and this minimum number is the *Eulerization number* $eu(G)$ of G. For example, if two adjacent vertices have odd degree, you could add a further edge joining them. This would mean that the inspector must travel the road between them twice. However, if the two vertices were not adjacent, one new edge will not suffice — it would be the same as requiring that a new road be built!

Sample Problem 7.7 *Consider the multigraph G of Figure* 7.9(a). *What is $eu(G)$? Find an Eulerization of the road network represented by G that uses the minimum number of edges.*

Solution. Look at the representation of the multigraph on the left of Figure 7.10. The black vertices have odd degree, so they need additional edges. As there are four black vertices, at least two new edges are needed; but

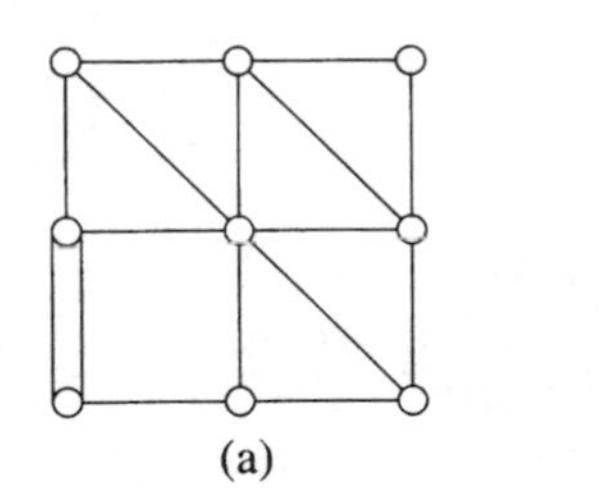
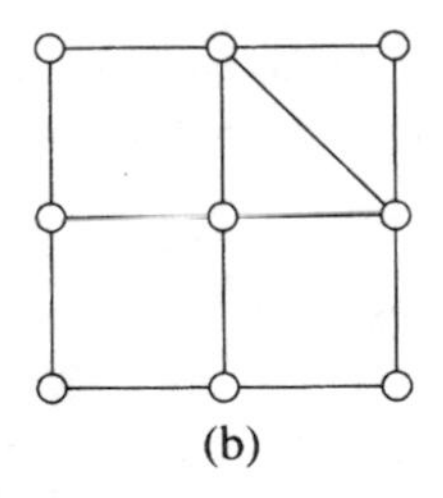

Figure 7.9: Find Eulerizations

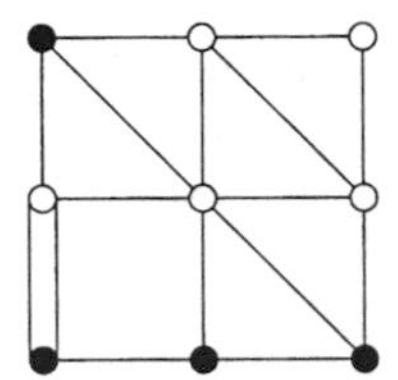
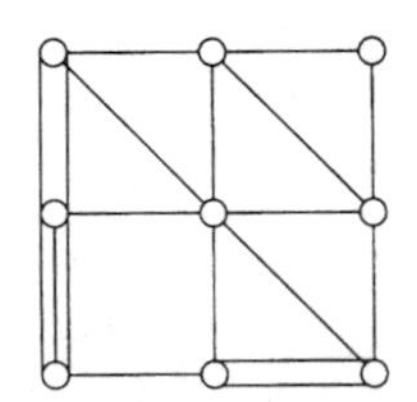
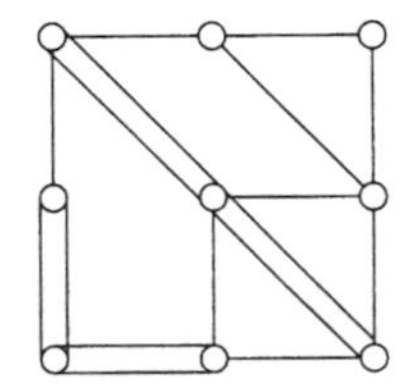

Figure 7.10: Finding an Eulerization

obviously no two edges will suffice. However, there are solutions with three added edges — two examples are shown in Figure 7.10 — so $eu(G) = 3$.

Practice Exercise. Consider the multigraph H of Figure 7.9(b). What is $eu(H)$? Find an Eulerization of the road network represented by H that uses the minimum number of edges.

Exercises 7.2

In Exercises 1 *to* 16, *decide whether the graph shown contains an Euler walk. If the graph contains an Euler walk, find one. Is it an Euler circuit?*

1.

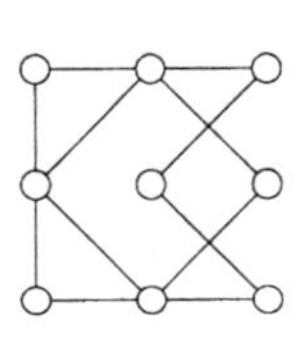

2.

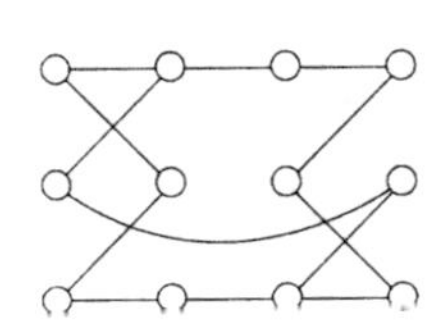

3.

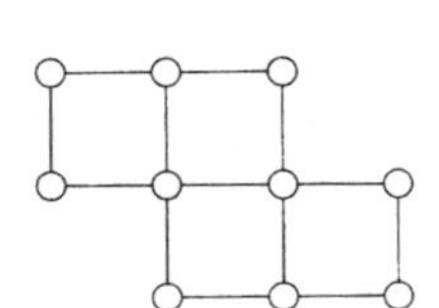

4.

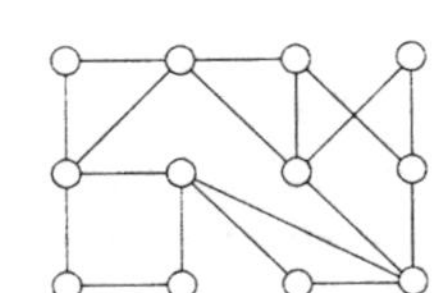

5.

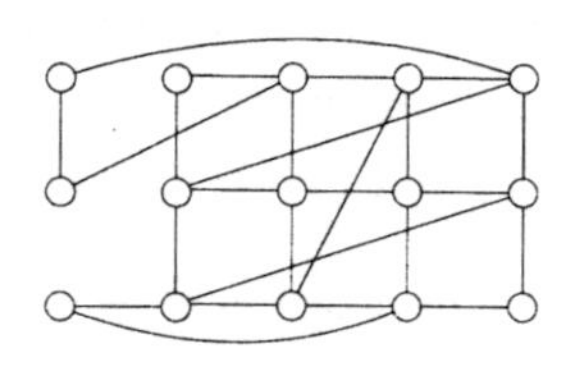

6.

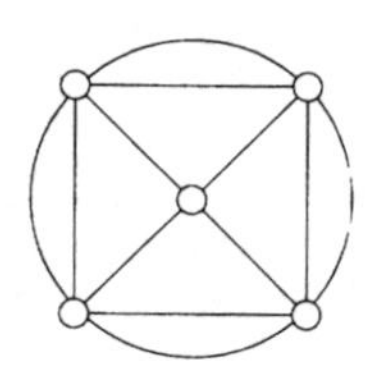

7.

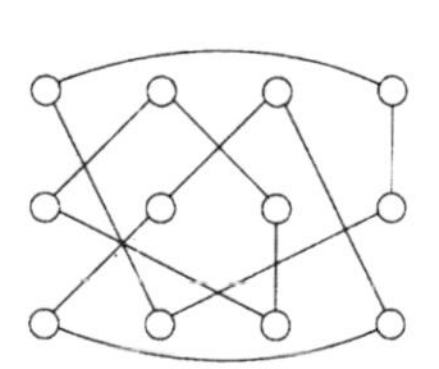

8.

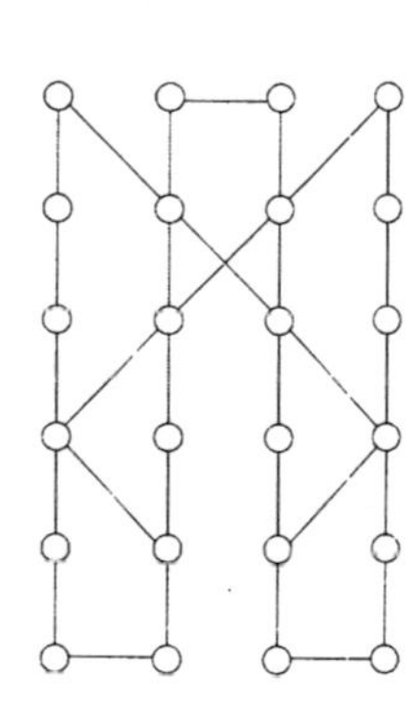

9.

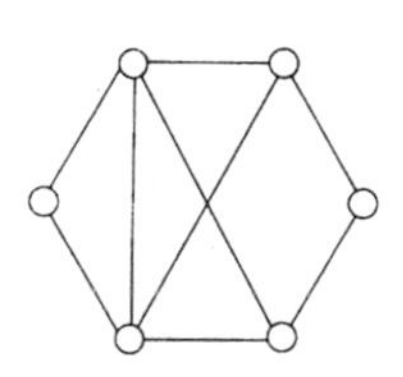

10.

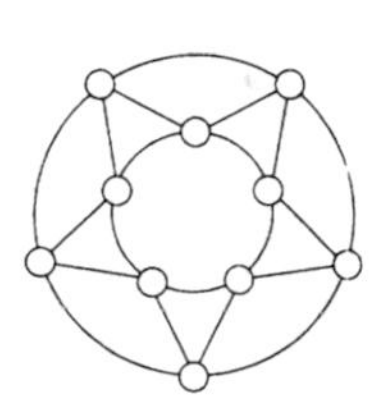

11.

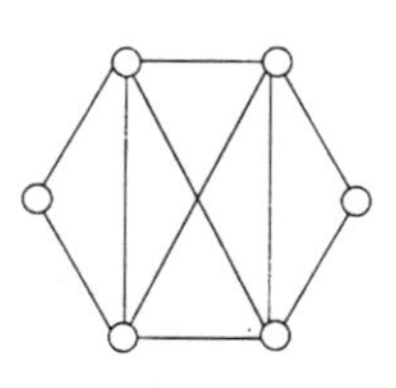

12.

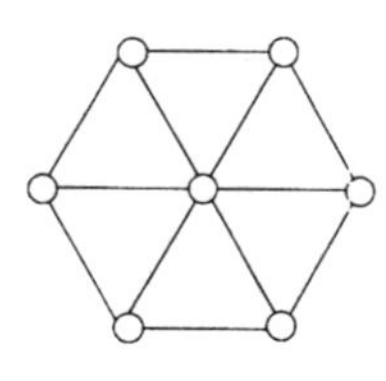

13.

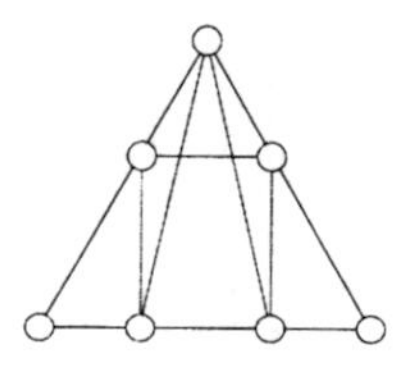

14.

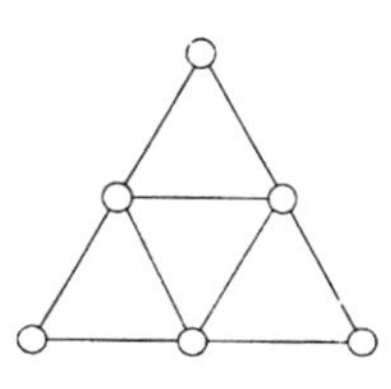

15.

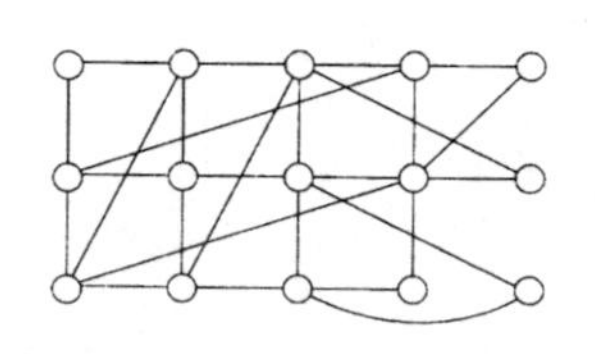

16.

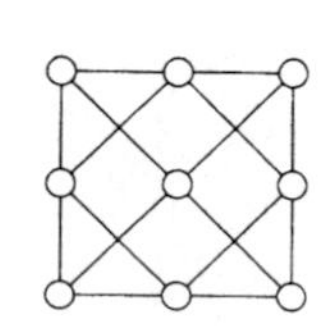

17. Suppose G is a connected graph with k vertices of odd degree. Show that G can be decomposed into $k/2$ edge-disjoint simple walks.

18. Find a graph G such that both G and its complement $\overline{G}$ have Euler circuits.

In Exercises 19 *to* 28, *state the Eulerization number of the graph and find a good Eulerization.*

19.

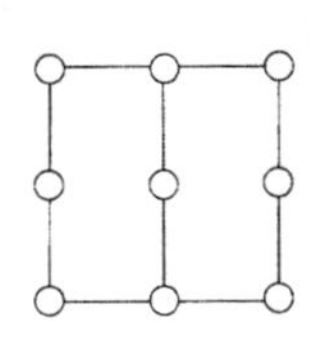

20.

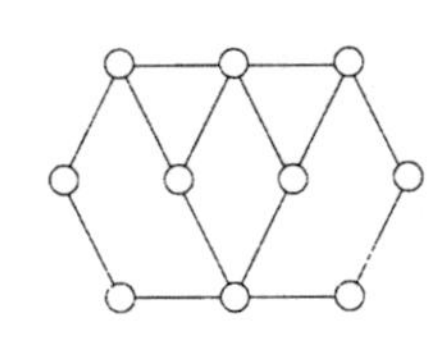

21.

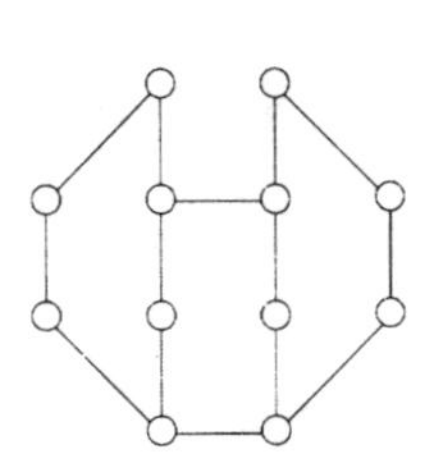

22.

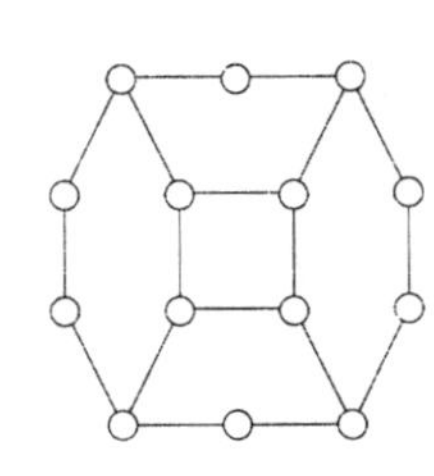

23.

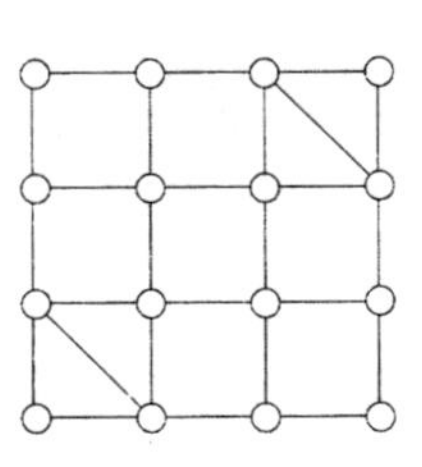

24.

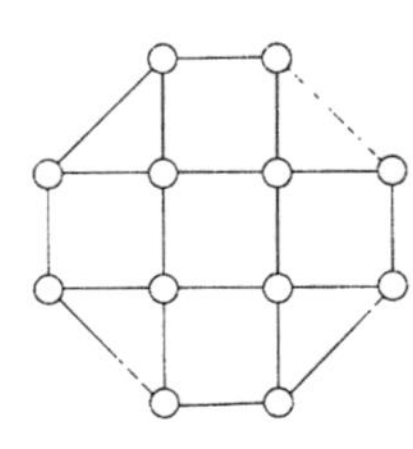

25.

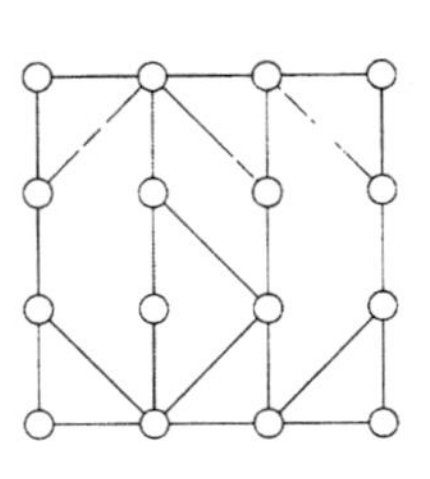

26.

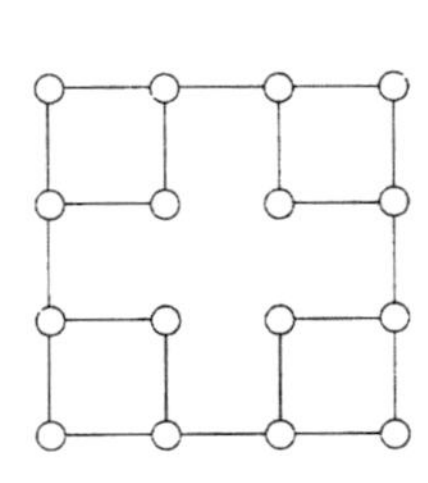

27.

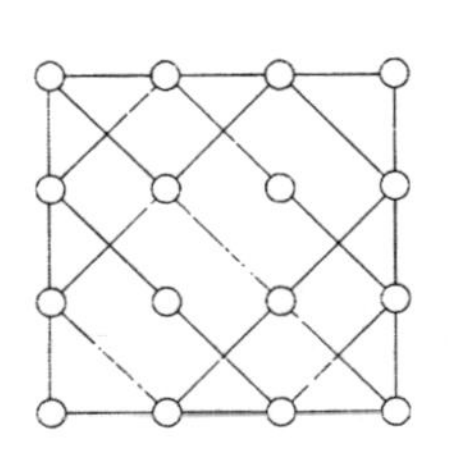

28.

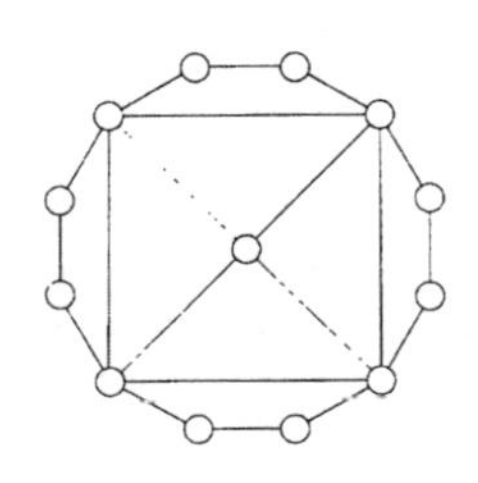

7.3 Walks, Paths and Cycles

To generalize the idea of an Euler walk, we shall define a *walk* in a graph G to be any finite sequence of vertices $x_0, x_1, \ldots, x_n$ and edges $a_1, a_2, \ldots, a_n$ of G:

$$x_0, a_1, x_1, a_2, \ldots, a_n, x_n,$$

where the endpoints of a_i are x_{i-1} and x_i for each i. The *length* of a walk is its number of edges. The walk shown above might be called an x_0x_n-*walk* of length n. A *closed* walk, sometimes called a *circuit*, is one in which the first and last vertices are the same.

Walks are called *simple* if there is no repeated edge. So an Euler walk is a simple walk that contains every edge. A walk in which no *vertex* is repeated is a *path*.

There is no such thing as a closed path — the last vertex is a repeat of the first — but we shall define a *cycle* to be a closed simple walk with no vertex repeats except for the endpoints.

We also use the word "path" for a graph that consists of a single n-vertex path, and write P_n for such a graph. A "cycle" C_n is defined similarly to mean a graph consisting of an n-vertex cycle. If multiple edges are allowed, C_2 would mean a multigraph with two vertices joined by a double edge, and if loops are allowed, then C_1 usually means a single vertex with a loop, but for graphs C_n is only defined when $n \geq 3$. C_n is a cycle of length n, while P_n is a path of length $n-1$.

Figure 7.11 shows P_5 and C_5.

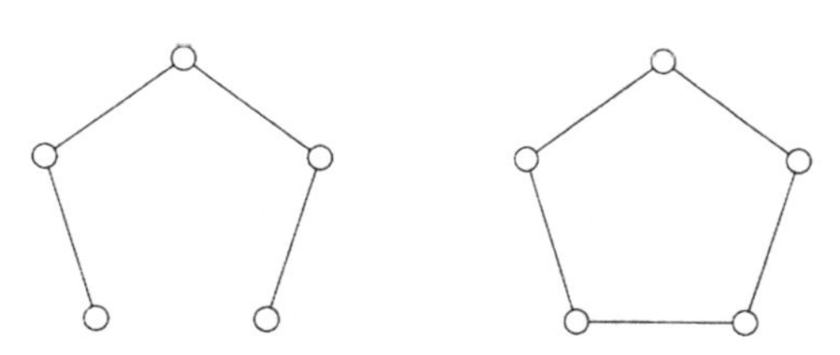

Figure 7.11: P_5 and C_5

Suppose there is a walk from vertex x to vertex y in a graph. Maybe there are repeated vertices in the walk. But suppose z is a repeated vertex. You could follow the walk from x up to the first time you come to z, then go from the last occurrence of z along the walk to y. In this way you replace the walk by one in which z is not a repeated vertex. This could be done until every repeat has been eliminated. If x and y are different, the end result is a path from x to y. (If $x = y$, *all* the edges are eliminated.)

Theorem 7.3 *If there is a walk from vertex x to vertex y in the graph G, where x is not equal to y, then there is a path in G whose first vertex is x and whose last vertex is y.*

Connectedness

We shall say that two vertices are *connected* when there is a walk joining them. (Theorem 7.3 tells us we could replace the word "walk" by "path.") Two vertices of G are connected if and only if they lie in the same component of G; G is a connected graph if and only if all pairs of its vertices are connected. (This definition of "connected graph" is consistent with the one given in Section 7.1.)

Among connected graphs, some are connected so slightly that removal of a single vertex or edge will disconnect them. Such vertices and edges are quite important. A vertex x is called a *cutpoint* in G if $G-x$ contains more components than does G; in particular if G is connected, then a cutpoint is a vertex x such that $G-x$ is disconnected. Similarly a *bridge* (or *cut-edge*) is an edge whose deletion increases the number of components.

Sample Problem 7.8 *If a graph represents a road network, what is the interpretation of a bridge? Why are such edges important?*

Solution. A bridge represents a road whose deletion would cut people in part of the network off from those in another part. (Often a real bridge, over a river etc., plays this role.) It is important to recognize such roads, so that extra effort can be taken to keep them open to traffic. For example, the highways department might give special priority to keeping them plowed in winter.

Practice Exercise. In such a graph, what is the interpretation of a cutpoint?

Connectivity

Generalizing the idea of a cutpoint, we define the *connectivity* $\kappa(G)$ of a graph G to be the smallest number of vertices whose removal from G results in either a disconnected graph or a single vertex. (The latter special case is included to avoid problems when discussing complete graphs.) If $\kappa(G) \geq k$, then G is called k-connected. The *edge-connectivity* $\kappa'(G)$ is defined to be the minimum number of edges whose removal disconnects G (no special case is needed). In other words, the edge-connectivity of G equals the size of the smallest cutset in G. From the definition, it is clear that the connectivity and edge-connectivity of a graph is at least as great as that of any of its subgraphs.

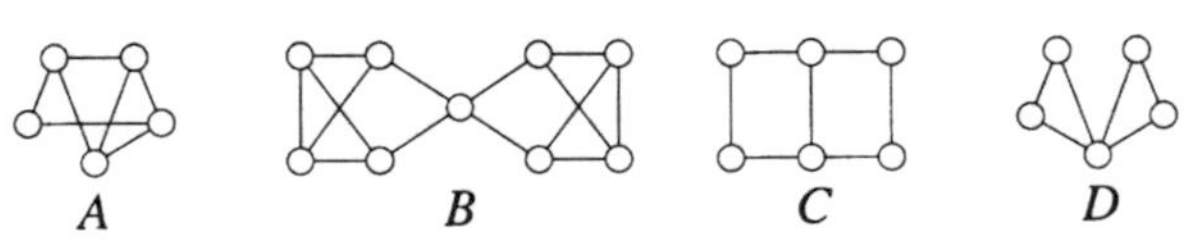

Figure 7.12: Examples for connectivity calculations

Sample Problem 7.9 *What are κ and κ' for the graphs A, B in Figure 7.12?*

Solution. A has no cutpoint or bridge, so $\kappa(A) \geq 2$ and $\kappa'(A) \geq 2$. Consider the leftmost vertex: deleting the two vertices adjacent to it, or the two edges containing it, will disconnect A. So $\kappa(A) = \kappa'(A) = 2$. The center vertex of B is a cutpoint, so $\kappa(B) = 1$, but there is no bridge; however, the graph can be disconnected by removing the two edges joining the center vertex from the left, so $\kappa'(B) = 2$.

Practice Exercise. Evaluate κ and κ' for the graphs C, D in Figure 7.12.

The following theorem was proven by Whitney in 1932. Recall that $\delta(G)$ denotes the minimum degree of vertices of G.

Theorem 7.4 *For any graph G,*

$$\kappa(G) \leq \kappa'(G) \leq \delta(G).$$

The proof is beyond our scope, but the easiest part is set as an exercise.

Sample Problem 7.10 *Calculate* δ *and verify Theorem* 7.4 *for the graphs A, B in Figure* 7.12.

Solution. A has minimum degree 2, so $\kappa(A), \kappa'(A), \delta(A) = 2,2,2$. B has minimum degree 3, so $\kappa(B), \kappa'(B), \delta(B) = 1,2,3$.

Practice Exercise. Do the same calculation for the graphs C, D.

It is easy to see that all combinations of strictness are possible in Theorem 7.4: both of the inequalities can be strict, or one of them, or neither. For example, graphs A and B in Figure 7.12 illustrate the possibilities of κ, κ' and δ being all equal or all different.

Exercises 7.3

1. Find all paths from s to t in the graph shown below.

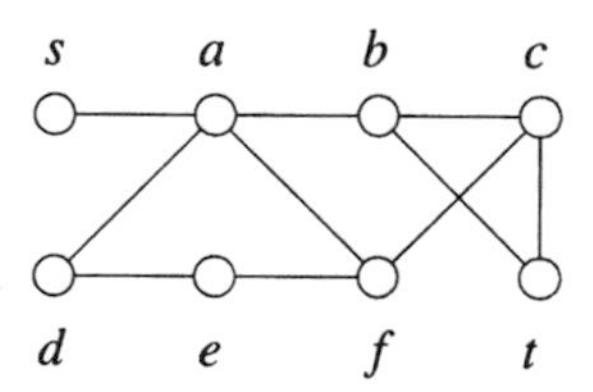

2. Figure 7.13 shows the *Petersen graph,* which arises in several contexts in the study of graphs. Find cycles of lengths 5, 6, 8 and 9 in this graph.

3. A graph contains at least two vertices. There is exactly one vertex of degree 1, and every other vertex has degree 2 or greater. Prove that the graph contains a cycle. Is this always true if we allowed graphs to have infinite vertex sets?

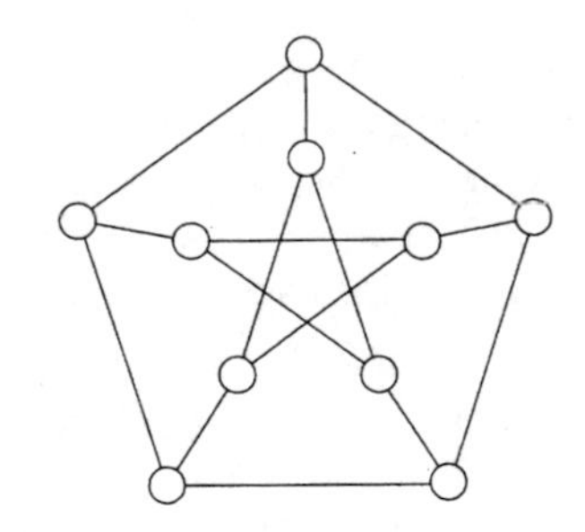

Figure 7.13: The Petersen graph

4. Find all bridges in the following graphs.

(i) 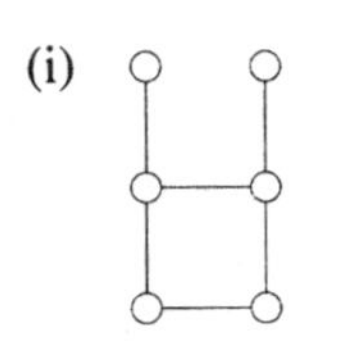 (ii) 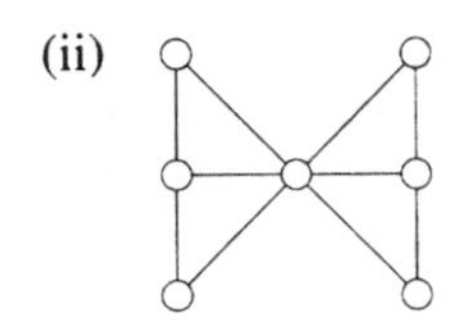 (iii) 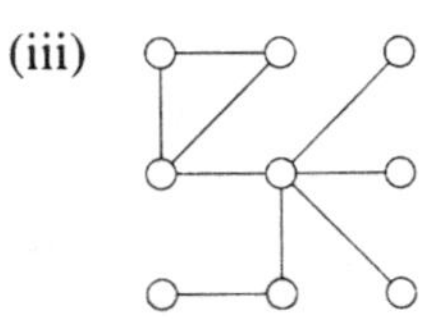

5. What are the connectivity and edge-connectivity of the graphs in Exercise 4?

In Exercises 6 *to* 14, *find the connectivity and edge-connectivity of the given graph.*

6. 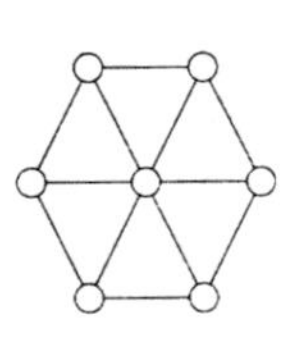

7. 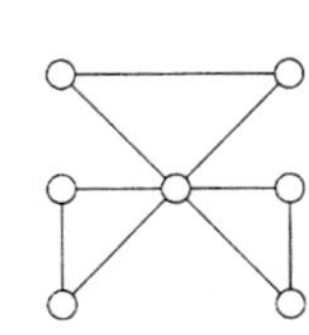

8. 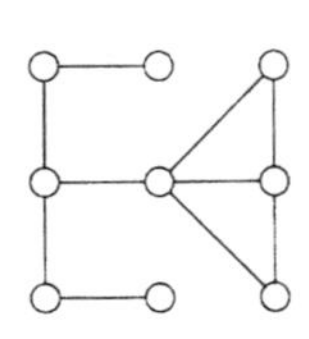

9. 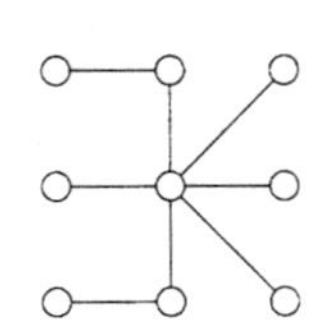

10. 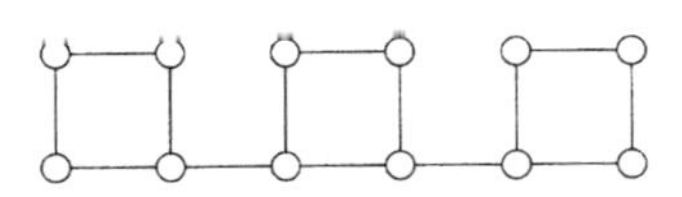

11. 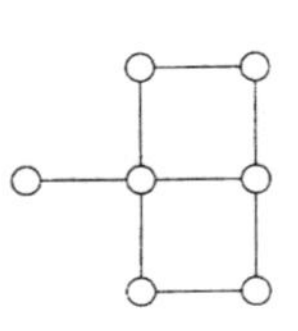

12. 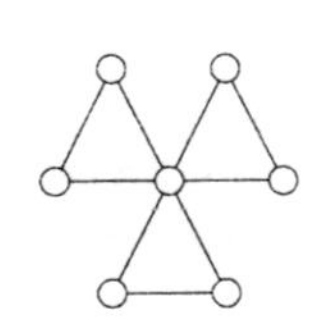

13.

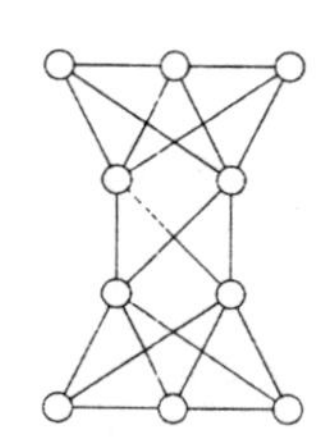

14.

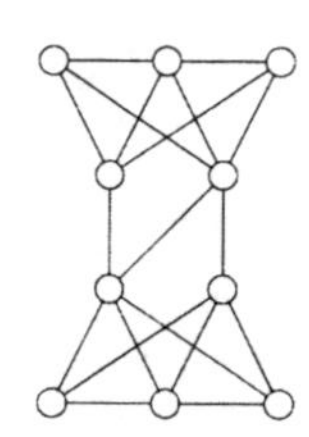

15. What are the connectivity and edge-connectivity of the n-vertex cycle C_n?

In Exercises 16 *to* 19, *find examples (other than the examples in Figure* 7.12*) of graphs with the specified relationship between minimum degree, connectivity and edge-connectivity.*

16. $\kappa(G) = \kappa'(G) = \delta(G)$.

17. $\kappa(G) < \kappa'(G) = \delta(G)$.

18. $\kappa(G) = \kappa'(G) < \delta(G)$.

19. $\kappa(G) < \kappa'(G) < \delta(G)$.

20. Show that $\kappa'(G) \le \delta(G)$ in any graph G.

7.4 Distances and Shortest Paths

Distance

If vertices x and y are adjacent, then there is a path of length 1 from x to y, and we shall say that the *distance* from x to y is 1. If x and y are connected, there may be several paths joining them, and we define the distance between them to be the length of the shortest path joining them; we denote it $D(x,y)$. For completeness, we define $D(x,x)$ to be 0.

Suppose x, y and z are vertices in a connected graph. One could travel from x to y by first going from x to z to y (distance $D(x,z)$), then from x to z to y (distance $D(z,y)$), a total of $D(x,z)$ + $D(z,y)$. There may be a shorter path from x to y, but in any event

$$D(x,y) \le D(x,z) + D(z,y)$$

for any vertices x, y and z in a connected graph.

Sample Problem 7.11 *Find the distance between every pair of vertices in the graph* (a) *below.*

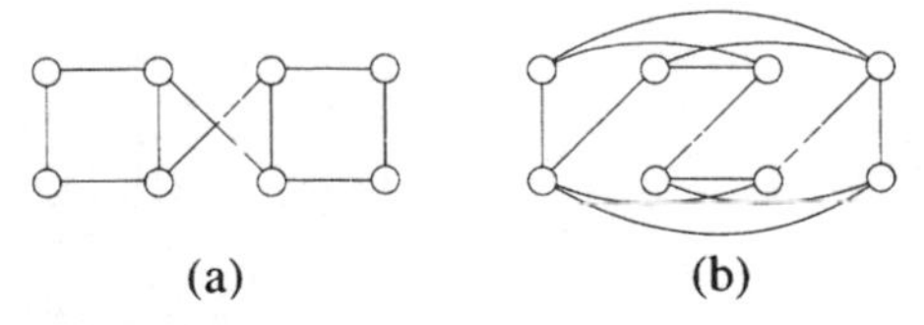

Solution. There are two different kinds of vertex in the graph. In each part of the following figure one vertex is highlighted and the number at a vertex shows the distance from the highlighted one.

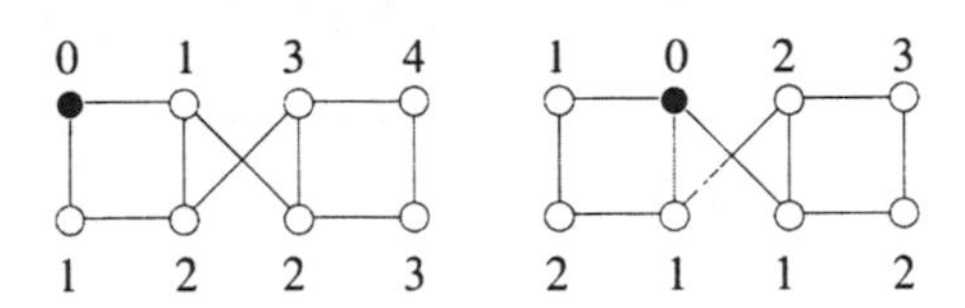

Practice Exercise. Find the distance between every pair of vertices in the graph of graph (b).

Distance arises in the example of a communications network. The vertices of a graph might represent computers. If two computers are directly linked, the corresponding vertices are adjacent, so that edges represent direct links. Often messages from one computer to another are sent by way of another machine — e-mail messages are often relayed through several intermediate computers. Two computers can communicate if the corresponding vertices are connected.

The physical separation between the computers is irrelevant. The number of times the signal is relayed may well be significant: the more relays, the greater the chance of a delay. (Direct communication between computers, by media such as telephone cable, is practically instantaneous, but delays may occur when an intermediate link is busy or down.) So the distance between vertices is a measure of the possibility of delay.

In this example, we are often interested in the *worst case* — what is the worst delay that can occur? To measure this, we define the *eccentricity* $\varepsilon(x)$ of vertex x in a connected graph to be the largest value of $D(x,y)$, where y ranges through all the vertices. The *diameter* $D = D(G)$ of G is the maximum value of $\varepsilon(x)$ for all vertices x of G, while the *radius* $R = R(G)$ is the smallest value of $\varepsilon(x)$.

Sample Problem 7.12 *Find the diameter and radius of: graph* (a) *of Sample Problem* 7.11; *the Petersen graph (see Figure* 7.13*); the cycle* C_{2n}.

Solution. In the first graph, four vertices have eccentricity 4, and the others have eccentricity 3. So $D = 4$, $R = 3$. Every vertex of the Petersen graph has three neighbors and is distance 2 from all other vertices, so $D = R = 2$. In C_{2n}, every vertex is distance n from its opposite vertex, and the distance is smaller for all other vertices, so $D = R = n$.

Practice Exercise. Find the diameter and radius of: the graph (b) of Sample Problem 7.11; the wheel W_n (defined in Exercise 7.1.20); the cycle C_{2n+1}.

Theorem 7.5 *In any graph G,*

$$R(G) \leq D(G) \leq 2R(G).$$

Proof. Clearly $D(G)$ equals the maximum distance between any two vertices in G. Say x and y attain this maximum distance, so that $D(x,y) = D(G)$, and say z is a vertex for which $\varepsilon(z) = R(G)$. As R is a distance between two vertices, necessarily $R \leq D$. But by definition $D(z,t) \leq \varepsilon(z) = R(G)$ for every vertex t, so $D(G) = D(x,y) \leq D(x,z) + D(z,y) \leq 2R(G)$.

Weighted distance

In many applications it is appropriate to define a positive function called a *weight* $w(x,y)$ associated with each edge (x,y). For example, if a graph represents a road system, a common weight is the length of the corresponding stretch of road. Weights also often represent costs or durations. The weight of a *path* P is the sum of the weights of the edges in P. Similarly, we could define the weights of walks, cycles, subgraphs and graphs.

The routes traveled by an airline can conveniently be shown in a graph, with vertices representing cities and edges representing services. A wide range of different weights might be used: the distance, the flying time and the airfare are all possibilities.

If x and y are connected vertices, then the *weighted distance* from x to y is the minimum among the weights of all the paths from x to y, and is denoted $W(x,y)$. Weighted versions of eccentricity, radius and diameter are defined in the obvious way.

Shortest paths

In many applications it is desirable to know the path of least weight between two vertices. This is usually called the *shortest path* problem, primarily because a common application is one in which weights represent physical distances. We shall describe an algorithm due to Dijkstra that finds the shortest path from vertex s to vertex t in a finite connected graph G. Informally stated, the algorithm arranges the vertices of G in order of increasing weighted distance from s. This algorithm is most easily described if w is defined for all pairs of vertices, so we write $w(x,y) = \infty$ if x and y are not adjacent. (In a computer implementation, ∞ can be replaced by some very large number.)

The algorithm actually orders the vertices of G as $s_0 (= s), s_1, s_2, \ldots$ so that the weighted distances $W(s,s_1), W(s,s_2), \ldots$ are in non-decreasing order. To do so, it attaches a temporary label $\ell(x)$ to each member x of $V(G)$. $\ell(x)$ is an upper bound on the weighted length of the shortest path from s to x. To start the algorithm, write $s_0 = s$ and $S_0 = \{s\}$. Define $\ell(s_0) = 0$ and for every other vertex y, $\ell(y) = \infty$. Call this step 0. After step k the set

$$S_k = \{s_0, s_1, \ldots, s_k\}$$

has been defined. In the next step, for each x not in S_k, the algorithm changes the value $\ell(x)$ to the weighted length of the shortest (s,x) path that has only one edge not in S_k, if this is an improvement. Then s_{k+1} is chosen to be a vertex x such that $\ell(x)$ is minimized. (In practice, it does not matter whether the new value of $\ell(x)$ is used or the old value is retained for those vertices x other than s_{k+1}. Moreover, for each member s_i of S_k, it is only necessary to consider one new vertex, a vertex x such that $w(s_i,x)$ is minimal.)

- s_0 is s;
- s_1 is the vertex x for which $w(s,x)$ is smallest, the vertex "closest" to s;
- to find s_2 one selects, for every vertex x other than s or s_1, either the lowest weight edge sx or the lowest weight path ss_1x, whichever is shorter, and then chooses the shortest of all these to be s_2;

and so on. After a vertex has been labeled s_i, its ℓ-value never changes, and this final value of $\ell(x)$ equals $W(s,x)$. Eventually the process must stop, because a new vertex is added at each step and the vertex set is finite. So $W(s,t)$ must eventually be found. Finally, whenever a vertex is labeled s_{k+1}, one can define a unique vertex s_i that "preceded" it in the process — the vertex before s_{k+1} in the ss_{k+1} path that had length $\ell(s_{k+1})$. By working back from t to the vertex before it, then to the one before that and so on, one eventually reaches s, and reversing the process gives a shortest st path. For convenience, the predecessor of each new vertex s_k is recorded at the step when s_k is selected.

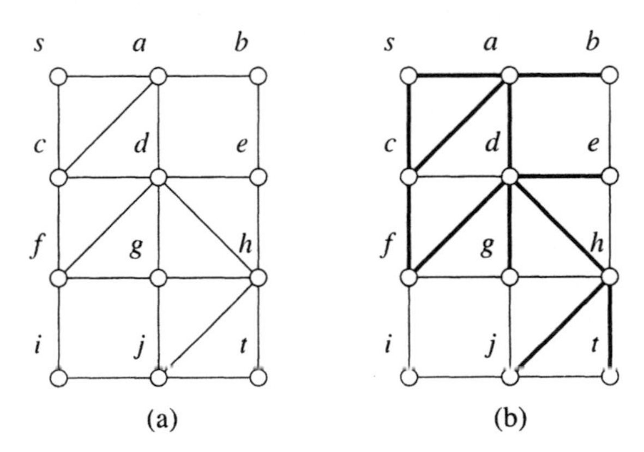

Figure 7.14: Find the path of minimum weight from s to t

Sample Problem 7.13 *Figure* 7.14(a) *shows a graph with weighted edges. Find the minimum weight path from s to t.*

Solution. Initially set $s_0 = s$, $S_0 = \{s\}$ and $\ell(s) = 0$. s has no predecessor.

The nearest vertex to s is a ($w(s,a) = 5$, $w(s,c) = 6$). So $s_1 = a$, $S_1 = \{s,a\}$ and $\ell(a) = 5$. a has predecessor s.

We consider each member of S_1. The nearest vertex in $V \backslash S_1$ to s (in fact, the only one) is c, and $w(s,c) = 6$. The nearest vertex to a is b ($w(a,b) = 2, w(a,c) = 4, w(a,d) = 3$). The candidate values are $\ell(c) = 6$ (through s) and $\ell(b) = 7$ (through a). The smaller is chosen. So $s_2 = c$, $S_2 = \{s,a,c\}$ and $\ell(c) = 6$. c has predecessor s.

We now process S_2. There is no vertex in $V \backslash S_1$ adjacent to s, so s can be ignored in this and later iterations. The nearest vertex to a is b; $w(a,b) = 2$ so $\ell(a) + w(a,b) = 7$. The nearest vertex to c is d; $w(c,d) = 4$ so $\ell(c) + w(c,d) = 10$. So $s_3 = b$, $S_3 = \{s,a,c,b\}$ and $\ell(b) = 7$. b has predecessor a.

The nearest vertex to a is d; $w(a,d) = 3$ so $\ell(a) + w(a,d) = 8$. The nearest vertex to c is d; $w(c,d) = 4$ so $\ell(c) + w(c,d) = 10$. The nearest vertex to b is e; $w(b,e) = 5$ so $\ell(b) + w(b,e) = 12$. So $s_4 = d$, $S_4 = \{s,a,c,b,d\}$ and $\ell(d) = 8$. d has predecessor a.

a need not be considered, as all its neighbors are in S_4. The nearest vertex to c is f; $w(c,f) = 5$ so $\ell(c) + w(c,f) = 11$. The nearest vertex to b is e; $w(b,e) = 5$ so $\ell(b) + w(b,e) = 12$. The nearest vertex to d is e; $w(d,e) = 2$ so $\ell(d) + w(d,e) = 10$. So $s_5 = e$, $S_5 = \{s,a,c,b,d,e\}$ and $\ell(e) = 10$. e has predecessor d.

From now on b need not be considered. The nearest vertex to c is f; $w(c,f)$ is 5 so $\ell(c) + w(c,f) = 11$. The nearest vertex to d is g; $w(d,g) = 3$ so $\ell(d) + w(d,g) = 11$. The nearest vertex to e is h; $w(e,h) = 2$ so $\ell(e) + w(e,h) = 12$. Either f or g could be chosen. For convenience, suppose the earlier member of the alphabet is always chosen when equal ℓ-values occur. Then $s_6 = f$, $S_6 = \{s,a,c,b,d,e,f\}$ and $\ell(f) = 11$. f has predecessor c.

Now c can be ignored. The nearest vertex to d is g; $w(d,g) = 3$ so $\ell(d) + w(d,g) = 11$. The nearest vertex to e is h; $w(e,h) = 2$ so $\ell(e) + w(e,h) = 12$. The nearest vertex to f is g; $w(f,g) = 2$ so $\ell(f) + w(f,g) = 13$. So $s_7 = g$, $S_7 = \{s,a,c,b,d,e,f,g\}$ and $\ell(g) = 11$. g has predecessor d.

The nearest vertex to d is h; $w(d,g) = 4$ so $\ell(d) + w(d,g) = 12$. The nearest vertex to e is h; $w(e,h) = 2$ so $\ell(e) + w(e,h) = 12$. The nearest vertex to f is i; $w(f,i) = 4$ so $\ell(f) + w(f,i) = 15$. The nearest vertex to g is h; $w(g,h) = 3$ so $\ell(g) + w(g,h) = 14$. So $s_8 = h$, $S_8 = \{s,a,c,b,d,e,f,g,h\}$ and $\ell(h) = 12$. We shall say h has predecessor d (e could also be used, but d is earlier in alphabetical order).

d and e are now eliminated. The nearest vertex to f is i; $w(f,i) = 4$ so $\ell(f) + w(f,i) = 15$. The nearest vertex to g is j; $w(g,j) = 6$ so $\ell(g) + w(g,j) = 17$. The nearest vertex to h is t; $w(h,t) = 3$ so $\ell(h) + w(h,t) = 15$. We always choose t when it has the equal-smallest ℓ-value. So $s_9 = t$, and the algorithm stops with $\ell(t) = 15$. Since t has predecessor h, the minimum weight path is *sadht*, with weight 15. In Figure 7.14(b), all links to predecessors are emphasized. It is easy to read off the minimum weight path from this figure, as well as the minimum weight paths from s to a, b, c, d, e, f, g and h.

Exercises 7.4

1. In Exercise 7.3.1 you found all paths from s to t in the graph shown in a given graph. What is the distance from s to t in each path?

2. Suppose the graph G has diameter greater than 3. Show that its complement, $\overline{G}$, has diameter less than 3.

In Exercises 3 *to* 14, *what are the diameter and radius of the graph?*

3. The complete graph K_v.

4. The n-vertex path P_n.

5. The complete bipartite graph $K_{m,n}$.

6. The *ladder* L_n, a $2n$-vertex graph constructed by taking two copies of P_n and joining corresponding vertices.

7.

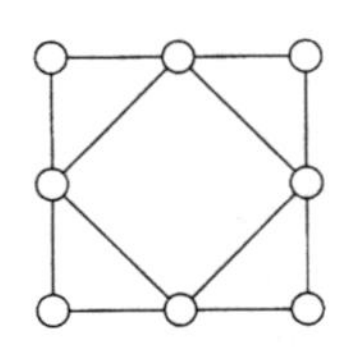

8.

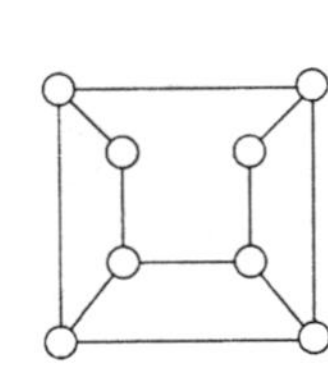

9.

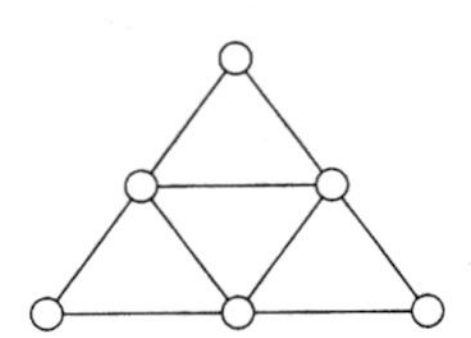

10.

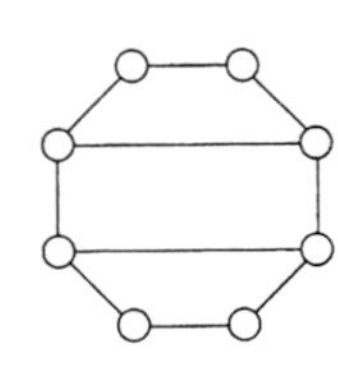

11.

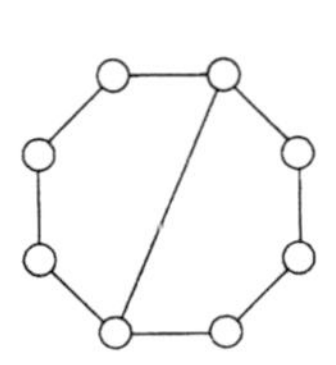

12.

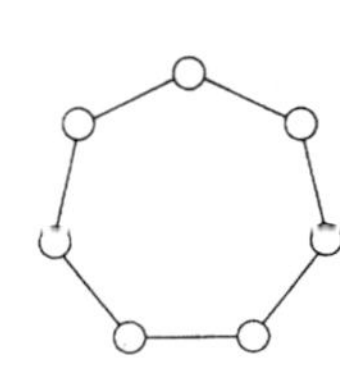

13.

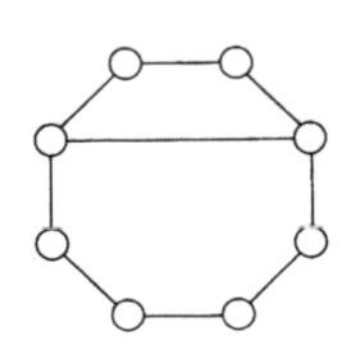

14.

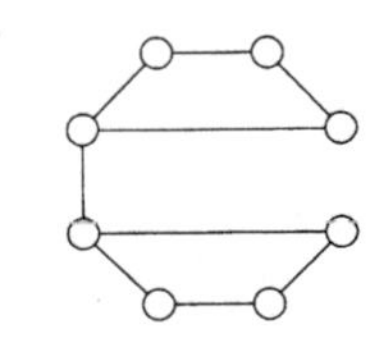

15. The *square* G^2 of a graph G has the same vertex set as G; x is adjacent to y in G^2 if and only if $D(x,y) = 1$ or 2 in G. What are the squares of $K_{1,n}$, P_5 and C_5?

16. Refer to the preceding exercise. What are the squares of $K_{3,4}$, C_6 and the Petersen graph?

In Exercises 17 *to* 22, *find the minimum weight paths from s to t in the weighted graph shown.*

17.

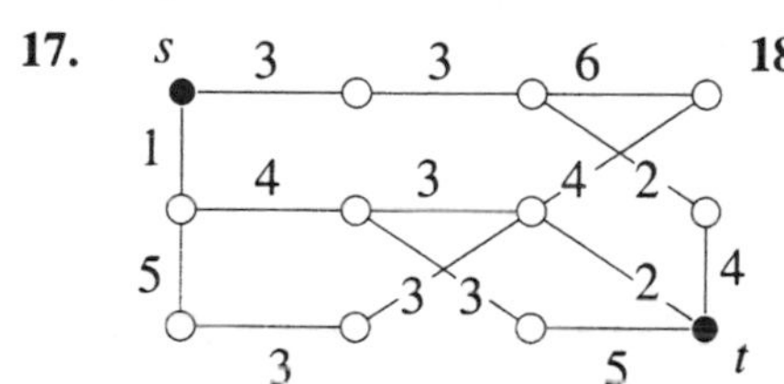

18.

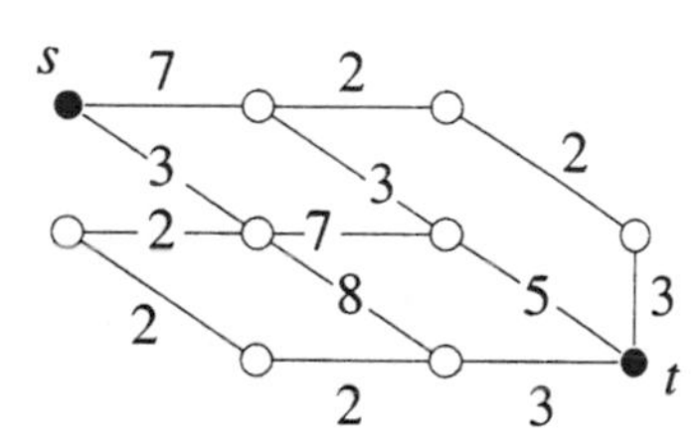

19.

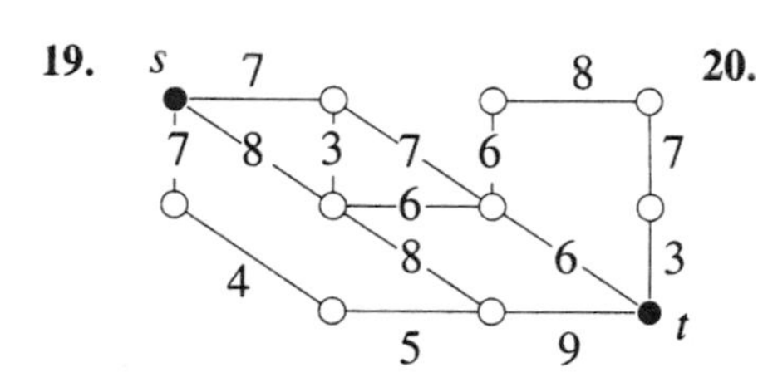

20.

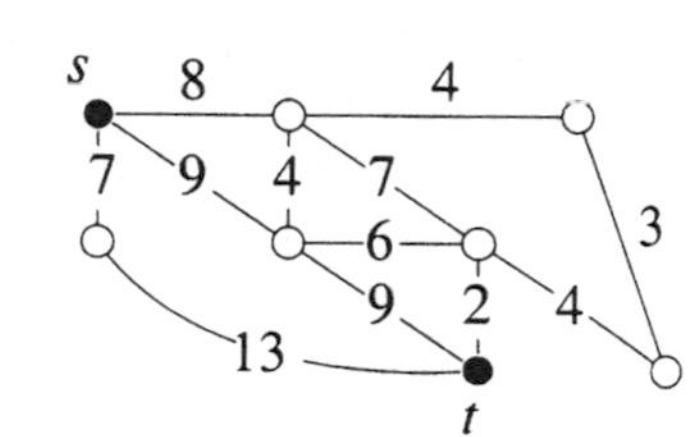

21.

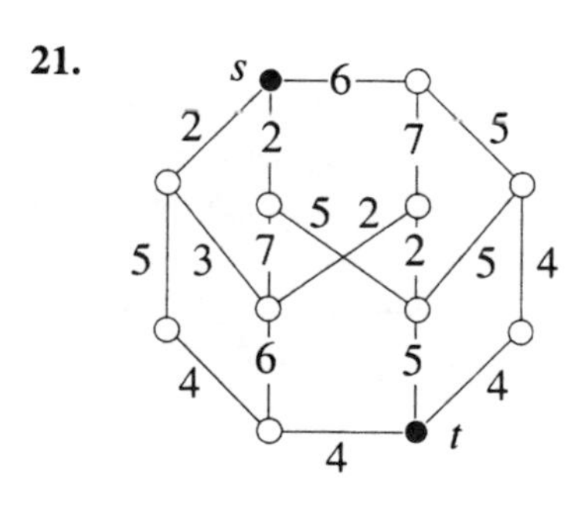

22.

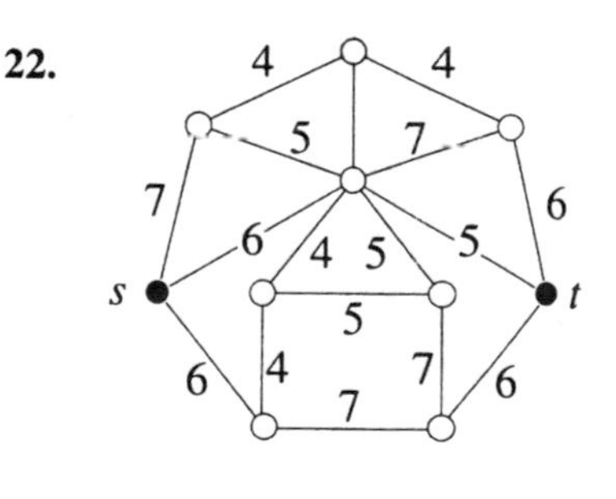

7.5 Trees

Sample Problem 7.14 *Suppose an oil field contains five oil wells* w_1, w_2, w_3, w_4, w_5, *and a depot* d. *It is required to build pipelines so that oil can be pumped from the wells to the depot. Oil can be relayed from one well to another, at very small cost. The only major expense is building the pipelines. Figure* 7.15(a) *shows which pipelines are feasible to build, represented as a graph in the obvious way, with the cost (in hundreds of thousands of dollars)* *shown (a missing edge might mean that the cost would be very high). Which pipelines should be built?*

Solution. Your first impulse might be to connect d to each well directly, as shown in Figure 7.15(b). This would cost \$2,600,000. However, the cheapest solution is shown in Figure 7.15(c), and costs \$1,600,000.

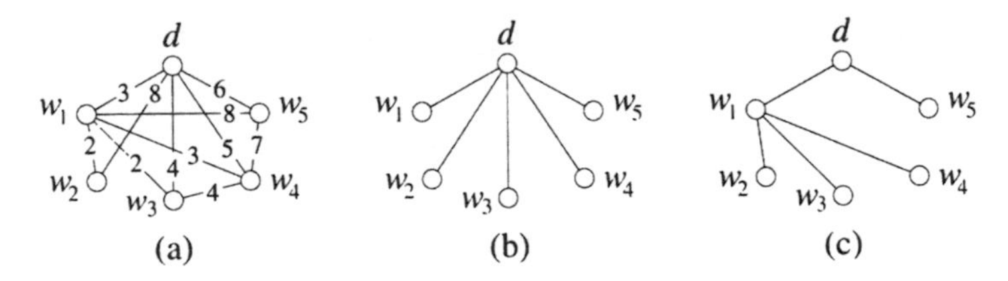

Figure 7.15: An oil pipeline problem

Once the pipelines in case (c) have been built, there is no reason to build a direct connection from d to w_3. Any oil being sent from w_3 to the depot can be relayed through w_1. Similarly, there would be no point in building a pipeline joining w_3 to w_4. The conditions of the problem imply that the graph does not need to contain any cycle. However, it must be connected.

A connected graph that contains no cycle is called a *tree*. So the solution of the problem is to find a subgraph of the underlying graph that is a tree, and among those to find the one that is cheapest to build. We shall return to this topic shortly. But first we look at trees as graphs.

We have already seen trees. The tree diagrams used in Chapter 6 are just trees laid out in a certain way. To form a tree diagram from a tree, select any vertex as a starting point (called the *root*). Draw edges to all of its neighbors. These are the first generation. Then treat each neighbor as if it were the starting point; draw edges to all its neighbors except the start. Continue in this way; to each vertex, attach all of its neighbors except for those that have already appeared in the diagram.

Figure 7.16 contains three examples of trees. It is also clear that every path is a tree, and the star $K_{1,n}$ is a tree for every n.

Sample Problem 7.15 *Draw the first tree in Figure 7.16 as a tree diagram. How many different-looking diagrams can be made?*

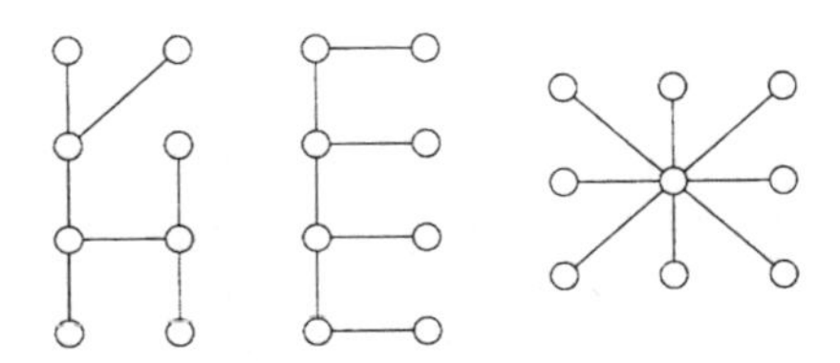

Figure 7.16: Some trees

Solution. There are four different-looking diagrams. In the following illustration, the vertices of the original tree are labelled 1, 2, 3, 4. The diagrams obtained using each type of vertex as root are shown.

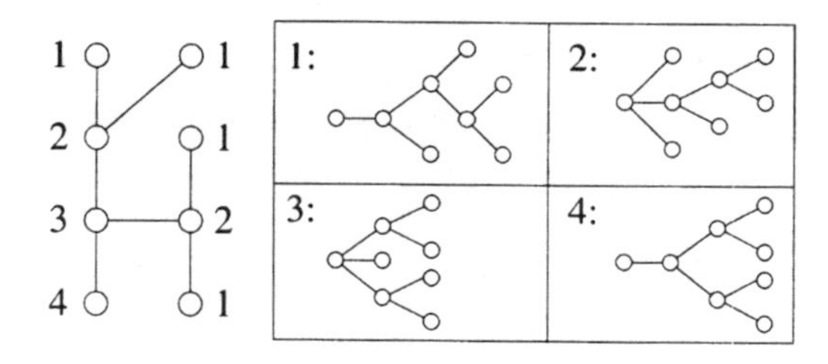

Practice Exercise. Repeat this problem for the other two trees in Figure 7.16.

A tree is a minimal connected graph in the following sense: if any vertex of degree at least 2, or any edge, is deleted, then the resulting graph is not connected. In fact it is easy to prove that a connected graph is a tree if and only if every edge is a bridge.

Trees can also be characterized among connected graphs by their number of edges.

Theorem 7.6 *A finite connected graph G with v vertices is a tree if and only if it has exactly $v-1$ edges.*

Proof. (i) Suppose that G is a tree with v vertices. We proceed by induction on v. The theorem is true for $v = 1$, since the only graph with one vertex is K_1, which is a tree. Suppose it is true for $w < v$, and suppose G is a tree with v vertices. Select an edge (G must have an edge, or it will be the unconnected graph $\bar{K}_v$) and delete it. The result is a union of two disjoint components, each of which is a tree with fewer than v vertices; say the first component has v_1 vertices and the second has v_2, where $v_1 + v_2 = v$. By the induction hypothesis, these graphs have $v_1 - 1$ and $v_2 - 1$ edges, respectively. Adding one edge for the one that was deleted, we find that the number of edges in G is

$$(v_1 - 1) + (v_2 - 1) + 1 = v - 1.$$

(ii) Conversely, suppose G is not a tree. Select an edge that is *not* a bridge, and delete it. If the resulting graph is not a tree, repeat the process. Eventually there will be only bridges left, and the graph is a tree. From what we have just said it must have $v-1$ edges, and the original graph had more than $v-1$ edges. □

Corollary 7.6.1 *Every tree other than K_1 has at least two vertices of degree* 1.

Proof. Suppose the tree has v vertices. It then has $v-1$ edges. So, by Theorem 7.1, the sum of all degrees of the vertices is $2(v-1)$. There can be no vertex of degree 0, since the tree is connected and is not K_1; if $v-1$ of the vertices have degree at least 2, then the sum of the degrees is at least $1+2(v-1)$, which is impossible. □

Sample Problem 7.16 *Find all trees with five vertices.*

Solution. If a tree has five vertices, then the largest possible degree is 4. Moreover there are four edges (by Theorem 7.6), so from Theorem 7.1 the sum of the five degrees is 8. As there are no vertices of degree 0 and at least two vertices of degree 2, the list of degrees must be one of

$$4,1,1,1,1 \quad 3,2,1,1,1 \quad 2,2,2,1,1.$$

In the first case, the only solution is the star $K_{1,4}$. If there is one vertex of degree 3, no two of its neighbors can be adjacent (this would form a cycle), so the fourth edge must join one of those three neighbors to the fifth vertex. The only case with the third degree list is the path P_5. These three cases are

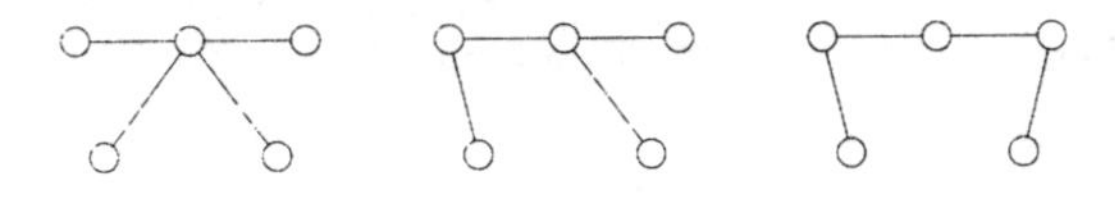

Practice Exercise. Find all trees with four vertices.

Spanning trees; the minimal spanning tree

A *spanning* subgraph of a graph G is one that contains every vertex of G. A *spanning tree* in a graph is a spanning subgraph that is a tree when considered as a graph in its own right. For example, it was essential that the tree used to plan the oil pipelines should be a spanning tree.

It is clear that any connected graph G has a spanning tree. If G is a tree, then the whole of G is itself the spanning tree. Otherwise G must contain a cycle; delete one edge from the cycle. The resulting graph is still a connected subgraph of G; and, as no vertex has been deleted, it is a spanning subgraph. Find a cycle in this new graph and delete it; repeat the process. Eventually the remaining graph will contain no cycle, so it is a tree. So when the process stops, we have found a spanning tree. A given graph might have many different spanning trees.

In many applications — such as the oil pipeline problem — each edge of a graph has a weight associated with it, such as distance or cost. It is sometimes desirable to find a spanning tree such that the weight of the tree — the total of the weights of its edges — is minimum. Such a tree is called a *minimal spanning tree*.

A finite graph can contain only finitely many spanning trees, so it is possible in theory to list all spanning trees and their weights, and to find a minimal spanning tree by choosing one with minimum weight. This process could take a very long time however, since the number of spanning trees in a graph can be very large. So efficient algorithms that find a minimal spanning tree are useful. We present here an example due to Prim.

We assume that G is a graph with vertex set V and edge set E, and suppose there is associated with G a map $w : E \to R$ called the *weight* of the edge; when xy is an edge of G we write $w(x,y)$ for the image of xy under w. We could quite easily modify the algorithm to allow for multiple edges, but the notation is slightly simpler in the graph case. The algorithm consists of finding a sequence of vertices $x_0, x_1, x_2, \ldots$, of G and a sequence of sets $S_0, S_1, S_2, \ldots$, where

$$S_i = \{x_0, x_1, \ldots, x_{i-1}\}.$$

We choose x_0 at random from V. When $n > 0$, we find x_n inductively using S_n as follows.

1. Given i, $0 \le i \le n-1$, choose y_i to be a member of $V \setminus S_n$ such that $w(x_i, y_i)$ is minimum, if possible. In other words:

   (a) if there is no member of $V \setminus S_n$ adjacent to x_i, then there is no y_i;

   (b) if $V \setminus S_n$ contains a vertex adjacent to x_i, then y_i is one of those vertices adjacent to x_i, and if $x_i \sim y$, then $w(x_i, y_i) \le w(x_i, y)$.

2. Provided that at least one y_i has been found in step (1), define x_n to be a y_i such that $w(x_i, y_i)$ is minimal, in other words, x_n is the y_i that satisfies

$$w(x_i, y_i) \le w(x_j, y_j) \text{ for all } j.$$

3. Put $S_{n+1} = S_n \cup \{x_n\}$.

This process stops only when there is no new vertex y_i. If there is no new vertex y_i, it must be true that no member of $V \setminus S_n$ is adjacent to a vertex of S_n. It is impossible to partition the vertices of a connected graph into two non-empty sets such that no edge joins one set to the other, and S_n is never empty, so the process stops only when $S_n = V$.

When we reach this stage, so that $S_n = V$, we construct a graph T as follows:

(i) T has vertex set V;

(ii) if x_k arose as y_i, then x_i is adjacent to x_k in T;

(iii) no edges of T exist other than those that may be found using (ii).

It is not hard to verify that T is a tree and that it is minimal.

Observe that X_n may not be defined uniquely at step (2) of the algorithm, and indeed y_i may not be uniquely defined. This is to be expected: after all, there may be more than one minimal spanning tree.

Sample Problem 7.17 *Find a minimal spanning tree in the graph G shown in Figure* 7.17(a).

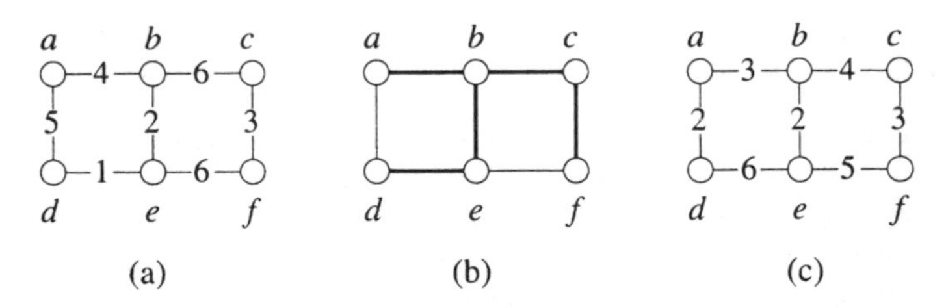

Figure 7.17: An example of Prim's algorithm

Solution. Select $x_0 = a$. Then $S_1 = \{a\}$. Now $y_0 = b$, and this is the only choice for x_1. So $S_2 = \{a, b\}$. The tree will contain edge ab.

Working from S_2, we get $y_0 = d$ and $y_1 = e$. Since be $(= x_1y_1)$ has smaller weight than ad $(= x_0y_0)$, we select $x_2 = e$. Then $S_3 = \{a, b, e\}$ and edge be goes into the tree. Similarly, from S_3, we get $x_3 = d$ and $S_4 = \{a, b, d, e\}$, and the new edge is de.

Now there is a choice. Working from S_4, $y_1 = c$ and $y_2 = f$. In both cases the weight of the edge to be considered is 6. So either may be used. Let us choose c, and use edge bc.

The final vertex is f, and the edge is cf. So the tree has edges ab, bc, be, cf, de and weight 16. It is shown in Figure 7.17(b).

Practice Exercise. Find a minimal spanning tree in the graph G shown in Figure 7.17(c).

The algorithm might be described as follows. First, choose a vertex x_0. Trivially the minimum weight tree with vertex-set $\{x_0\}$ — the *only* tree with vertex-set $\{x_0\}$ — is the K_1 with vertex x_0. Call this the *champion*. Then find the smallest weight tree with two vertices, one of which is x_0; in other words, find the minimum weight tree that can be formed by adding just one edge to the current champion. This tree is the new champion. Continue in this way: each time a champion is found, look for the cheapest tree that can be formed by adding one edge to it. One can consider each new tree to be an approximation to the final minimal spanning tree, with successive approximations having more and more edges.

Prim's algorithm was a refinement of an earlier algorithm due to Kruskal. In that algorithm, one starts by listing all edges in order of increasing weight. The first approximation is the K_2 consisting of the edge of least weight. The second approximation is formed by appending the next edge in the ordering. At each stage the next approximation is formed by adding on the smallest edge that has not been used, provided only that it does not form a cycle with the edges already chosen. In this case the successive approximations are not necessarily connected, until the last one. The advantage of Prim's algorithm is that, in large graphs, the initial sorting stage of Kruskal's algorithm can be very time consuming.

Exercises 7.5

1. Find all different trees on six vertices.

2. For each $n = 2, 3, 4, 5, 6$, find a tree on seven vertices with n vertices of degree 1.

3. Prove that every bridge in a connected graph lies on every spanning tree of the graph.

In Exercises 4 *to* 10, *find the number of spanning trees in the graph, and sketch all the trees.*

4.

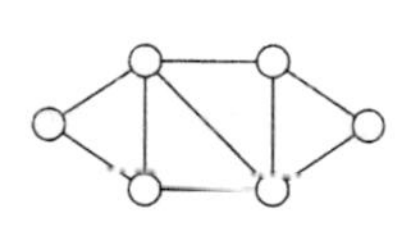

5.

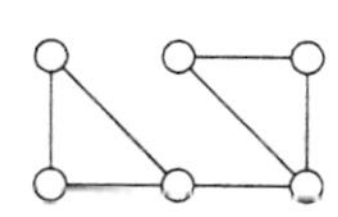

6.

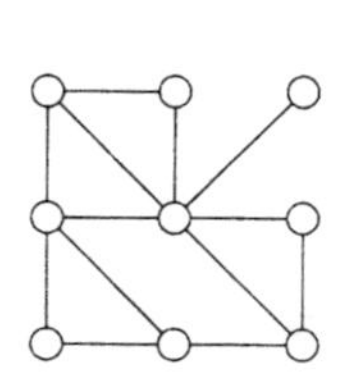

7.

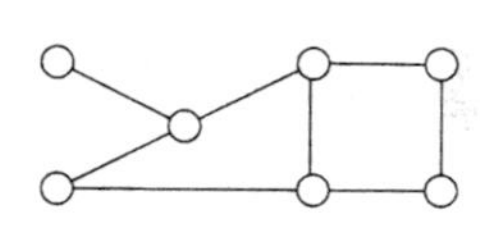

8.

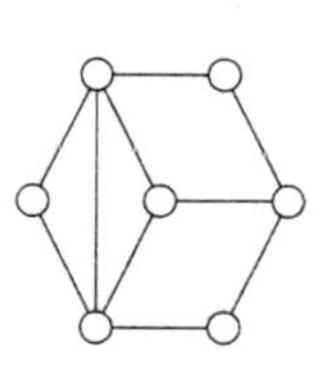

9.

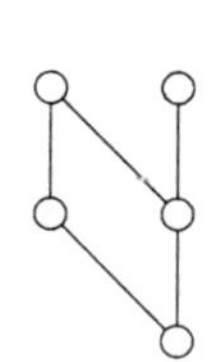

10.

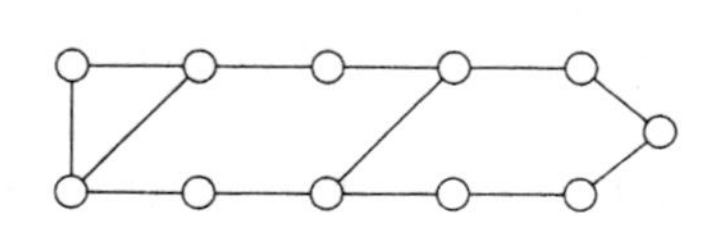

In Exercises 11 *to* 16, *find all different ways to draw the tree as a tree diagram.*

11.

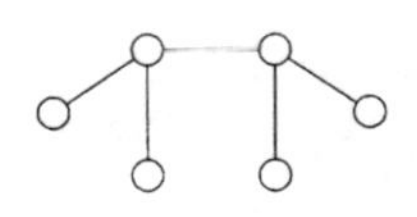

12.

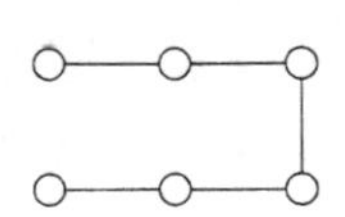

13.

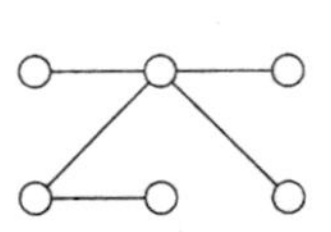

14.

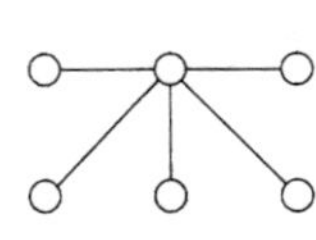

15.

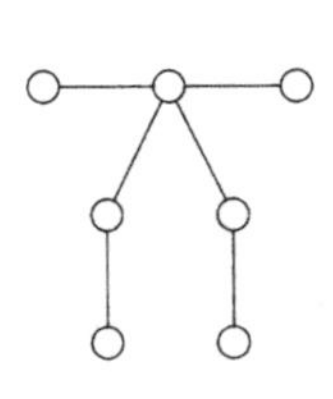

16.

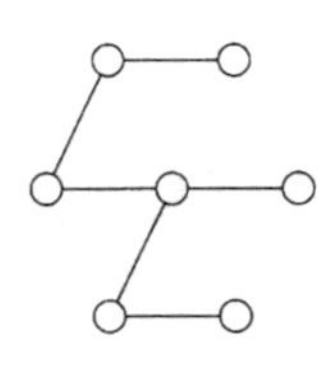

17. (i) On graph A below, a weight function is shown. Find a minimal spanning tree in A.

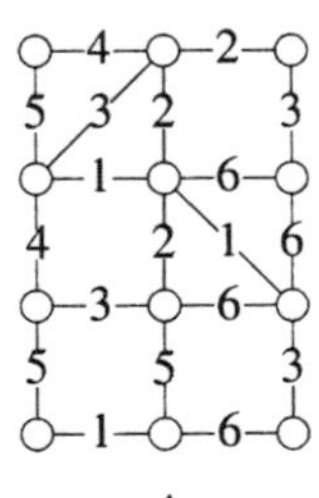

A

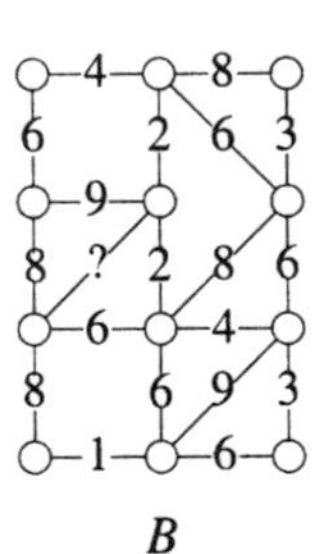

B

(ii) B is similar, except that no weight is specified for one edge. Find a minimal spanning tree in B if that edge has weight

(a) 1; (b) 4; (c) 7.

In Exercises 18 *to* 25, *find minimal spanning trees in the graph, using both Kruskal's and Prim's methods.*

18.

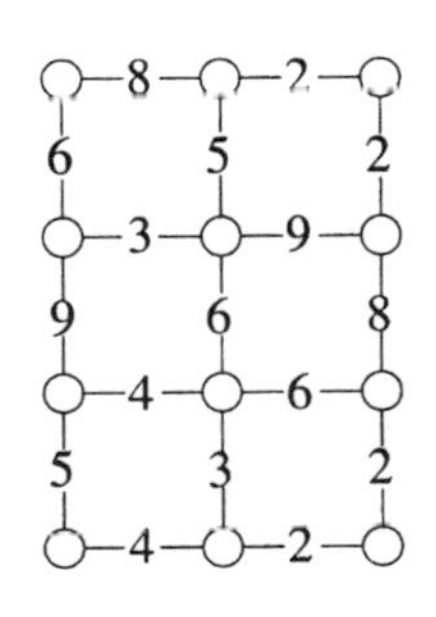

19.

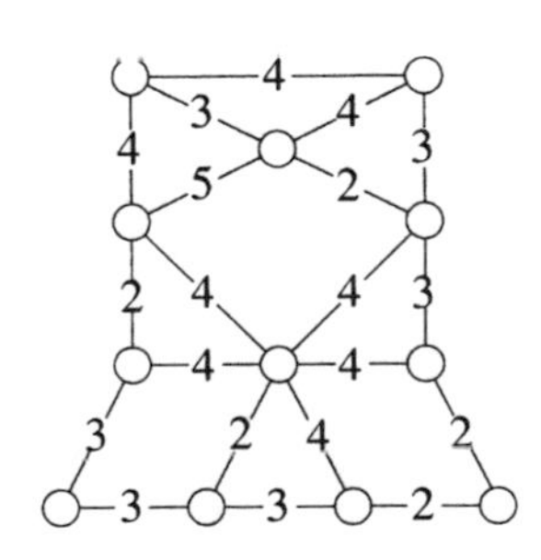

20.

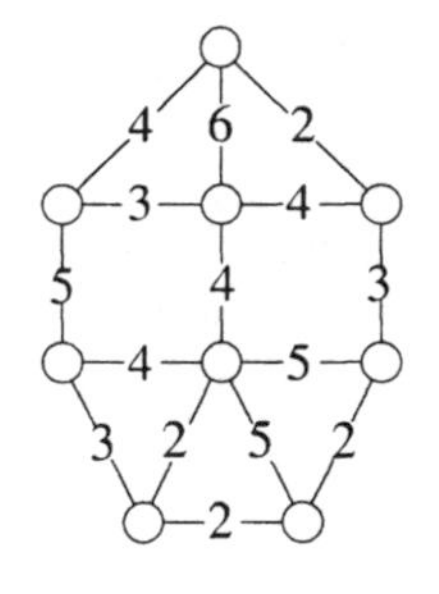

21.

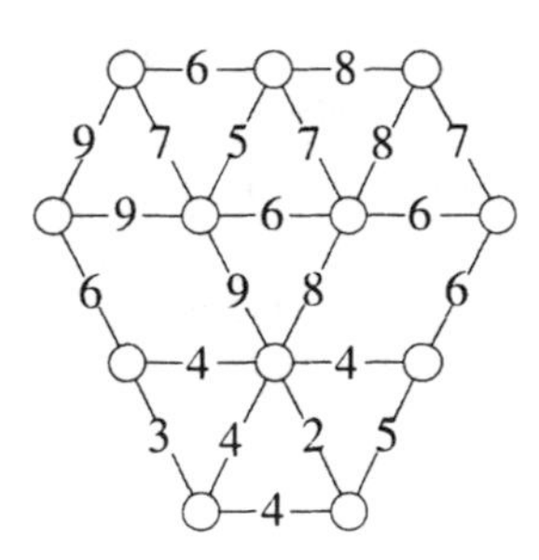

22.

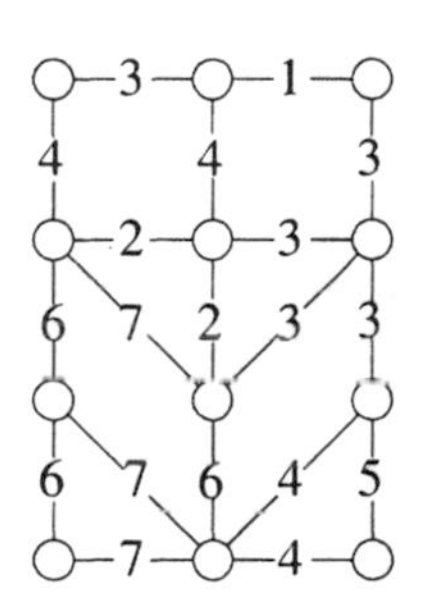

23.

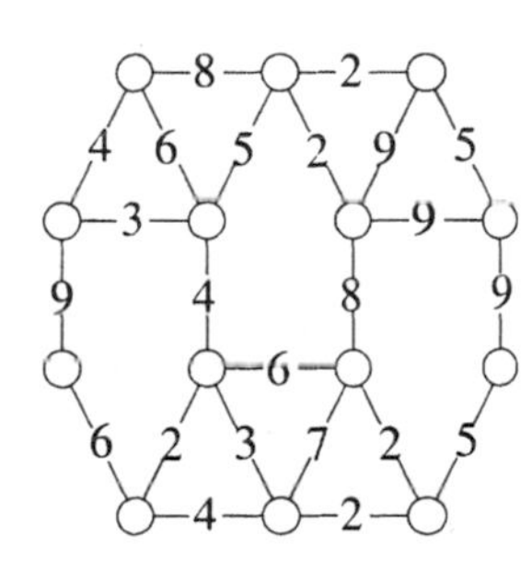

24.

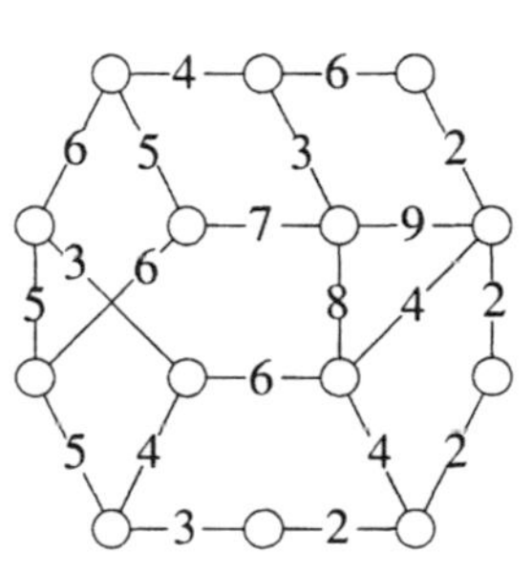

25.

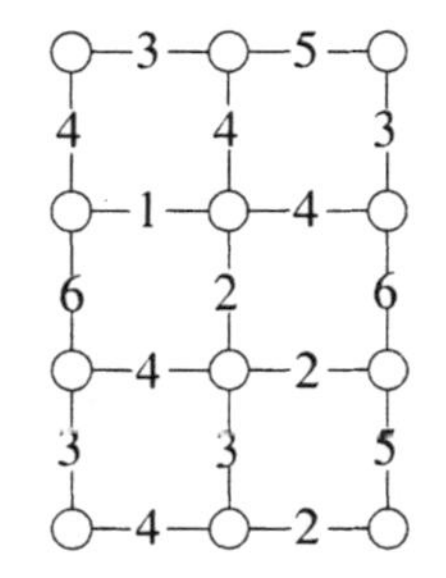

7.6 Hamiltonian Cycles

Modeling road systems again

When we discussed modeling road systems in Section 7.2, the emphasis was on visiting all the roads: crossing the Königsberg bridges, or inspecting a highway. Another viewpoint is that of the traveling salesman or tourist who wants to visit the towns. The salesman travels from one town to another, trying not to pass through any town twice on his trip. In terms of the underlying graph, the salesman plans to follow a cycle.

A cycle that passes through every vertex in a graph is called a *Hamiltonian cycle* and a graph with such a cycle is called *Hamiltonian.* The idea of such a spanning cycle was simultaneously developed by Hamilton in 1859 in a special case, and more generally by Kirkman in 1856.

A *Hamiltonian path* is a path that contains every vertex. If you take a Hamiltonian cycle and delete an edge you obtain a Hamiltonian path, but the reverse is not always true. Some graphs contain a Hamiltonian path but no Hamiltonian cycle; an example is the Petersen graph.

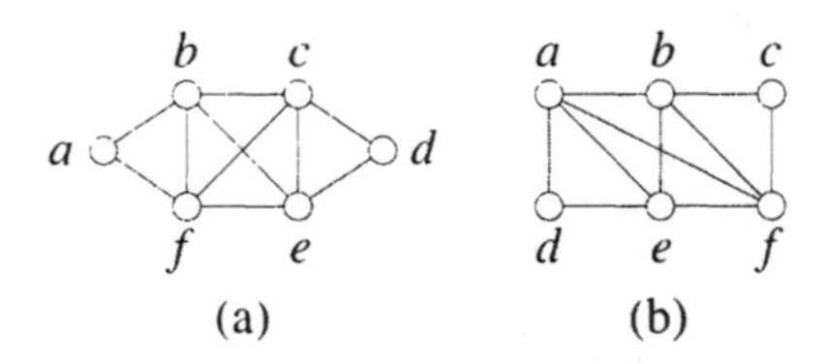

Figure 7.18: Sample graphs for discussing Hamiltonian cycles

Sample Problem 7.18 *Consider the graph in Figure* 7.18(a). *Which of the following are Hamiltonian cycles?*

(i) (a,b,e,d,c,f,a)

(ii) (a,b,e,c,d,e,f,a)

(iii) (a,b,c,d,e,f,a)

(iv) (a,b,c,e,f,a)

Solution. (i) and (iii) are Hamiltonian. (ii) is not; it contains a repeat of e. (iv) is not; vertex d is omitted.

Practice Exercise. Repeat this problem for the graph in Figure7.18(b) and the following cycles.

(i) (a,b,c,f,e,d,a)

(ii) (a,b,f,c,b,e,d,a)

(iii) (a,b,c,d,e,f,a)

(iv) (a,d,e,b,c,f,a)

Which graphs are Hamiltonian?

At first, the problem of deciding whether a graph is Hamiltonian sounds similar to the problem of Euler circuits. However, the two problems are strikingly different in one regard. We found a very easy test for the Eulerian property, but no nice necessary and sufficient conditions are known for the existence of Hamiltonian cycles.

It is easy to see that the complete graphs with three or more vertices are Hamiltonian, and any ordering of the vertices gives a Hamiltonian cycle. On the other hand, no tree is Hamiltonian, because they contain no cycles at all. We can discuss Hamiltonicity in a number of other particular cases, and there are a number of small theorems.

One useful sufficient condition for a graph to be Hamiltonian is the following, due to Ore. We include the proof, but most students should skip it on a first reading.

Theorem 7.7 *If G is a graph with v vertices, $v \geq 3$, and $d(x) + d(y) \geq v$ whenever x and y are non-adjacent vertices of G, then G is Hamiltonian.*

Proof. Suppose the theorem is false. There must be at least one counterexample — a graph that satisfies the conditions but is not Hamiltonian. Suppose there is a counterexample with v vertices, for some particular v. There may be more than one counterexample with v vertices; if there are, select one that has the largest possible number of edges among counterexamples, and call it G. Choose two nonadjacent vertices p and q of G. $G + pq$ must be Hamiltonian, because it has more edges than G. Moreover, pq must be an edge in every Hamiltonian cycle of $G + pq$, because any Hamiltonian cycle that does not contain pq would be Hamiltonian in G. Since G satisfies the conditions, $d(p) + d(q) \geq v$.

Consider any Hamiltonian cycle in $G + pq$. Since it contains edge pq, it will look like

$$p, x_1, x_2, \ldots, x_{v-2}, q, p.$$

If x_i is any vertex adjacent to p, then x_{i-1} cannot be adjacent to q, because if it were, then

$$p, x_1, x_2, \ldots, x_{i-1}, q, x_{v-2}, x_{v-3}, \ldots, x_i, p$$

would be a Hamiltonian cycle in G. So each of the $d(p)$ vertices adjacent to p in G must be preceded in the cycle by vertices not adjacent to q, and none of these vertices can be q itself. So there are at least $d(p) + 1$ vertices in G that are not adjacent to q. So there are at least $d(q) + d(p) + 1$ vertices in G, whence

$$d(p) + d(q) \leq v - 1,$$

which is a contradiction. □

As a consequence, it follows that if a graph has v vertices ($v \geq 3$) and every vertex has degree at least $\frac{v}{2}$, then the graph is Hamiltonian.

On the other hand, suppose a graph G contains a Hamiltonian cycle

$$x_1, x_2, \ldots, x_v, x_1.$$

Since x_i occurs only once in the cycle, only two of the edges touching x_i can be in the cycle. One can sometimes use this fact to prove that a graph contains no Hamiltonian cycle. For example, consider the graph of Figure 7.19.

Suppose the graph contains a Hamiltonian cycle.

The vertices on the outer circuit are each of degree 3, and only two of the edges touching any given vertex can be in a Hamiltonian cycle. Figure 7.20(a) shows as

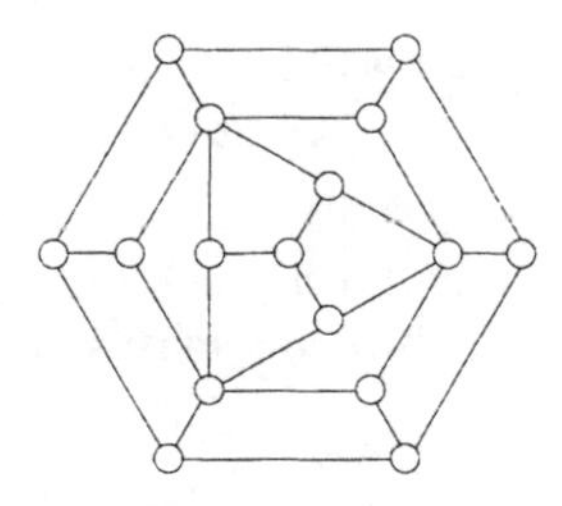

Figure 7.19: A graph with no Hamiltonian cycle

heavy lines all the edges touching three of those vertices; of the nine edges, three are not in the cycle. Similarly, Figure 7.20(b) shows the fifteen edges touching the three vertices of degree 5; nine of these are out of the cycle. These sets of edges are disjoint, so there are at least twelve edges not in the cycle. (If the sets were not disjoint, but had k common elements, only $12-k$ edges would definitely be eliminated.) Similarly, one of the edges touching the central vertex must be deleted in forming the cycle. So thirteen edges are barred, and the Hamiltonian cycle must be chosen from the remaining fourteen edges. Since the graph has sixteen vertices, a Hamiltonian cycle in it must contain sixteen edges, which is impossible.

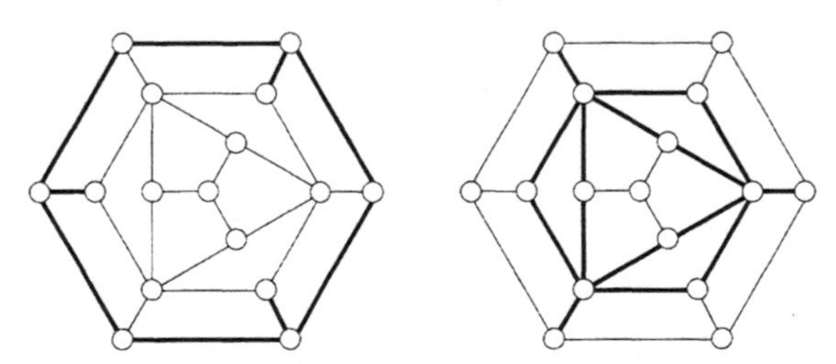

Figure 7.20: Steps in proving there is no Hamiltonian cycle

A similar argument could be used to prove the impossibility of a Hamiltonian path in this graph.

Another test is applicable only to bipartite graphs. As a bipartite graph is a subgraph of some $K_{m,n}$, its vertices can be partitioned into two subsets, of sizes m and n, such that the graph contains no edge that joins two vertices in the same subset.

Theorem 7.8 *A bipartite graph with vertex sets of sizes m and n can contain a Hamiltonian cycle only if $m = n$, and can contain a Hamiltonian path only if m and n differ by at most* 1.

Proof. Suppose a bipartite graph G has vertex sets V_1 and V_2, and suppose it contains a Hamiltonian path:

$$x_1, x_2, \ldots, x_v.$$

Suppose that x_1 belongs to V_1. Then x_2 must be in V_2, x_3 in V_1, and so on. Since the path contains every vertex, it follows that

$$V_1 = \{x_1, x_3, \ldots\},$$
$$V_2 = \{x_2, x_4, \ldots\}.$$

If v is even, then V_1 and V_2 each contain $v/2$ elements; if v is odd, then $|V_1| = (v+1)/2$ and $|V_2| = (v-1)/2$. In either case, the difference in orders is at most 1. If G contains a Hamiltonian cycle

$$x_1, x_2, \ldots, x_v, x_1,$$

and x_1 is in V_1, then x_v must belong to V_2; so $|V_1| = |V_2| = v/2$.

It should be realized that neither of these necessary conditions is sufficient; in particular, the following bipartite graph has four vertices in each subset (the vertices in the two sets are colored black and white respectively). It contains no Hamiltonian cycle; this cannot be proved using the above methods.

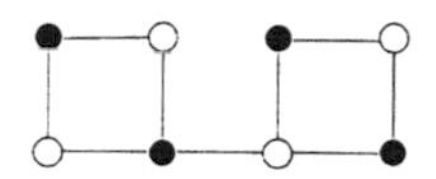

Exercises 7.6

In Exercises 1 *to* 6, *find a Hamiltonian cycle in the graph shown.*

1.

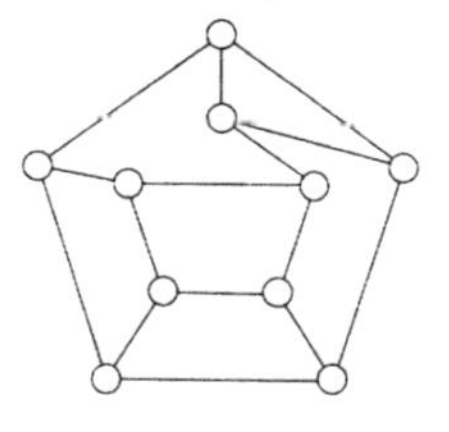

2.

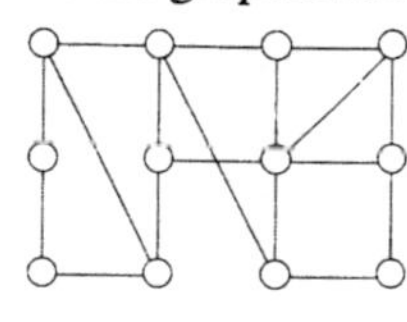

3.

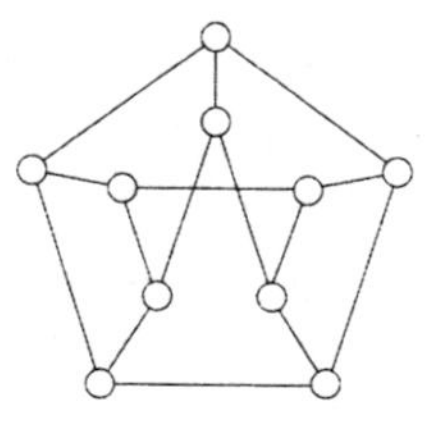

4.

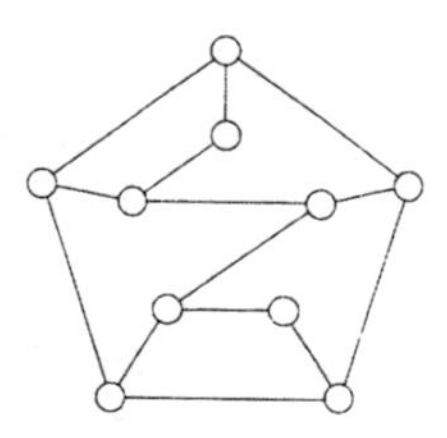

5.

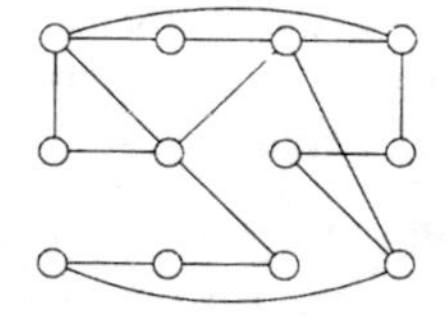

6.

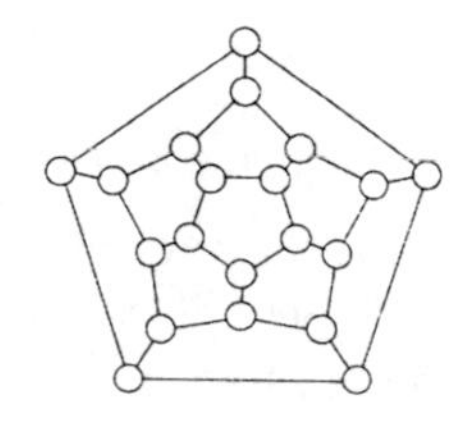

7. Prove that the following graph contains no Hamiltonian cycle.

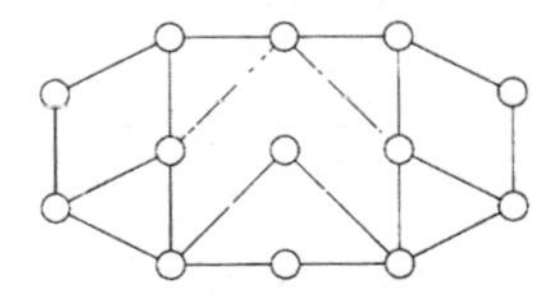

8. Prove that the following graphs contain no Hamiltonian paths.

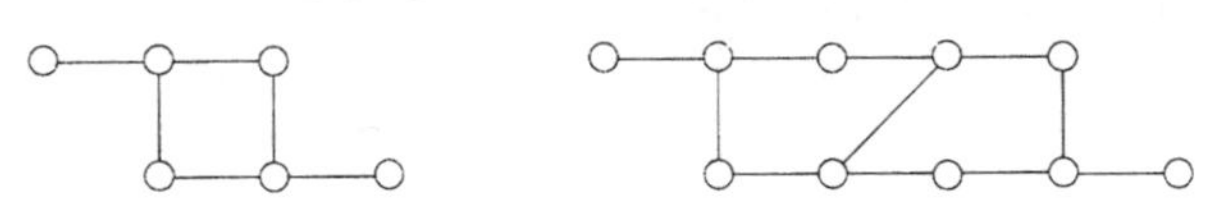

9. Prove that the following graph contains no Hamiltonian cycle.

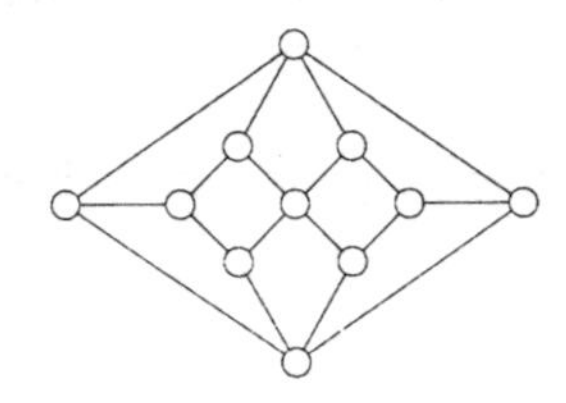

10. Prove that a spanning subgraph of $K_{m,n}$ can have a Hamiltonian cycle only if $m = n$.

11. Use Theorem 7.8 to prove that the graph of Figure 7.19 contains no Hamiltonian path.

12. Prove that the Petersen graph contains a Hamiltonian path, but no Hamiltonian cycle.

7.7 The Traveling Salesman Problem

Remember our example of a traveling saleman who wishes to visit several cities? If the cities are represented as vertices and the possible routes between them as edges, then the salesman's preferred itinerary is a Hamiltonian cycle in the graph.

In most cases there is a cost associated with every edge. Depending on the salesman's priorities, the cost might be a dollar cost such as airfare, or the number of miles to be driven, or the number of hours the trip will take. The cost of the trip is the sum of the costs of the edges, and the most desirable itinerary will be the one for which this sum of costs is a minimum. The problem of finding this cheapest Hamiltonian cycle is called the Traveling Salesman Problem.

We shall continue to speak in terms of a salesman, but these problems have many other applications. They arise in airline and delivery routing and in telephone routing. More recently they have been important in manufacturing integrated circuits and computer chips, and for internet routing.

In order to solve a Traveling Salesman Problem on n vertices, your first instinct might be to list all possible arrangements of n vertices and then delete those with two consecutive vertices that are not adjacent in the graph. This process can then be made more efficient by observing that the n lists $a_1a_2\ldots a_{n-1}a_n$, $a_2a_3\ldots a_n - a_1, \ldots, a_na_1\ldots a_{n-2}a_{n-1}$ all represent the same cycle (written with a different starting point), and also $a_n\ldots a_3a_2a_1$ is the same cycle as $a_1a_2\ldots a_{n-1}a_n$ (traversed in the opposite direction). If there is a vertex of degree 2, then any Hamiltonian cycle must contain the two edges touching it. If x has neighbors y and z, then xy and xz are in each cycle, and the edge yz, if it exists, is not in any cycle. So it suffices to delete x, add an edge yz if there is not one already,, find all Hamiltonian cycles in the new graph, and delete any that do not contain the edge yz. The Hamiltonian cycles in the original are then formed by inserting x between y and z.

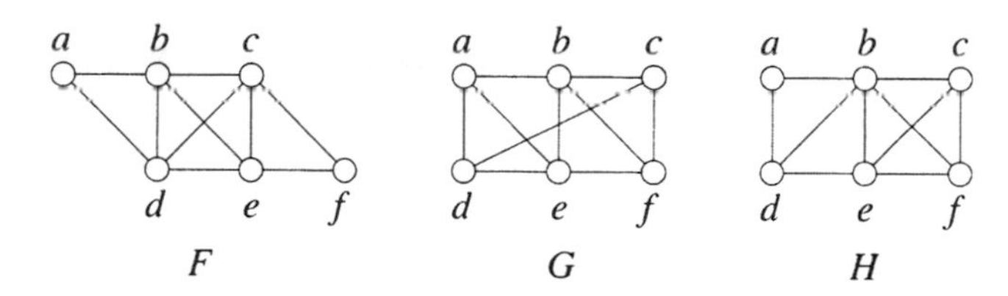

Figure 7.21: Find all Hamiltonian cycles

Sample Problem 7.19 *Find all Hamiltonian cycles in the graph F of Figure* 7.21.

Solution. The graph F has two vertices of degree 2, namely a and f. When we replace these vertices by edges bd and ce, we obtain the complete graph with vertices a,b,c,d (multiple edges can be ignored). This complete graph has three Hamiltonian cycles: there are six arrangements starting with b, namely $bcde$, $bced$, $bdce$, $bedc$, $bdec$, $becd$, and the latter three are just the former three written in reverse. $bcde$ yields a cycle that does not contain edges bd and ce. The other two give the cycles $bcfeda$ and $badcfe$, so these are the two Hamiltonian cycles in F.

Practice Exercise. Repeat this problem for graphs G and H of Figure 7.21.

However, this technique does not reduce the problem enough to make it manageable, because of the large number of Hamiltonian cycles in complete graphs.

Theorem 7.9 *The complete graph K_n contains $(n-1)!/2$ Hamiltonian cycles.*

Proof. There are $n!$ arrangements of the vertices. Each Hamiltonian cycle gives rise to n arrangements in the same cyclic order, and a further n are obtained by reversing them. So there are $n!/(2n) = (n-1)!/2$ Hamiltonian cycles. □

This number grows very quickly. For $n = 3,4,5,6,7$ the value of $(n-1)!/2$ is 1, 3, 12, 60, 360; in K_{10}, there are 181,440, and in K_{24}, there are about 10^{23}

Hamiltonian cycles. Twenty-four vertices is not an unreasonably large network, but performing so many summations and comparing them ·vould be impossible in practice. To give you some idea of the times involved, if you had a computer capable of evaluating and sorting through a million ten-vertex cycles per second, a complete search solution of the Traveling Salesman Problem for K_{10} would take about 0.18 seconds. Assuming the computer took about twice as much time to process a 24-vertex cycle, the complete search for K_{24} would take about a billion years.

Some fast methods of solution have been developed for *sparse* graphs — graphs where the average degree is low. This is a reasonable assumption in some cases, but very often the graph is complete or nearly so. In that case there is no known algorithm for solving the Traveling Salesman Problem that is substantially better than listing all Hamiltonian cycles. However, fast methods of reaching a "good solution" have been developed. Although they are not guaranteed to give the optimal answer, these approximation algorithms often give a route that is significantly cheaper than the average.

Approximation algorithms

The *nearest neighbor* method works as follows. Starting at some vertex x, one first chooses the edge incident with x whose cost is least. Say that edge is xy. Then an edge incident with y is chosen in accordance with the following rule: if y is the vertex most recently reached, then eliminate from consideration all edges incident with y that lead to vertices that have already been chosen (including x), and then select an edge of minimum cost from among those remaining. This rule is followed until every vertex has been chosen. The cycle is completed by going from the last vertex chosen back to the starting position x. This algorithm produces a directed cycle in the complete graph, but not necessarily the cheapest one, and different solutions may come from different choices of initial vertex x.

The *sorted edges* method does not depend on the choice of an initial vertex. One first produces a list of all the edges in ascending order of cost. At each stage, the cheapest edge is chosen with the restriction that no vertex can have degree 3 among the chosen edges, and the collection of edges contains no cycle of length less than v, the number of vertices in the graph. This method always produces an undirected cycle, and it can be traversed in either direction.

Sample Problem 7.20 *Suppose the costs of travel between St. Louis, Nashville, Evansville and Memphis are as shown in dollars on the left in Figure 7.22. You wish to visit all four cities, returning to your starting point. What routes do the two algorithms recommend?*

Solution. The nearest neighbor algorithm, applied starting from Evansville, starts by selecting the edge EM, because it has the least cost of the three edges incident with E. The next edge must have M as an endpoint, and ME is not allowed (one cannot return to E, it has already been used), so the cheaper of the remaining edges is chosen, namely MN. The cheapest edge originating at N is NE, with cost $110, but inclusion of this edge would lead back to E, a vertex that has already been visited, so NE is not allowed, and similarly NM is not available. It follows that NS must be chosen. So the algorithm finds route EMNSE, with cost $520.

A different result is achieved if one starts at Nashville. Then the first edge selected is NE, with cost $110. The next choice is EM, then MS, then SN, and the resulting cycle NEMSN costs $530.

If you start at St. Louis, the first stop will be Evansville ($120 is the cheapest flight from St. Louis), then Memphis, then Nashville, the same cycle as the Evansville case (with a different starting point), costing $520. From Memphis, the cheapest leg is to Evansville, then Nashville, and finally St. Louis, for $530 — the same cycle as from Nashville, in the opposite direction.

To apply the sorted edges algorithm, first sort the edges in order of increasing cost: EM($100), EN($110), ES($120), MN($130), MS ($150), NS($170). Edge EM is included, and so is EN. The next choice would be ES, but this is not allowed because its inclusion would give degree 3 to E. MN would complete a cycle of length 3 (too short), so the only other choices are MS and NS, forming route EMSNE (or ENSME) at a cost of $530.

However, in this example, the best route is ENMSE, with cost $510, and it does not arise from the nearest neighbor algorithm, no matter which starting vertex is used, or from the sorted edges algorithm.

Practice Exercise. A new cut-rate airline offers the fares shown on the right side of Figure 7.22. What do the algorithms say now?

In the real world, the cost of an itinerary will not necessarily be the sum of the individual legs. For example, airlines often offer fares for a flight with a stopover that are cheaper than the sum of the fares for the individual legs. However, the additive model applies in the majority of applications. In any case, the considerations discussed in this section still apply in the more general case — even if the fare is not additive, the time needed to check all Hamiltonian cycles is still prohibitive.

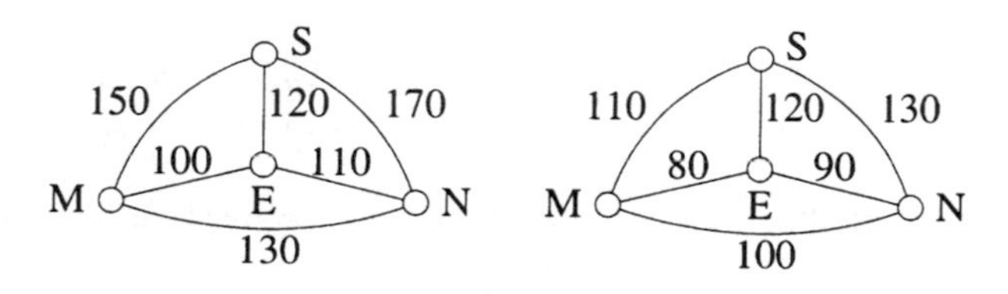

Figure 7.22: Traveling Salesman Problem example

Exercises 7.7

In Exercises 1 to 6, *list all Hamiltonian cycles in the graph shown.*

1.

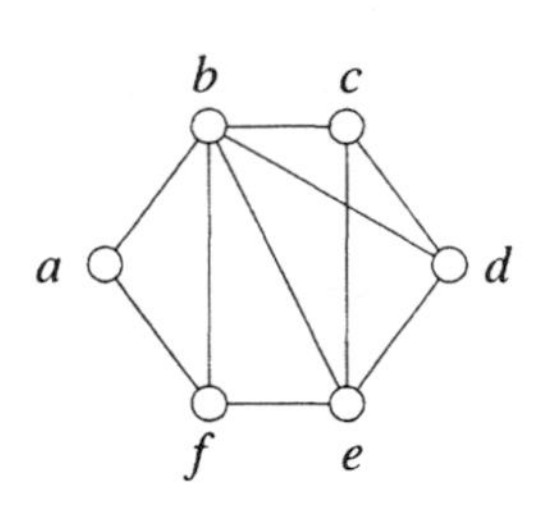

2.

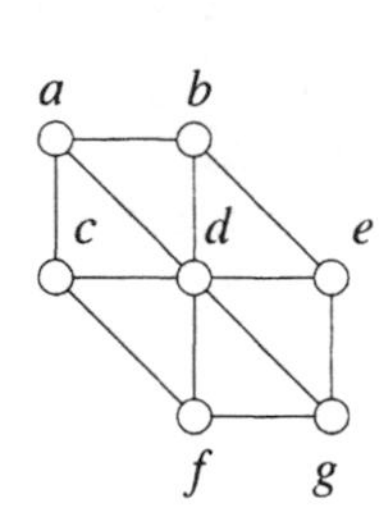

3.

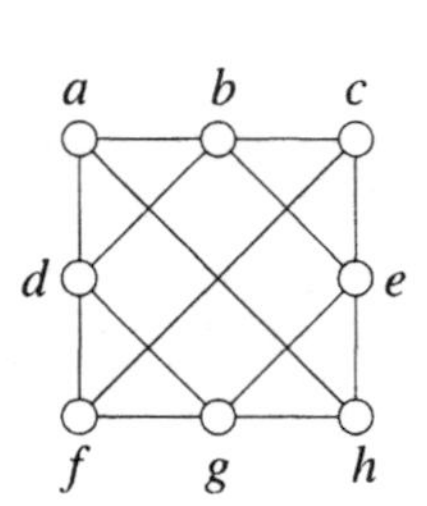

4.

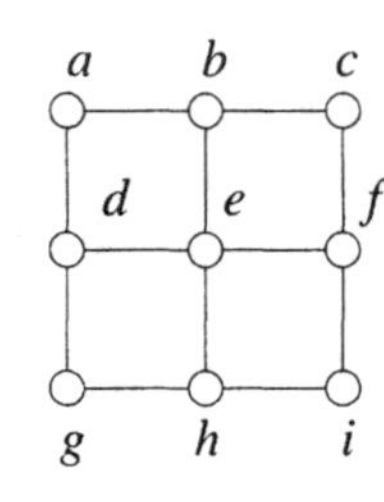

5.

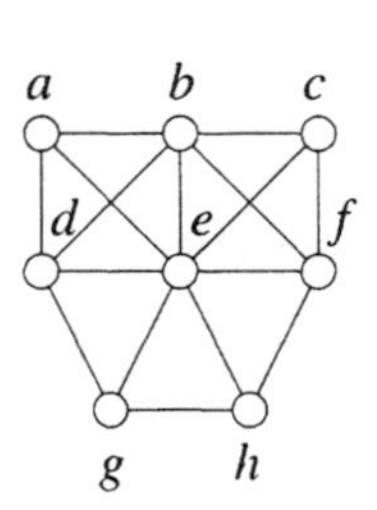

6.

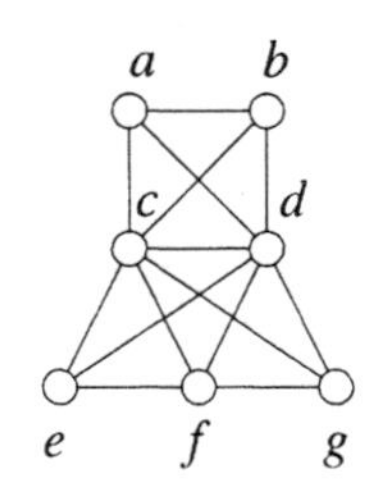

7. Solve the Traveling Salesman Problem for the graphs G and H in Figure 7.21 by finding all their Hamiltonian cycles, if the costs associated with the edges are as follows:

For G:

$ab = 18$ $ad = 12$ $ae = 30$
$bc = 24$ $be = 14$ $bf = 28$
$cd = 30$ $cf = 16$
$de = 21$ $ef = 25$

For H:

$ab = 19$ $ad = 16$ $bc = 22$
$bd = 17$ $be = 21$ $bf = 25$
$ce = 24$ $cf = 19$
$de = 18$ $ef = 21$

8. Consider a complete graph K_5 with vertices *abcde*, the costs associated with the edges being as follows.

$ab = 1$ $ac = 2$ $ad = 3$ $ae = 4$ $bc = 7$
$bd = 8$ $be = 9$ $cd = 5$ $ce = 6$ $de = 10$

Find the cost of the cheapest Hamiltonian cycle in this graph by a complete search. Then find the costs of the routes generated by the nearest neighbor algorithm starting at each of the five vertices in turn and show that the nearest neighbor solution is never the cheapest.

In Exercises 9 *to* 18, *you are given the costs associated with the edges of a complete graph with vertices abcde. Find the costs of the routes generated by the nearest neighbor algorithm starting at each of the five vertices in turn and by the sorted edges algorithm.*

9. $ab = 24 \quad ac = 23 \quad ad = 20 \quad ae = 21 \quad bc = 27$
$bd = 22 \quad be = 30 \quad cd = 26 \quad ce = 27 \quad de = 28$

10. $ab = 24 \quad ac = 26 \quad ad = 20 \quad ae = 21 \quad bc = 33$
$bd = 29 \quad be = 30 \quad cd = 25 \quad ce = 27 \quad de = 28$

11. $ab = 24 \quad ac = 22 \quad ad = 30 \quad ae = 29 \quad bc = 17$
$bd = 19 \quad be = 30 \quad cd = 18 \quad ce = 21 \quad de = 25$

12. $ab = 59 \quad ac = 69 \quad ad = 60 \quad ae = 58 \quad bc = 56$
$bd = 69 \quad be = 54 \quad cd = 58 \quad ce = 66 \quad de = 61$

13. $ab = 12 \quad ac = 18 \quad ad = 14 \quad ae = 19 \quad bc = 22$
$bd = 16 \quad be = 15 \quad cd = 18 \quad ce = 17 \quad de = 11$

14. $ab = 16 \quad ac = 24 \quad ad = 30 \quad ae = 48 \quad bc = 27$
$bd = 29 \quad be = 44 \quad cd = 16 \quad ce = 46 \quad de = 51$

15. $ab = 27 \quad ac = 22 \quad ad = 28 \quad ae = 17 \quad bc = 33$
$bd = 14 \quad be = 23 \quad cd = 24 \quad ce = 26 \quad de = 19$

16. $ab = 91 \quad ac = 79 \quad ad = 75 \quad ae = 82 \quad bc = 87$
$bd = 64 \quad be = 78 \quad cd = 68 \quad ce = 81 \quad de = 88$

17. $ab = 45 \quad ac = 28 \quad ad = 50 \quad ae = 36 \quad bc = 21$
$bd = 42 \quad be = 34 \quad cd = 44 \quad ce = 39 \quad de = 25$

18. $ab = 11 \quad ac = 15 \quad ad = 13 \quad ae = 18 \quad bc = 20$
$bd = 12 \quad be = 16 \quad cd = 14 \quad ce = 19 \quad de = 17$

8

Matrices

A matrix is a two-dimensional array of numbers; matrices are useful for representing data that is naturally classified in two ways. Matrices arise in many parts of mathematics; usually they are studied as part of linear algebra.

There are many applications of matrices in discrete mathematics. They are particularly important in the study of relations and graphs, so we have included a section that briefly introduces those applications. Many other properties of graphs and relations can be deduced using further algebraic properties of matrices.

8.1 Vectors and Matrices

Vectors

A *vector* is an ordered finite set of numbers. We say *real vector* if all the elements are real numbers; we can also talk about *integer vectors*, and so on. If it is required that all elements belong to some number set, those numbers are called the *scalars* for those vectors.

The number of elements in a vector is called its *dimension* or *length*. The set of all real vectors of dimension n will be denoted $\mathbb{R}^n$. The set $\mathbb{R}^2$ is familiar from geometry, because the coordinates of points in the plane are just members of $\mathbb{R}^2$. Similarly, $\mathbb{R}^3$ is the set of coordinates of points in three-dimensional space.

When we talk about the elements of a vector and specify their positions, we usually call them *entries*. The usual notation for a vector is to denote its i-th entry by subscript i. The vector $\boldsymbol{v}$ has entries $v_1, v_2, \ldots$, and we write $\boldsymbol{v} = (v_1, v_2, \ldots)$. Two vectors are equal if and only if they are equal in every position (*component-wise equal*): $\boldsymbol{u} = \boldsymbol{v}$ if and only if $u_i = v_i$ for every i. It follows that equal vectors must be of the same length.

There are two standard operations on vectors: one may *multiply by a scalar*, and one may *add vectors*. They are defined as follows.

If $\boldsymbol{v} = (v_1, v_2, \ldots, v_n)$, and k is any scalar, then $k\boldsymbol{v} = (kv_1, kv_2, \ldots, kv_n)$. If $\boldsymbol{u} = (u_1, u_2, \ldots, u_n)$, and $\boldsymbol{v} = (v_1, v_2, \ldots, v_n)$, then $\boldsymbol{u} + \boldsymbol{v} = ((u_1 + v_1), (u_2 + v_2), \ldots, (u_n + v_n)$. If $\boldsymbol{u}$ and $\boldsymbol{v}$ are vectors of different lengths, then $\boldsymbol{u} + \boldsymbol{v}$ is not defined.

It is easy to see that vector addition satisfies the commutative and associative laws: if $\boldsymbol{u}$ and $\boldsymbol{v}$ are any two vectors of the same length, then

$$\boldsymbol{u} + \boldsymbol{v} = \boldsymbol{v} + \boldsymbol{u},$$

and any three vectors $\boldsymbol{t}, \boldsymbol{u}$ and $\boldsymbol{v}$ of equal length satisfy

$$\boldsymbol{t} + (\boldsymbol{u} + \boldsymbol{v}) = (\boldsymbol{t} + \boldsymbol{u}) + \boldsymbol{v}.$$

We usually omit brackets when writing the sum of three or more vectors, so this common value is simply denoted $\boldsymbol{t} + \boldsymbol{u} + \boldsymbol{v}$.

There are also two laws involving multiplication by a scalar. If a and b are any numbers and $\boldsymbol{u}$ and $\boldsymbol{v}$ are any two equal-length vectors, then

$$a(b\boldsymbol{u}) = (ab)\boldsymbol{u}$$

and

$$a(\boldsymbol{u} + \boldsymbol{v}) = (a\boldsymbol{u}) + (a\boldsymbol{v}).$$

Again we follow the same convention as ordinary arithmetic and omit brackets, writing the last expression as simply $a\boldsymbol{u} + a\boldsymbol{v}$.

Sample Problem 8.1 *Calculate* $3(1, -1, 3)$ *and* $(2, 2) + (-1, 3)$.

Solution. $3(1, -1, 3) = (3, -3, 9)$; $(2, 2) + (-1, 3) = (1, 5)$.

Practice Exercise. Calculate $4(2, 0, -1) + (1, 4, -3)$.

Subtraction is defined in the natural way: $\boldsymbol{u} - \boldsymbol{v} = ((u_1 - v_1), (u_2 - v_2), \ldots, (u_n - v_n))$. If we write $-\boldsymbol{v} = (-v_1, -v_2, \ldots, -v_n)$, then $\boldsymbol{u} - \boldsymbol{v} = \boldsymbol{u} + (-\boldsymbol{v})$, and moreover $-\boldsymbol{v} = (-1)\boldsymbol{v}$.

It is natural to define a *zero vector* $\boldsymbol{0} = (0, 0, \ldots, 0)$ (in fact, a family of zero vectors, one for each possible dimension). Then $\boldsymbol{v} + (-\boldsymbol{v}) = 0$, as one would expect.

Lines; the scalar product

The equation of a straight line in coordinate geometry has the form

$$ax + by = c,$$

where a, b and c are numbers and x, y are the usual variables. The equation involves two vectors, the vector (a, b) of coefficients and the vector (x, y) of variables. For this reason it is natural to associate $ax + by$ with the two vectors (a, b) and (x, y).

We define the *scalar product* of two vectors $\boldsymbol{u} = (u_1, u_2, \ldots, u_n)$ and $\boldsymbol{v} = (v_1, v_2, \ldots, v_n)$ to be

$$\boldsymbol{u} \cdot \boldsymbol{v} = (u_1 v_1 + u_2 v_2 + \cdots + u_n v_n) = \sum_{k=1}^{n} u_k v_k.$$

This is also called the *dot product* of the two vectors. In this notation a typical straight line in two-dimensional geometry has an equation of the form

$$\boldsymbol{a} \cdot \boldsymbol{x} = c,$$

where $\boldsymbol{a}$ is some vector of two real numbers, $\boldsymbol{x}$ is the vector of variables (x, y), and c is a constant. If $\boldsymbol{a}$ and $\boldsymbol{x}$ are of length three, then $\boldsymbol{a} \cdot \boldsymbol{x} = c$ could be the equation of a plane in three-dimensional space.

Sample Problem 8.2 *Suppose* $\boldsymbol{t} = (1,2,3), \boldsymbol{u} = (-1,3,0)$ *and* $\boldsymbol{v} = (2, -2, 2)$. *Calculate* $\boldsymbol{u} \cdot \boldsymbol{v}, (\boldsymbol{t} - \boldsymbol{u}) \cdot \boldsymbol{v}$ *and* $3(\boldsymbol{v} \cdot \boldsymbol{t})$.

Solution. $\boldsymbol{u} \cdot \boldsymbol{v} = -2 - 6 + 0 = -8$; $(\boldsymbol{t} - \boldsymbol{u}) \cdot \boldsymbol{v} = (2, -1, 3) \cdot (2, -2, 2) = 4 + 2 + 6 = 12$; $3(\boldsymbol{v} \cdot \boldsymbol{t}) = 3 \times 8 = 24$.

Practice Exercise. Calculate $\boldsymbol{u} \cdot \boldsymbol{t}$ and $(2\boldsymbol{u} - 3\boldsymbol{v}) \cdot \boldsymbol{t}$.

Sample Problem 8.3 *Suppose* $(x, 3) = (4, x + y)$. *What are* x *and* y*?*

Solution. Two vectors are equal if and only if their components are equal. So we have the two equations $x = 4$ and $3 = x + y$. So $x = 4$ and $y = -1$.

Practice Exercise. Suppose $(2x, 2) = (y, x)$. What are x and y?

It is not hard to see that the scalar product is commutative. There is no need to discuss the associative law, because scalar products involving three vectors are not defined. For example, consider $\boldsymbol{t} \cdot (\boldsymbol{u} \cdot \boldsymbol{v})$. Since $(\boldsymbol{u} \cdot \boldsymbol{v})$ is a scalar, not a vector, we cannot calculate its dot product with anything.

Matrices

Just as one linear equation can be written as one vector equation, a set of linear equations can be written as a set of vector equations. For example, the equations

$$\begin{aligned} 2x+y &= 3, \\ 3x-2y &= 1 \end{aligned}$$

could be written as

$$\big((2,1)\cdot(x,y),\ (3,-2)\cdot(x,y)\big) = \big(3,1\big)$$

or, a little easier to read,

$$\begin{bmatrix} (2,1)\cdot(x,y) \\ (3,-2)\cdot(x,y) \end{bmatrix} = \begin{bmatrix} 3 \\ 1 \end{bmatrix}.$$

This can be seen as the two-dimensional array of numbers

$$\begin{bmatrix} 2 & 1 \\ 3 & -2 \end{bmatrix}$$

acting on the vector (x,y) to produce the vector $(3,1)$. We shall call this array a *matrix*. We usually denote matrices by capital letters, to distinguish them from scalars (numbers).

Just as a vector generalizes a number, a matrix generalizes a vector. A matrix is any rectangular (two-dimensional) array of numbers. A typical matrix is

$$M = \begin{bmatrix} 1 & -1 & 4 & 3 \\ 2 & 3 & -4 & 1 \\ 4 & -4 & 4 & 8 \end{bmatrix}.$$

This example is a 3×4 real matrix (and it is in fact a 3×4 integer matrix). We call 3×4 the *shape* or *size* of the matrix, and the two numbers 3 and 4 are its *dimensions*.

A vector can be interpreted as a matrix with one of its dimensions 1. A $1\times n$ matrix is called a *row* vector and an $n\times 1$ matrix is called a *column* vector. An $m\times n$ matrix can be viewed as a vertical stack of m row vectors, and those vectors are called the *rows of the matrix.* Alternatively we could see it as a horizontal array of column vectors, called the *columns of the matrix.* Our example M could be viewed as

$$\begin{bmatrix} (1 & -1 & 4 & 3) \\ (2 & 3 & -4 & 1) \\ (4 & -4 & 4 & 8) \end{bmatrix} \text{ or } \begin{bmatrix} \begin{bmatrix} 1 \\ 2 \\ 4 \end{bmatrix} \begin{bmatrix} -1 \\ 3 \\ -4 \end{bmatrix} \begin{bmatrix} 4 \\ -4 \\ 4 \end{bmatrix} \begin{bmatrix} 3 \\ 1 \\ 8 \end{bmatrix} \end{bmatrix}.$$

The element in the ith row and jth column is called the (i,j) *element* of the matrix. It is standard to denote the (i,j) element of a matrix A by a_{ij}, and write $A=(a_{ij})$. This is particularly useful when a formula is given for a_{ij}. For example, we might refer to the 3×3 matrix whose (i,j) element equals $i+j$, namely

$$\begin{bmatrix} 2 & 3 & 4 \\ 3 & 4 & 5 \\ 4 & 5 & 6 \end{bmatrix},$$

as "the 3×3 matrix $(i+j)$."

Sums and products

The *sum* of two matrices and the *product* of a scalar with a matrix are defined analogously to the vector case. We again define negatives by $-A = (-1)A$. Addition satisfies the commutative and associative laws, and the scalar product laws $a(bC) = (ab)C$ and $a(C+D) = aC + aD$.

Sample Problem 8.4 *Suppose A, B and C are the matrices*

$$A = \begin{bmatrix} 1 & 3 \\ -1 & 2 \end{bmatrix}, B = \begin{bmatrix} -2 & 0 \\ 1 & 4 \end{bmatrix}, C = \begin{bmatrix} -1 & 4 & 3 \\ -1 & -2 & -1 \end{bmatrix}.$$

Find $A+B$, $2A-3B$, $3A+C$, $-C$.

Solution.

$$A+B = \begin{bmatrix} -1 & 3 \\ 0 & 6 \end{bmatrix}, \quad 2A-3B = \begin{bmatrix} -4 & 6 \\ -5 & -8 \end{bmatrix},$$

$$-C = \begin{bmatrix} 1 & -4 & -3 \\ 1 & 2 & 1 \end{bmatrix}.$$

$3A+C$ is not defined, as A and C are of different sizes.

Practice Exercise. Calculate $-A$, $3A-B$, $B+C$.

We define the *product* of two matrices as a generalization of the scalar product of vectors. Suppose the rows of the matrix A are $\boldsymbol{a}_1, \boldsymbol{a}_2, \ldots$, and the columns of the matrix B are $\boldsymbol{b}_1, \boldsymbol{b}_2, \ldots$. Then AB is the matrix with (i, j) entry $\boldsymbol{a}_i \cdot \boldsymbol{b}_j$. These entries will only exist if the number of columns of A equals the number of rows of A, so this is a necessary condition for the product AB to exist.

Sample Problem 8.5 *Suppose*

$$A = \begin{bmatrix} 1 & 2 \\ 1 & -1 \end{bmatrix}, B = \begin{bmatrix} -1 & 1 \\ 2 & 0 \end{bmatrix}.$$

Find AB and BA.

Solution. First we find AB. The rows of A are $\boldsymbol{a}_1 = (1, 2)$ and $\boldsymbol{a}_2 = (1, -1)$. The columns of B are $\boldsymbol{b}_1 = (-1, 2)$ and $\boldsymbol{b}_2 = (1, 0)$. (Since we are treating them as vectors, it doesn't matter whether we write them as row or column vectors.) Then $\boldsymbol{a}_1 \cdot \boldsymbol{b}_1 = -1 + 4 = 3$, and similarly $\boldsymbol{a}_1 \cdot \boldsymbol{b}_2 = 1$, $\boldsymbol{a}_2 \cdot \boldsymbol{b}_1 = -3$ and $\boldsymbol{a}_2 \cdot \boldsymbol{b}_2 = 1$. So

$$AB = \begin{bmatrix} \boldsymbol{a}_1 \cdot \boldsymbol{b}_1 & \boldsymbol{a}_1 \cdot \boldsymbol{b}_2 \\ \boldsymbol{a}_2 \cdot \boldsymbol{b}_1 & \boldsymbol{a}_2 \cdot \boldsymbol{b}_2 \end{bmatrix} = \begin{bmatrix} 3 & 1 \\ -3 & 1 \end{bmatrix}.$$

Similarly we find

$$BA = \begin{bmatrix} 0 & -3 \\ 2 & 4 \end{bmatrix}.$$

The entries in AB will only exist if the number of columns in A equals the number of rows in B. For example, if A were 3×2 and B were 4×4, the product would not exist. In general, we can say the following.

Theorem 8.1 *Suppose A is an $m \times n$ matrix and B is an $r \times s$ matrix. If $n = r$, then AB exists and is an $n \times s$ matrix. If $n \neq r$, then AB does not exist.*

If A were 2×3 and B were 3×4, then AB would be a 2×4 matrix but BA would not exist. It is also possible that AB and BA might both exist but might be of different shapes; for example, if A and B have shapes 2×3 and 3×2 respectively, then AB is 2×2 and BA is 3×3. And we observe from the preceding example that, even when AB and BA both exist and are the same shape, they need not be equal. *There is no commutative law for matrix multiplication.*

We shall look at some more properties of matrix multiplication in the next section.

Transposition

If A is an $m \times n$ matrix, then we can form an $n \times m$ matrix whose (i, j) entry equals the (j, i) entry of A. This new matrix is called the *transpose* of A, and written A^T. A matrix A is called *symmetric* if $A = A^T$.

Sample Problem 8.6 *What is the transpose of the matrix A from Sample Problem* 8.5*?*

Solution.

$$A^T = \begin{bmatrix} 1 & 1 \\ 2 & -1 \end{bmatrix}.$$

Practice Exercise. What is the transpose of the matrix B from Sample Problem 8.5?

The transposition symbol is applied only to the matrix nearest to it. For example, the expression AB^T means $A(B^T)$, not $(AB)^T$.

Exercises 8.1

1. Verify that real matrices satisfy the commutative and associative laws.

Carry out the vector computations in Exercises 2 *to* 23.

2. $4(2, -2)$

3. $-(2, -2)$

4. $2(5,1,-1)$

5. $3(3,6,1)$

6. $(2,3)+(1,4)$

7. $3(2,3)-2(1,4)$

8. $(1,0,3)+3(4,4,4)$

9. $2(-1,-1,2)-2(2,-1,-1)$

10. $3(-1,2,3)+2(1,1,-1)$

11. $3(1,-2,2)+2(2,3,-1)$

12. $3(1,0,1,0)-4(2,0,-1,-1)$

13. $2(4,-1,2,3)-3(1,6,-2,-3)$

14. $(2,3)\cdot(1,-1)$

15. $(1,-1)\cdot(2,3)$

16. $(1,1,-1)\cdot(2,0,3)$

17. $(1,3,3)\cdot(1,0,-2)$

18. $(4,2,1)\cdot(1,2,4)$

19. $(1,-1,-2)\cdot(3,2,-1)$

20. $(0,2,1,1)\cdot(3,4,2,1)$

21. $(-1,2,-1,3)\cdot(2,4,-3,-1)$

22. $(3,-1,3,2)\cdot(-1,-1,2,1)$

23. $(1,-2,5,2)\cdot(2,2,3,1)$

A is a 2×4 matrix; B is 2×4; C is 1×3; D is 4×2; E is 3×4; F is 4×3; G is 4×4. In Exercises 24 to 41, say whether the indicated matrix exists. If it does exist, what is its shape?

24. $A+B$

25. $2A-B$

26. CE

27. AD

28. $D(A+B)$

29. CF

30. F^T

31. CF^T

32. $2FC$

33. DA

34. $AD+DA$

35. BFE

36. GG

37. FF

38. $DA+3G$

39. CEF

40. CFE

41. AGF

42. Suppose a,b are any two real numbers and C,D are any two real matrices of the same shape. Prove the following equalities.

(i) $a(bC)=(ab)C$.

(ii) $a(C+D)=aC+aD$.

(iii) $(a+b)C=aC+bC$.

In Exercises 43 to 48, carry out the matrix computations.

43. $\begin{bmatrix} 4 & -1 \\ -2 & 0 \end{bmatrix}+\begin{bmatrix} 3 & -1 \\ -1 & 2 \end{bmatrix}$

44. $3\begin{bmatrix} 1 & -1 & -1 \\ -2 & 0 & 10 \end{bmatrix}$

45. $3\begin{bmatrix} 10 & -1 \\ 2 & 7 \end{bmatrix} - 2\begin{bmatrix} 1 & -1 \\ -1 & 1 \end{bmatrix}$

46. $\begin{bmatrix} 3 & -4 \\ 3 & 0 \end{bmatrix}\begin{bmatrix} 6 & -2 \\ -7 & 4 \end{bmatrix}$

47. $\begin{bmatrix} 2 & 0 & 1 \\ 3 & -1 & 1 \end{bmatrix}\begin{bmatrix} 2 & -1 \\ 1 & -1 \\ -1 & 1 \end{bmatrix}$

48. $\begin{bmatrix} 2 & -1 \\ 1 & -1 \\ -1 & 1 \end{bmatrix}\begin{bmatrix} 2 & 0 & 1 \\ 3 & -1 & 1 \end{bmatrix}$

In Exercises 49 *to* 64,

$$A = \begin{bmatrix} 1 & -1 \\ -2 & 3 \end{bmatrix}, \quad B = \begin{bmatrix} 3 & 0 & -1 \\ -1 & 4 & 4 \end{bmatrix}, \quad C = \begin{bmatrix} 6 \\ 2 \end{bmatrix},$$

$$D = \begin{bmatrix} -1 & 1 & -1 \\ 1 & 3 & 3 \\ -2 & 2 & 0 \end{bmatrix}, \quad E = \begin{bmatrix} 1 & -1 \\ 1 & 0 \\ 2 & 2 \end{bmatrix}, \quad F = \begin{bmatrix} 2 \\ -1 \\ 2 \end{bmatrix},$$

$$G = \begin{bmatrix} 2 & 1 & -1 \end{bmatrix}, \quad H = \begin{bmatrix} 2 & 2 \end{bmatrix}, \quad K = \begin{bmatrix} -1 & 2 \end{bmatrix}.$$

Carry out the matrix computations, or explain why they are impossible.

49. $2A$

50. BD

51. AC

52. EF

53. BF

54. B^T

55. $2H - 3K$

56. $HA + 2K$

57. CF

58. $D - EB$

59. BG

60. KB

61. GD

62. CE^T

63. $EK + KB$

64. $2F + G$

65. Suppose

$$\begin{bmatrix} x & -1 \\ -1 & 2 \end{bmatrix} = \begin{bmatrix} y+1 & -1 \\ -1 & x \end{bmatrix}.$$

What are the values of x and y?

66. Find x, y and z so that

$$\begin{bmatrix} x-2 & 3 & z \\ y & x & 2y \end{bmatrix} = \begin{bmatrix} y & z & 3 \\ 3z & y+2 & 6z \end{bmatrix}.$$

67. A and B are any two matrices such that AB exists. Prove that B^TA^T exists, and that

$$B^TA^T = (AB)^T.$$

8.2 Properties of the Matrix Product

Identity elements

Suppose A is an $r \times s$ matrix. Adding 0 to every entry will not change A. So the $r \times s$ matrix with every entry 0 will act like a zero element, or additive identity, for A. We shall denote this matrix by $O_{r,s}$ and call it a *zero matrix*. Usually the subscripts can be omitted, because the shape can be deduced from the context. For any matrix A, the appropriate zero matrix satisfies

$$A + O = O + A = A \text{ and } A + (-A) = (-A) + A = O.$$

Writing $-A$ for $(-1)A$ is an extension of the notation we used for zero vectors, and is consistent with the usual arithmetical notations for zero and negatives..

The zero matrix also behaves under multiplication the way you would expect: provided zero matrices of appropriate size are used,

$$OA = O \text{ and } AO = O.$$

This is not just one rule, but an infinite set of rules. If we write in the subscripts, then the full statement is

If A is any $r \times s$ matrix, then $O_{m,r}A = O_{m,s}$ for any positive integer m, and $AO_{s,n} = O_{r,n}$ for any positive integer n.

There are also multiplicative identity elements. We define I_n to be the $n \times n$ matrix with its $(1,1), (2,2), \ldots, (n,n)$ entries 1 and all other entries 0. For example,

$$I_3 = \begin{bmatrix} 1 & 0 & 0 \\ 0 & 1 & 0 \\ 0 & 0 & 1 \end{bmatrix}.$$

If A is any $r \times s$ matrix, then $I_r A = A = AI_s$.

We call I_n an *identity matrix* of order n.

Commutativity

We saw that the commutative law does not hold for matrices in general. Even if AB and BA are both defined and are the same size, it is possible for the two products to be different (see Sample Problem 8.5). On the other hand, some pairs of matrices have the same product in either order. If $AB = BA$ we say that A and B *commute*, or A *commutes with* B. For example, any 3×3 matrix commutes with I_3. There are many other examples.

Sample Problem 8.7 *Show that the following matrices commute.*

$$A = \begin{bmatrix} 1 & -2 \\ 1 & -1 \end{bmatrix}, \quad B = \begin{bmatrix} 1 & 2 \\ -1 & 3 \end{bmatrix}.$$

Solution.

$$AB = BA = \begin{bmatrix} 3 & -4 \\ 2 & -1 \end{bmatrix}.$$

Practice Exercise. Show that the following matrices commute.

$$C = \begin{bmatrix} 3 & 1 \\ -1 & 1 \end{bmatrix}, \quad D = \begin{bmatrix} 2 & 1 \\ -1 & 0 \end{bmatrix}.$$

Suppose A has shape $m \times n$ and B is $r \times s$. If both AB and BA exist, then necessarily $n = r$ and $m = s$; then AB is $m \times m$ and BA is $n \times n$. In order for A and B to commute, we must have $m = n$. Both A and B must have the same number of rows as columns. Such a matrix is called *square*, and the common dimension is called its *order*.

If A is square, we can evaluate the product AA. We call this A *squared*, and write it as A^2, just as with powers of numbers. We define other positive integer powers similarly: $A^3 = AAA = AA^2$, and in general $A^{n+1} = AA^n$.

Inverses

If the matrices A and B satisfy $AB = BA = I$, we say that B is an *inverse* of A.

In the real numbers, everything but 0 has an inverse. In the integers, only 1 and -1 have integer inverses, but we know that we can obtain inverses of other non-zero integers by going to the rational numbers. The situation is obviously more complicated for matrices, because only a square matrix can have an inverse. Moreover, there are non-zero square matrices without inverses, even if we restrict our attention to the 2×2 case.

Sample Problem 8.8 *Show that the matrix*

$$A = \begin{bmatrix} 2 & 1 \\ 2 & 1 \end{bmatrix}$$

has no inverse.

Solution. Suppose A has inverse

$$B = \begin{bmatrix} x & y \\ z & t \end{bmatrix}.$$

Then $AB = I$, so

$$\begin{bmatrix} 2 & 1 \\ 2 & 1 \end{bmatrix} \begin{bmatrix} x & y \\ z & t \end{bmatrix} = \begin{bmatrix} 2x+z & 2y+t \\ 2x+z & 2y+t \end{bmatrix} = \begin{bmatrix} 1 & 0 \\ 0 & 1 \end{bmatrix}.$$

The $(1,1)$ entries of the two matrices must be equal, so $2x + z = 1$; but the $(2,1)$ entries must also be equal, so $2x + z = 0$. This is impossible.

Practice Exercise. Show that the matrices

$$C = \begin{bmatrix} 1 & 0 \\ 0 & 0 \end{bmatrix} \text{ and } D = \begin{bmatrix} 0 & 1 \\ 0 & 0 \end{bmatrix}$$

have no inverses.

A matrix that has an inverse will be called *invertible* or *non-singular*; a square matrix without an inverse is called *singular*.

We used the phrase "an inverse" above. However, we shall prove that, if a matrix has an inverse, it is unique.

Theorem 8.2 *If matrices A, B, C satisfy $AB = BA = I$ and $AC = CA = I$, then $B = C$.*

Proof. Suppose A, B and C satisfy the given equations. Then

$$C = CI = C(AB) = (CA)B = IB = B$$

so B and C are equal. □

In fact, it can be shown that either of the conditions $AB = I$ or $BA = I$ is enough to determine that B is the inverse of A. However, this requires more algebra than we shall cover in this book.

Sample Problem 8.9 *Suppose*

$$A = \begin{bmatrix} 3 & 2 \\ 4 & 3 \end{bmatrix}, \quad B = \begin{bmatrix} 2 & 1 \\ 1 & 1 \end{bmatrix}.$$

What is the inverse of A if it exists?

Solution. Suppose the inverse is

$$C = \begin{bmatrix} x & z \\ y & t \end{bmatrix}.$$

Then $AC = I$ means

$$\begin{bmatrix} 3 & 2 \\ 4 & 3 \end{bmatrix} \begin{bmatrix} x & z \\ y & t \end{bmatrix} = \begin{bmatrix} 1 & 0 \\ 0 & 1 \end{bmatrix}$$

which is equivalent to the four equations

$$\begin{aligned} 3x + 2y &= 1, \qquad & 3z + 2t &= 0, \\ 4x + 3y &= 0, \qquad & 4z + 3t &= 1. \end{aligned}$$

The left-hand pair of equations is easily solved to give $x = 3$ and $y = -4$, while the right-hand pair give $z = -2$ and $t = 3$. So the inverse exists, and is

$$A^{-1} = C = \begin{bmatrix} 3 & -2 \\ -4 & 3 \end{bmatrix}.$$

Practice Exercise. What is the inverse of B if it exists?

The above procedure can be used to invert square matrices of any order; if there is no inverse, then the equations will have no solution. In the next section we shall show how to reduce the number of computations required.

The usual notation for the inverse of A, if it exists, is A^{-1}. If we define $A^0 = I$ whenever A is square, then powers of matrices satisfy the usual index laws

$$A^m A^n = A^{m+n}, (A^m)^n = A^{mn}$$

for all non-negative integers m and n, and for negative values also provided that A^{-1} exists. If x and y are non-zero reals, then $(xy)^{-1} = x^{-1}y^{-1}$. The fact that matrices do not necessarily commute means that we have to be a little more careful:

Theorem 8.3 *If A and B are invertible matrices of the same order, then AB is invertible, and*

$$(AB)^{-1} = B^{-1}A^{-1}.$$

Proof. We need to show that both $(B^{-1}A^{-1})(AB)$ and $(AB)(B^{-1}A^{-1})$. equal the identity. But $(B^{-1}A^{-1})(AB) = B^{-1}(A^{-1}A)B = B^{-1}IB = B^{-1}B = I = AA^{-1} = AIA^{-1} = A(BB^{-1})A^{-1} = (AB)(B^{-1}A^{-1})$. □

There are two cancellation laws for matrix multiplication. If A is an invertible $r \times r$ matrix and B and C are $r \times s$ matrices such that $AB = AC$, then

$$\begin{aligned} AB = AC \quad &\Rightarrow \quad A^{-1}(AB) = A^{-1}(AC) \\ &\Rightarrow \quad (A^{-1}A)B = (A^{-1}A)C \Rightarrow IB = IC \Rightarrow B = C \end{aligned}$$

so $B = C$. Similarly, if A is an invertible $s \times s$ matrix and B and C are $r \times s$ matrices such that $BA = CA$, then $B = C$.

The requirement that A be invertible is necessary. We can find matrices A, B and C such that AB and AC are the same size, A is non-zero and $AB = AC$, but B and C are different. One very easy example is

$$\begin{bmatrix} 1 & 0 \\ 0 & 0 \end{bmatrix}\begin{bmatrix} 1 & -1 \\ 2 & 3 \end{bmatrix} = \begin{bmatrix} 1 & 0 \\ 0 & 0 \end{bmatrix}\begin{bmatrix} 1 & -1 \\ 1 & 4 \end{bmatrix}.$$

Some other examples are given in the exercises.

Moreover we can only cancel on one side of an equation; we cannot mix the two sides. Even if A is invertible it is possible that $AB = CA$ but $B \neq C$ (see Exercises 8.2.21 and 8.2.22).

Exercises 8.2

1. Is it correct to say that, for any matrix A, $AO = O = OA$? Why, or why not?

2. Prove: if A is an invertible $s \times s$ matrix and B and C are $r \times s$ matrices such that $BA = CA$, then $B = C$.

In Exercises 3 *to* 8, *find the products* AB *and* BA. *Do the two matrices commute?*

3. $A = \begin{bmatrix} 0 & 1 \\ 2 & 3 \end{bmatrix}$, $B = \begin{bmatrix} 1 & 2 \\ -1 & 1 \end{bmatrix}$.

4. $A = \begin{bmatrix} 1 & -1 \\ 1 & 1 \end{bmatrix}$, $B = \begin{bmatrix} 2 & 1 \\ -1 & 2 \end{bmatrix}$.

5. $A = \begin{bmatrix} 2 & 0 \\ 1 & -1 \end{bmatrix}$, $B = \begin{bmatrix} -2 & 4 \\ 3 & 2 \end{bmatrix}$.

6. $A = \begin{bmatrix} 1 & 2 \\ 3 & 4 \end{bmatrix}$, $B = \begin{bmatrix} 4 & 3 \\ 2 & 1 \end{bmatrix}$.

7. $A = \begin{bmatrix} 3 & 1 \\ -1 & 3 \end{bmatrix}$, $B = \begin{bmatrix} 1 & 3 \\ -3 & 1 \end{bmatrix}$.

8. $A = \begin{bmatrix} 3 & 1 & -2 \\ 1 & 2 & -1 \\ -1 & -1 & 3 \end{bmatrix}$, $B = \begin{bmatrix} 2 & 0 & 1 \\ -1 & 3 & 0 \\ 0 & 1 & 2 \end{bmatrix}$.

In Exercises 9 *to* 14, A *is given. Find* A^2 *and* A^3.

9. $\begin{bmatrix} 1 & 1 \\ -1 & 1 \end{bmatrix}$

10. $\begin{bmatrix} 2 & -1 \\ -1 & 0 \end{bmatrix}$

11. $\begin{bmatrix} -1 & -1 \\ -1 & -1 \end{bmatrix}$

12. $\begin{bmatrix} 1 & 3 \\ -1 & 1 \end{bmatrix}$

13. $\begin{bmatrix} 2 & 0 & 3 \\ 0 & 3 & -1 \\ -1 & -1 & 1 \end{bmatrix}$

14. $\begin{bmatrix} 1 & 2 & 1 \\ 0 & -1 & 1 \\ 0 & 0 & -2 \end{bmatrix}$

15. Consider the matrix

$$A = \begin{bmatrix} 1 & 3 \\ 5 & 3 \end{bmatrix}.$$

(i) Find A^2 and A^3.

(ii) Evaluate $A^3 - 2A - I$.

(iii) Show that $A^2 - 4A - 12I = O$.

In Exercises 16 *to* 19, *show that the matrices are inverses.*

16. $\begin{bmatrix} 2 & 5 \\ 1 & 3 \end{bmatrix}$ and $\begin{bmatrix} 3 & -5 \\ -1 & 2 \end{bmatrix}$

17. $\begin{bmatrix} \frac{3}{2} & -1 \\ 0 & 1 \end{bmatrix}$ and $\begin{bmatrix} \frac{2}{3} & \frac{2}{3} \\ 0 & 1 \end{bmatrix}$

18. $\begin{bmatrix} \frac{1}{2} & -1 \\ 1 & -1 \end{bmatrix}$ and $\begin{bmatrix} -2 & 2 \\ -2 & 1 \end{bmatrix}$

19. $\begin{bmatrix} 2 & -3 \\ -1 & 2 \end{bmatrix}$ and $\begin{bmatrix} 2 & 3 \\ 1 & 2 \end{bmatrix}$

20. Find a matrix A such that

$$A\begin{bmatrix} 2 & 1 \\ 3 & 2 \end{bmatrix} = \begin{bmatrix} 1 & -1 \\ 1 & 4 \end{bmatrix}.$$

In Exercises 21 *and* 22, *show that* $AB = AC$.

21. $A = \begin{bmatrix} 1 & -1 \\ 2 & -2 \end{bmatrix}$, $B = \begin{bmatrix} 2 & 1 \\ 1 & 4 \end{bmatrix}$, $C = \begin{bmatrix} 4 & 0 \\ 3 & 3 \end{bmatrix}$;

22. $A = \begin{bmatrix} 1 & 0 & 1 \\ 1 & 1 & 1 \\ 1 & 2 & 1 \end{bmatrix}$, $B = \begin{bmatrix} 1 & 1 & 2 \\ 1 & 1 & 0 \\ 1 & 2 & 2 \end{bmatrix}$, $C = \begin{bmatrix} 2 & 2 & 1 \\ 1 & 1 & 0 \\ 0 & 1 & 3 \end{bmatrix}$.

In Exercises 23 *and* 24, *show that* A^{-1} *exists, but* $AB = CA$, *even though* $B \neq C$.

23. $A = \begin{bmatrix} -1 & 3 \\ 1 & -2 \end{bmatrix}$, $B = \begin{bmatrix} 1 & 2 \\ 1 & 1 \end{bmatrix}$, $C = \begin{bmatrix} 5 & 7 \\ -2 & -3 \end{bmatrix}$;

24. $A = \begin{bmatrix} 2 & -1 \\ 1 & 1 \end{bmatrix}$, $B = \begin{bmatrix} 3 & 0 \\ 4 & 1 \end{bmatrix}$, $C = \begin{bmatrix} 1 & 0 \\ 2 & 3 \end{bmatrix}$.

25. The (leading) *diagonal* of a matrix is the set of entries in positions $(1,1)$, $(2,2), (3,3), \ldots$. A *diagonal* matrix is a square matrix with all of its elements zero except those on the diagonal. Prove that any two diagonal matrices of the same order commute.

26. Suppose M_x denotes the 2×2 matrix

$$\begin{bmatrix} 1 & x \\ 0 & 1 \end{bmatrix},$$

where x may be any real number.

(i) Compute M_xM_y, and show that the matrices M_x and M_y commute for any real numbers x and y.

(ii) Find $M_x{}^2$, $M_x{}^3$ and $M_x{}^4$.

(iii) Find a formula for $M_x{}^n$, where n is any positive integer.

(iv) What is $M_x{}^{-1}$?

27. A square matrix A is called *idempotent* if it satisfies $A^2 = A$.

(i) Which of the following matrices are idempotent?

$$\begin{bmatrix} 1 & 0 & 0 \\ 0 & 1 & 0 \\ 0 & 0 & 0 \end{bmatrix}, \begin{bmatrix} 1 & -1 & 0 \\ 0 & 0 & -2 \\ -2 & 1 & 1 \end{bmatrix}, \begin{bmatrix} 2 & -2 & -4 \\ -1 & 3 & 4 \\ 1 & -2 & -3 \end{bmatrix}.$$

(ii) Prove that, if A is an idempotent matrix, then $I-A$ is idempotent.

8.3 Systems of Linear Equations

Matrix representation of equations

The solution set of a system of equations such as

$$\begin{aligned} 2x+4y-4z &= 4 \\ -2y+4z &= 6 \\ x-y+4z &= 10 \end{aligned}$$

is the set of all assignments of values to the variables that make all the equations true. For example, the above system has solution $x=3, y=1, z=2$. We often write the solutions as vectors with the variables taken in standard (alphabetical) order, and would say the solution set is $\{(3,1,2)\}$. Other systems have infinite solution sets; or the set could be empty, in which case we say the system is *inconsistent.*

We introduced matrices in Section 8.1 by pointing out their relationship to sets of linear equations. The above system of equations can be written as

$$\begin{bmatrix} (2,4,-4)\cdot(x,y,z) \\ (0,-2,4)\cdot(x,y,z) \\ (1,-1,4)\cdot(x,y,z) \end{bmatrix} = \begin{bmatrix} 4 \\ 6 \\ 10 \end{bmatrix},$$

which can be seen as the matrix

$$\begin{bmatrix} 2 & 4 & -4 \\ 0 & -2 & 4 \\ 1 & -1 & 4 \end{bmatrix}$$

acting on the vector (x,y,z) to produce the vector $(4,6,10)$.

The typical set of m linear equations in n unknowns can be written

$$A\boldsymbol{x} = \boldsymbol{b}$$

where A is an $m \times n$ matrix of coefficients, $\boldsymbol{x}$ is an $n \times 1$ matrix (column vector) of unknowns, and $\boldsymbol{b}$ is the $m \times 1$ matrix (column vector) of right-hand sides of the equations.

The usual way of solving such a system of equations is to use one equation to express one variable in terms of the other variables and substitute for that variable in other equations. Repeat this process in the new set of $m-1$ equations. Continue until only one equation remains. This equation is used to evaluate one variable (either as a constant, or in terms of those other variables not yet eliminated), and the remaining variables are evaluated from the equations used earlier. If this process fails (for example, if two contradictory equations result), the equations were *inconsistent*, and there are no solutions.

We shall formalize this process. We define the *augmented matrix* of the system $Ax = \boldsymbol{b}$ to be the matrix $[A \mid \boldsymbol{b}]$ formed by adjoining the vector of constants $\boldsymbol{b}$ to the matrix A of coefficients. The vertical line indicates the division between the two types of element. The augmented matrix of the system given above is

$$\left[\begin{array}{rrr|r} 2 & 4 & -3 & 4 \\ 0 & -2 & 4 & 6 \\ 1 & -1 & 4 & 10 \end{array}\right].$$

The first step is to select a variable to eliminate. This is equivalent to choosing a column in the augmented matrix and selecting a row — an equation — to use for substitution. The only requirement is that the matrix has a non-zero entry in that row and column. Let us choose row 3, column 1, representing variable x in the first equation. It will be convenient to interchange rows 1 and 3, so that we are operating on the $(1,1)$ entry. This is equivalent to rewriting the equations in a different order. The matrix is now

$$\left[\begin{array}{rrr|r} 1 & -1 & 4 & 10 \\ 0 & -2 & 4 & 6 \\ 2 & 4 & -3 & 4 \end{array}\right] \quad \begin{array}{l} R1 \leftarrow R3 \\ \\ R3 \leftarrow R1, \end{array}$$

where the annotations mean *the new row 1 is the old row 3* and *the new row 3 is the old row 1*.

Now substitute for x in the other equations. No action is required in the second equation, but x must be eliminated from the third. So we subtract twice the first row from the third row. This yields precisely the equation we would get if we used equation 1 to substitute for x in equation 3, but for consistency we have kept all the variables on the left-hand side of the equation. The augmented matrix becomes

$$\left[\begin{array}{rrr|r} 1 & -1 & 4 & 10 \\ 0 & -2 & 4 & 6 \\ 0 & 6 & -11 & -16 \end{array}\right] \quad \begin{array}{l} \\ \\ R3 \leftarrow R3 - 2R1, \end{array}$$

where the legend means *the new row* 3 *is* (*the old row* 3) -2(*the old row* 1). (When we say *old* we are referring to the preceding augmented matrix, not to the original one.)

Now multiply row 2 by $-\frac{1}{2}$. Then eliminate the $6y$ from the third equation. The result is

$$\left[\begin{array}{rrr|r} 1 & -1 & 4 & 10 \\ 0 & 1 & -2 & -3 \\ 0 & 0 & 1 & 2 \end{array}\right] \quad \begin{array}{l} \\ R2 \leftarrow \frac{1}{2}R2 \\ R3 \leftarrow R3 - 6(\frac{1}{2}R2). \end{array}$$

This could have been broken into two steps.

So far we have done the equivalent of substituting in the later equations. Now we substitute back to find the values. We know from the third equation that $z = 3$. To substitute this in the earlier equations, we add twice row 3 to row 2 and subtract four times row 3 from row 1:

$$\left[\begin{array}{ccc|c} 1 & -1 & 0 & 2 \\ 0 & 1 & 0 & 1 \\ 0 & 0 & 1 & 2 \end{array}\right] \quad \begin{array}{l} R1 \leftarrow R1 - 4R3 \\ R2 \leftarrow R2 + 2R3 \end{array} .$$

Next add row 2 to row 1:

$$\left[\begin{array}{ccc|c} 1 & 0 & 0 & 3 \\ 0 & 1 & 0 & 1 \\ 0 & 0 & 1 & 2 \end{array}\right] \quad R1 \leftarrow R1 + R2 .$$

The resulting array can be translated into the equations

$$x = 3, \ y = 1, \ z = 2.$$

Elementary operations

In our example, we used three operations:

E1: *permute the rows of the matrix;*

E2: *multiply a row by a (non-zero) constant;*

E3: *add a multiple of one row to another row.*

We shall call them *elementary row operations*. Their importance comes from the following theorem.

Theorem 8.4 *Suppose P is the augmented matrix of a system of linear equations, and Q is obtained from P by a sequence of elementary row operations. Then the system of equations corresponding to Q has the same solutions as the system corresponding to P.*

Proof. Operation E1 does not change the system.

The equations $\boldsymbol{a} \cdot \boldsymbol{x} = b$ and $k\boldsymbol{a} \cdot \boldsymbol{x} = kb$ have the same solutions when k is non-zero. So E2 does not change the solution set.

Finally, suppose x_0 is a solution of the system of two equations $\boldsymbol{a} \cdot \boldsymbol{x} = b, \boldsymbol{a}' \cdot \boldsymbol{x} = b'$. Then $\boldsymbol{a} \cdot \ x_0 = b$ and $\boldsymbol{a}' \cdot \boldsymbol{x}_0 = b'$, so $(\boldsymbol{a} + k\boldsymbol{a}') \cdot \boldsymbol{x}_0 = \boldsymbol{b} + kb'$, and x_0 is a solution of the system $\boldsymbol{a} \cdot \boldsymbol{x} = b, \boldsymbol{a} + k\boldsymbol{a}' \cdot \boldsymbol{x} = \boldsymbol{b} + kb'$; conversely any solution of the second system is a solution of the first set. If the same further set of equations is appended to each system, the two resulting systems still have the same solutions, so E3 does not change the solution set. □

The solution algorithm

It is clear that repeated application of the three steps to the augmented matrix will provide a solution. The algorithm for solving systems of linear equations works as follows.

Stage 1.

1. Find the leftmost column in the matrix of coefficients that contains a non-zero element, say column j. Use E1 to make the row containing this element into the first row, and E2 to convert its leftmost non-zero element to 1. This is called a *leading* 1. Then use E3 to change all entries below the leading 1 to zero. That is, if the (i, j) entry is a_{ij}, then subtract $a_{ij} \times$ (row 1) from row j.

At this stage we say column j is *processed*. Processed rows are not disturbed in the first stage.

2. Find the leftmost unprocessed column in the augmented matrix that contains a non-zero element, say column k. Use E1 to make this row the first row under the processed row(s), and E2 to convert its leftmost non-zero element to 1, another leading 1. Use E3 to change all entries below the leading 1 to zero (but do not change the processed row or rows). Now column k is also processed.

3. If you have not either reached the last column of coefficients (the vertical line) or the bottom of the matrix, go back to step 2, make another leading 1 and proceed from there.

Stage 2.

4. Choose the bottom-most leading 1 and eliminate all elements above it in its column by use of E3. Do the same to the next leading 1 up, then the next, until you reach the top.

The process is now finished. We shall illustrate the interpretation of the results with an example.

Sample Problem 8.10 *Solve the system*

$$\begin{aligned} 2x + 2y + 4z &= 0 \\ 3x - \ y + 2z &= 1 \\ 8x \qquad + 8z &= 2 \end{aligned}$$

by row operations.

Solution. The augmented matrix is

$$\left[\begin{array}{ccc|c} 2 & 2 & 4 & 0 \\ 3 & -1 & 2 & 1 \\ 8 & 0 & 8 & 2 \end{array}\right].$$

At step 1 we choose the element in the $(1, 1)$ position and divide row 1 by 2:

$$\left[\begin{array}{rrr|r} 1 & 1 & 2 & 0 \\ 3 & -1 & 2 & 1 \\ 8 & 0 & 8 & 2 \end{array}\right] \quad R1 \leftarrow \tfrac{1}{2}R1 \text{ (using E2)}.$$

Then we eliminate the rest of column 1:

$$\left[\begin{array}{rrr|r} 1 & 1 & 2 & 0 \\ 0 & -4 & -4 & 1 \\ 0 & -8 & -8 & 2 \end{array}\right] \quad \begin{array}{l} R2 \leftarrow R2 - 3 \times R1 \text{ (using E3)} \\ R3 \leftarrow R3 - 8 \times R1 \text{ (using E3)} \end{array}.$$

In step 2 we choose the $(2,2)$ position and divide by -4, then eliminate the entries below the $(2,2)$ position, obtaining successively

$$\left[\begin{array}{rrr|r} 1 & 1 & 2 & 0 \\ 0 & 1 & 1 & -\frac{1}{4} \\ 0 & -8 & -8 & 2 \end{array}\right] \quad R2 \leftarrow -\tfrac{1}{4} \times R2 \text{ (using E2)}$$

$$\left[\begin{array}{rrr|r} 1 & 1 & 2 & 0 \\ 0 & 1 & 1 & -\frac{1}{4} \\ 0 & 0 & 0 & 0 \end{array}\right] \quad R3 \leftarrow R3 - 8 \times R2 \text{ (using E3)}.$$

There are no further numbers available for leading 1's, so we move to step 4. We use the $(2,2)$ element:

$$\left[\begin{array}{rrr|r} 1 & 0 & 1 & \frac{1}{4} \\ 0 & 1 & 1 & -\frac{1}{4} \\ 0 & 0 & 0 & 0 \end{array}\right] \quad R1 \leftarrow R1 - R2 \text{ (using E3)}.$$

The process is finished. There is no restriction on z. The final augmented matrix converts to the system

$$\begin{array}{rcr} x \quad + z & = & \frac{1}{4} \\ y + z & = & -\frac{1}{4} \end{array}$$

(the third equation can be ignored) and the final solution could be expressed as

$$x = \tfrac{1}{4} - z, \; y = -\tfrac{1}{4} - z, \text{ any real number } z.$$

It is important to notice that the sequence of calculations is completely determined by the matrix of coefficients, the left-hand part of the augmented matrix.

If the column corresponding to a variable receives a leading 1, we shall call that variable *dependent*; the others are *independent*. One standard way of recording the answer is to give an equation for each dependent variable, with a constant and the independent variables on the right; the independent variables take any real number value. Another way to express the above solution would be to use set notation $\{\frac{1}{4} - z, -\frac{1}{4} - z, z) \mid z \in \mathbb{R}\}$, or perhaps $\{t + \frac{1}{4}, t, -\frac{1}{4}, -t) \mid t \in \mathbb{R}\}$. In this case t is a *parameter*.

Sometimes there will be no solution to a system of equations. As we said above, the equations are then *inconsistent*.

Sample Problem 8.11 *Solve the system*

$$\begin{aligned} 2x+2y+4z &= 0 \\ 3x-\ y+2z &= 1 \\ 8x \qquad +8z &= 3. \end{aligned}$$

Solution. The augmented matrix is

$$\left[\begin{array}{rrr|r} 2 & 2 & 4 & 0 \\ 3 & -1 & 2 & 1 \\ 8 & 0 & 8 & 3 \end{array}\right].$$

The left-hand part of this equation is the same as in Sample Problem 8.10, so we go through the same steps, making the appropriate changes to the right-hand column. At the end of Stage 1, we have

$$\left[\begin{array}{rrr|r} 1 & 1 & 2 & 0 \\ 0 & 1 & 1 & -\frac{1}{4} \\ 0 & 0 & 0 & 1 \end{array}\right].$$

When we convert back to equations, the third row gives the equation

$$0 = 1,$$

which is impossible. No values of x, y and z make this true, so the equations are inconsistent. There is no need to implement Stage 2.

In set-theoretic terms, we could report that the solution set is $\emptyset$.

Exercises 8.3

In Exercises 1 *to* 14, *the augmented matrix of a system of equations is shown. Assuming the variables are* x, y, z, *what is the solution of the system?*

1. $\left[\begin{array}{rrr|r} 1 & 0 & 0 & 3 \\ 0 & 1 & 0 & 1 \\ 0 & 0 & 0 & 0 \end{array}\right]$

2. $\left[\begin{array}{rrr|r} 1 & 0 & 0 & -1 \\ 0 & 1 & 0 & -1 \\ 0 & 0 & 1 & -1 \end{array}\right]$

3. $\left[\begin{array}{rrr|r} 1 & 0 & 1 & 2 \\ 0 & 1 & -1 & 1 \\ 0 & 0 & 0 & 0 \end{array}\right]$

4. $\left[\begin{array}{rrr|r} 1 & 1 & 0 & 2 \\ 0 & 0 & 1 & -1 \\ 0 & 0 & 0 & 0 \\ 0 & 0 & 0 & 0 \end{array}\right]$

5. $\left[\begin{array}{rrr|r} 1 & 0 & 1 & 2 \\ 0 & 1 & 2 & 1 \\ 0 & 0 & 0 & 2 \end{array}\right]$

6. $\left[\begin{array}{rrr|r} 1 & 0 & 0 & 3 \\ 0 & 1 & 0 & 1 \\ 0 & 0 & 1 & 2 \end{array}\right]$

7. $\left[\begin{array}{ccc|c} 1 & 0 & 1 & 2 \\ 0 & 1 & 1 & 1 \\ 0 & 0 & 0 & 1 \end{array}\right]$

8. $\left[\begin{array}{ccc|c} 1 & 0 & 0 & 2 \\ 0 & 1 & 0 & 1 \\ 0 & 0 & 1 & 1 \\ 0 & 0 & 0 & 1 \end{array}\right]$

9. $\left[\begin{array}{ccc|c} 1 & 1 & 1 & 4 \\ -1 & 2 & 1 & -1 \\ 3 & 2 & 4 & 6 \end{array}\right]$

10. $\left[\begin{array}{ccc|c} 4 & 2 & -3 & 1 \\ 3 & -1 & -1 & 1 \\ 1 & -7 & 3 & 1 \end{array}\right]$

11. $\left[\begin{array}{ccc|c} 4 & 3 & 1 & 11 \\ 2 & -2 & 4 & 2 \\ 1 & 3 & -2 & 5 \end{array}\right]$

12. $\left[\begin{array}{ccc|c} 0 & 1 & -1 & -1 \\ 1 & 0 & 1 & 1 \\ 1 & 2 & -1 & 0 \end{array}\right]$

13. $\left[\begin{array}{ccc|c} 2 & -1 & 1 & -2 \\ 3 & 2 & 3 & 8 \\ 1 & -1 & -1 & 0 \end{array}\right]$

14. $\left[\begin{array}{ccc|c} 2 & 3 & 1 & -1 \\ 1 & 2 & 1 & 0 \\ 3 & 2 & -1 & -4 \end{array}\right]$

In Exercises 15 *to* 36, *solve the system of equations.*

15. $\begin{aligned} 2x+6y &= 6 \\ 4x+11y &= 10 \end{aligned}$

16. $\begin{aligned} 3x-2y &= 4 \\ -6x+4y &= 2 \end{aligned}$

17. $\begin{aligned} 2x+3y &= 5 \\ 4x+6y &= 10 \end{aligned}$

18. $\begin{aligned} x-y &= 3 \\ 2x+y &= 3 \end{aligned}$

19. $\begin{aligned} 3x-y &= 4 \\ 6x-2y &= 2 \end{aligned}$

20. $\begin{aligned} 3x-2y &= -1 \\ -6x+4y &= 2 \end{aligned}$

21. $\begin{aligned} x-2y &= 4 \\ -3x-4y &= -2 \\ 2x+3y &= 1 \end{aligned}$

22. $\begin{aligned} 3x+2y &= 4 \\ 2x+3y &= 1 \\ 5x-4y &= 14 \end{aligned}$

23. $\begin{aligned} x+2y+z &= 3 \\ x+y-2z &= 2 \end{aligned}$

24. $\begin{aligned} x+y+3z &= 2 \\ 4x+2y+2z &= 10 \end{aligned}$

25. $\begin{aligned} x+y+z &= 3 \\ x+2y+2z &= 3 \\ x+y+2z &= 1 \end{aligned}$

26. $\begin{aligned} 2x+4z &= 6 \\ 2x+y+5z &= 7 \\ x-y+z &= 2 \end{aligned}$

27.
$$\begin{aligned} 2x+2z &= 2 \\ x+2y+6z &= 3 \\ 2x-2y &= 1 \end{aligned}$$

28.
$$\begin{aligned} x+z &= 4 \\ x+4y+z &= 7 \\ x-2y+z &= 3 \end{aligned}$$

29.
$$\begin{aligned} x+2y+z &= -1 \\ 2x+3y-2z &= 7 \\ -2x+2y-3z &= -2 \end{aligned}$$

30.
$$\begin{aligned} x+y-z &= 4 \\ 3x+4y-7z &= 8 \\ -y+4z &= 4 \end{aligned}$$

31.
$$\begin{aligned} x+y &= 2 \\ x-y+5z &= 3 \\ -3x-3y+2z &= -6 \end{aligned}$$

32.
$$\begin{aligned} x+2y+3z &= 4 \\ 4x+5y+6z &= 16 \\ 7x+8y+9z &= 28 \end{aligned}$$

33.
$$\begin{aligned} 3x-2y-8z+7t &= 1 \\ x+y-z-t &= 3 \\ x-y-3z+3t &= -1 \end{aligned}$$

34.
$$\begin{aligned} 2x-y+z-3t &= 2 \\ -4x-3y+t &= 1 \\ 2x-6y+3z-8t &= 4 \end{aligned}$$

35.
$$\begin{aligned} x+2y+3z+4t &= 8 \\ x-3y+4z+4t &= 8 \\ 2x-2y-z+t &= -3 \\ x-7y-7z-3t &= -11 \end{aligned}$$

36.
$$\begin{aligned} 2x+2y-2z+3t &= 2 \\ 4x-2y-z+t &= -4 \\ 6x-3z+4t &= -2 \\ 2x+8y-5z+8t &= 10 \end{aligned}$$

8.4 More About Linear Systems and Inverses

Classification of systems of equations

We say that a matrix is in *reduced row echelon form* if it has the following properties:

- All rows with every element zero lie at the bottom of the matrix;
- The leftmost non-zero element of each row is a 1 (called a *leading* 1);
- Every other entry in the same column as a leading 1 is zero;
- Each leading 1 is to the right of the leading 1 in the preceding row.

So the process of solving a system of equations consists of converting its augmented matrix to reduced row echelon form. For this reason we often speak of *reducing* the matrix (or reducing the system).

The system is inconsistent when the reduction process results in a leading 1 in the right-hand column of the augmented matrix. In this case there is no need to complete the algorithm.

Suppose the system is consistent. There are two possibilities.

Sometimes the reduced matrix will have a leading 1 in each column, so that it is an identity matrix, possibly with some zero rows appended. In this case there will be exactly one solution.

Alternatively, there will be some columns that contain no leading 1. The corresponding variables are independent, and can take any real value, so the system has infinitely many solutions.

We have a classification of systems of equations into three classes. A system can have no solutions, exactly one solution, or infinitely many solutions. Alternatively, we could say that the solution set is empty, a singleton set, or infinite. There can be no system whose finite solution set is finite but has more than one member.

Calculating the inverse

The inverse of a square matrix can be calculated by the algorithm used for solving equations. Suppose A is an $n \times n$ matrix with inverse B. Write $\boldsymbol{b}_1, \boldsymbol{b}_2, \ldots, \boldsymbol{b}_n$ for the columns of B, and write $\boldsymbol{u}_1, \boldsymbol{u}_2, \ldots, \boldsymbol{u}_n$ for the columns of the identity matrix of order n. Consider the equation $AB = I$. Column j of the left-hand side is $A\boldsymbol{b}_j$. So the equation is equivalent to the set of n systems

$$A\boldsymbol{b}_1 = \boldsymbol{u}_1, A\boldsymbol{b}_2 = \boldsymbol{u}_2, \ldots, A\boldsymbol{b}_n = \boldsymbol{u}_n.$$

If all these systems have solutions, then the inverse is formed by putting the solution vectors next to each other in order. If any system has no solution, there is no inverse.

To solve $A\boldsymbol{b}_j = \boldsymbol{u}_j$, we reduce the augmented matrix $[A \mid \boldsymbol{u}_j]$ to reduced row echelon form. The same steps will produce this result, no matter what vector is on the right-hand side. It follows that we can carry out the reduction simultaneously for all n systems of equations. So we have the following technique for inverting an $n \times n$ matrix A.

Row reduce the matrix $[A \mid I_n]$. If the resulting matrix has form $[I_n \mid B]$ then A is invertible, and B is A^{-1}. Otherwise, A is singular.

It follows from this that if a matrix has a row with every entry zero, it must be singular. This is also true if the matrix has a column with every entry zero. For example, if A has every entry of its first column zero, then BA has every element zero in *its* first column for any choice of B, so the equation $A^{-1}A = I$ cannot possibly be true — it must fail in the $(1,1)$ position.

Sample Problem 8.12 *For the following matrices, find the inverse or show that the matrix is singular:*

$$A = \begin{bmatrix} 1 & 2 & -1 \\ 0 & -2 & 1 \\ 1 & 1 & -1 \end{bmatrix}, \quad B = \begin{bmatrix} 2 & 3 & 2 \\ 3 & -1 & 2 \\ 1 & 7 & 2 \end{bmatrix}.$$

Solution. For A, we have

$$\left[\begin{array}{rrr|rrr} 1 & 2 & -1 & 1 & 0 & 0 \\ 0 & -2 & 1 & 0 & 1 & 0 \\ 1 & 1 & -1 & 0 & 0 & 1 \end{array}\right] \Rightarrow \left[\begin{array}{rrr|rrr} 1 & 2 & -1 & 1 & 0 & 0 \\ 0 & -2 & 1 & 0 & 1 & 0 \\ 0 & -1 & 0 & -1 & 0 & 1 \end{array}\right]$$

$$\Rightarrow \left[\begin{array}{rrr|rrr} 1 & 2 & -1 & 1 & 0 & 0 \\ 0 & 1 & 0 & 1 & 0 & -1 \\ 0 & -2 & 1 & 0 & 1 & 0 \end{array}\right] \Rightarrow \left[\begin{array}{rrr|rrr} 1 & 0 & -1 & -1 & 0 & 2 \\ 0 & 1 & 0 & 1 & 0 & -1 \\ 0 & 0 & 1 & 2 & 1 & -2 \end{array}\right]$$

$$\Rightarrow \left[\begin{array}{rrr|rrr} 1 & 0 & 0 & 1 & 1 & 0 \\ 0 & 1 & 0 & 1 & 0 & -1 \\ 0 & 0 & 1 & 2 & 1 & -2 \end{array}\right].$$

So A has inverse

$$\begin{bmatrix} 1 & 1 & 0 \\ 1 & 0 & -1 \\ 2 & 1 & -2 \end{bmatrix}.$$

For B, we get

$$\left[\begin{array}{rrr|rrr} 2 & 3 & 2 & 1 & 0 & 0 \\ 3 & -1 & 2 & 0 & 1 & 0 \\ 1 & 7 & 2 & 0 & 0 & 1 \end{array}\right] \Rightarrow \left[\begin{array}{rrr|rrr} 1 & 7 & 2 & 0 & 0 & 1 \\ 2 & 3 & 2 & 0 & 1 & 0 \\ 3 & -1 & 2 & 1 & 0 & 0 \end{array}\right]$$

$$\Rightarrow \left[\begin{array}{rrr|rrr} 1 & 7 & 2 & 0 & 0 & 1 \\ 0 & -11 & -2 & 0 & 1 & -2 \\ 0 & -22 & -4 & 1 & 0 & -3 \end{array}\right] \Rightarrow \left[\begin{array}{rrr|rrr} 1 & 7 & 2 & 0 & 0 & 1 \\ 0 & -11 & -2 & 0 & 1 & -2 \\ 0 & 0 & 0 & 1 & -2 & 1 \end{array}\right]$$

and the zero row on the left tells us that B is singular.

Practice Exercise. In the above calculations, identify the steps that have been taken at each stage. (For example, for A, the first step was

$$R3 \leftarrow R3 - R1.)$$

This method can be used to get a general solution for the inverse of a 2×2 matrix.

Theorem 8.5 *The matrix*

$$A = \begin{bmatrix} a & b \\ c & d \end{bmatrix}$$

is singular if $ad - bc = 0$. Otherwise it is invertible, with inverse

$$\frac{1}{(ad-bc)} \begin{bmatrix} d & -b \\ -c & a \end{bmatrix}. \tag{8.1}$$

Proof. If $a = c = 0$, then $ad - bc = 0$, and A has no inverse.

First, suppose a and c are both non-zero. The inverse procedure is

$$\left[\begin{array}{cc|cc} a & b & 1 & 0 \\ c & d & 0 & 1 \end{array}\right]$$

$$\Rightarrow \left[\begin{array}{cc|cc} ac & bc & c & 0 \\ ac & ad & 0 & a \end{array}\right] \quad \begin{array}{l} R1 \leftarrow cR1 \\ R2 \leftarrow aR2 \end{array}$$

$$\Rightarrow \left[\begin{array}{cc|cc} ac & bc & c & 0 \\ 0 & ad-bc & -c & a \end{array}\right] \quad R2 \leftarrow R2 - R1$$

$$\Rightarrow \left[\begin{array}{cc|cc} ac & 0 & (1+\frac{bc}{ad-bc})c & -\frac{abc}{ad-bc} \\ 0 & ad-bc & -c & a \end{array}\right] \quad R1 \leftarrow R1 - \tfrac{bc}{ad-bc}R2.$$

If $ad - bc = 0$, then we are finished and there is no inverse. Otherwise, notice that

$$1 + \frac{bc}{ad-bc} = \frac{ad-bc+bc}{ad-bc} = \frac{ad}{ad-bc},$$

so we have

$$\left[\begin{array}{cc|cc} ac & 0 & \frac{acd}{ad-bc} & -\frac{abc}{ad-bc} \\ 0 & ad-bc & -c & a \end{array}\right]$$

$$\Rightarrow \left[\begin{array}{cc|cc} 1 & 0 & \frac{d}{ad-bc} & -\frac{b}{ad-bc} \\ 0 & 1 & -\frac{c}{ad-bc} & \frac{a}{ad-bc} \end{array}\right] \quad \begin{array}{l} R1 \leftarrow \frac{1}{ac}R1 \\ R1 \leftarrow \frac{1}{ad-bc}R2 \end{array}$$

as required.

If $a \neq 0$ and $c = 0$ the calculations are simpler. We obtain the inverse

$$\begin{bmatrix} \frac{1}{a} & -\frac{b}{ad} \\ 0 & \frac{1}{d} \end{bmatrix}$$

and this is the form taken by (8.1) when $c = 0$.

The case where $a = 0, c \neq 0$ is left as an exercise. □

The number $ad - bc$ is called the *determinant* of the matrix A, written $\det(A)$. Determinants may be defined for square matrices of any order, and it is a general theorem that a matrix is invertible if and only if its determinant is non-zero. For more details, see books on linear algebra.

Sample Problem 8.13 *Find the determinants of the following matrices and use them to find their inverses, if possible:*

$$A = \begin{bmatrix} 3 & 1 \\ 2 & 2 \end{bmatrix}, \quad B = \begin{bmatrix} 2 & -2 \\ -1 & 1 \end{bmatrix}.$$

Solution. $\det(A) = 3 \cdot 2 - 1 \cdot 2 = 4$, so

$$A^{-1} = \frac{1}{4} \begin{bmatrix} 2 & -1 \\ -2 & 3 \end{bmatrix} = \begin{bmatrix} \frac{1}{2} & -\frac{1}{4} \\ -\frac{1}{2} & \frac{3}{4} \end{bmatrix}.$$

$\det(B) = 2 \cdot 1 - (-2) \cdot (-1) = 0$, so B has no inverse.

Practice Exercise. Repeat for

$$C = \begin{bmatrix} 1 & 1 \\ 3 & 4 \end{bmatrix}, \quad D = \begin{bmatrix} 2 & 4 \\ 1 & 2 \end{bmatrix}.$$

Using the inverse

Consider the system of equations $A\boldsymbol{x} = \boldsymbol{b}$, where A is an invertible matrix. Multiplying by A^{-1}, $\boldsymbol{x} = A^{-1}A\boldsymbol{x} = A^{-1}\boldsymbol{b}$, so the equations have the unique solution $A^{-1}\boldsymbol{b}$. This could be used to solve the equations. This technique is not usually helpful in practical situations, because the process of finding the inverse takes at least as long as solving the equations, but it is useful when there are several sets of equations with the same left-hand sides, or when the inverse is already known. It is also important in theoretical studies.

Sample Problem 8.14 *Solve the systems:*

$$\begin{array}{rcrcrcr} x & + & 2y & - & z & = & 3 \\ & - & 2y & + & z & = & 1 \\ x & + & y & - & z & = & 0 \end{array} \qquad \begin{array}{rcrcrcr} x & + & 2y & - & z & = & -1 \\ & - & 2y & + & z & = & -2 \\ x & + & y & - & z & = & 2 \end{array}$$

Solution. We saw in Sample Problem 8.12 that the matrix of coefficients has inverse

$$\begin{bmatrix} 1 & 1 & 0 \\ 1 & 0 & -1 \\ 2 & 1 & -2 \end{bmatrix}.$$

Now

$$\begin{bmatrix} 1 & 1 & 0 \\ 1 & 0 & -1 \\ 2 & 1 & -2 \end{bmatrix} \begin{bmatrix} 3 \\ 1 \\ 0 \end{bmatrix} = \begin{bmatrix} 4 \\ 3 \\ 7 \end{bmatrix}$$

so the first system has solution $x = 4, y = 3, z = 7$.

$$\begin{bmatrix} 1 & 1 & 0 \\ 1 & 0 & -1 \\ 2 & 1 & -2 \end{bmatrix} \begin{bmatrix} -1 \\ -2 \\ 2 \end{bmatrix} = \begin{bmatrix} -3 \\ -3 \\ -8 \end{bmatrix}$$

and the second has solution $x = -3, y = -3, z = -8$.

Practice Exercise. Solve the systems:

$$\begin{array}{rcrcrcr} x & + & 2y & - & z & = & 2 \\ & - & 2y & + & z & = & 2 \\ x & + & y & - & z & = & 1 \end{array} \qquad \begin{array}{rcrcrcr} x & + & 2y & - & z & = & 4 \\ & - & 2y & + & z & = & -1 \\ x & + & y & - & z & = & 3 \end{array}$$

Exercises 8.4

In Exercises 1 to 6, the reduced row echelon form of an augmented system is shown. Say whether its solution set is empty, singleton or infinite.

1. $\left[\begin{array}{ccc|c} 1 & 0 & 3 & 1 \\ 0 & 1 & 1 & 1 \\ 0 & 0 & 0 & 1 \end{array}\right]$

2. $\left[\begin{array}{ccc|c} 1 & 0 & 0 & 2 \\ 0 & 1 & 0 & 1 \\ 0 & 0 & 1 & -1 \end{array}\right]$

3. $\left[\begin{array}{ccc|c} 1 & 0 & 0 & 1 \\ 0 & 1 & 0 & 2 \\ 0 & 0 & 0 & 0 \end{array}\right]$

4. $\left[\begin{array}{ccc|c} 1 & 1 & 3 & 1 \\ 0 & 0 & 0 & 2 \\ 0 & 0 & 0 & 0 \end{array}\right]$

5. $\left[\begin{array}{ccc|c} 1 & 0 & 0 & 0 \\ 0 & 1 & 0 & 0 \\ 0 & 0 & 1 & 0 \end{array}\right]$

6. $\left[\begin{array}{ccc|c} 1 & 0 & 2 & 3 \\ 0 & 1 & 1 & 3 \\ 0 & 0 & 0 & 0 \end{array}\right]$

In Exercises 7 to 24, use row reduction either to find the inverse of the given matrix or to show that the matrix is singular.

7. $\begin{bmatrix} 2 & -2 \\ 4 & 0 \end{bmatrix}$

8. $\begin{bmatrix} 2 & -2 \\ 3 & -3 \end{bmatrix}$

9. $\begin{bmatrix} 3 & 7 \\ 2 & 5 \end{bmatrix}$

10. $\begin{bmatrix} 1 & 1 \\ 1 & .5 \end{bmatrix}$

11. $\begin{bmatrix} 4 & 2 \\ 2 & 1 \end{bmatrix}$

12. $\begin{bmatrix} 3 & 1 \\ -4 & -2 \end{bmatrix}$

13. $\begin{bmatrix} 0 & 0 & 1 \\ 0 & 1 & 0 \\ 1 & 0 & 1 \end{bmatrix}$

14. $\begin{bmatrix} 0 & 1 & 0 \\ 0 & 0 & 1 \\ 1 & 0 & 0 \end{bmatrix}$

15. $\begin{bmatrix} 2 & 3 & 2 \\ 1 & 2 & 2 \\ 3 & 1 & 3 \end{bmatrix}$

16. $\begin{bmatrix} 9 & 8 & 7 \\ 6 & 5 & 4 \\ 3 & 2 & 1 \end{bmatrix}$

17. $\begin{bmatrix} 0 & 1 & 1 \\ 5 & 1 & -2 \\ 2 & -3 & -3 \end{bmatrix}$ **18.** $\begin{bmatrix} 1 & -2 & 3 \\ 3 & 5 & 1 \\ 6 & 4 & 2 \end{bmatrix}$

19. $\begin{bmatrix} 1 & 3 & 3 \\ 1 & 4 & 3 \\ 1 & 3 & 4 \end{bmatrix}$ **20.** $\begin{bmatrix} 2 & 3 & -1 \\ 4 & 2 & 3 \\ 2 & 7 & -6 \end{bmatrix}$

21. $\begin{bmatrix} 1 & 1 & -1 \\ -2 & -1 & 7 \\ 3 & 2 & -8 \end{bmatrix}$ **22.** $\begin{bmatrix} 1 & 0 & 2 \\ 2 & -1 & 3 \\ 4 & 1 & 8 \end{bmatrix}$

23. $\begin{bmatrix} 1 & 2 & 3 \\ 2 & 5 & 1 \\ 3 & 7 & 4 \end{bmatrix}$ **24.** $\begin{bmatrix} 1 & 2 & 3 \\ 2 & 5 & 5 \\ 3 & 8 & 8 \end{bmatrix}$

In Exercises 25 *to* 30, *find the determinant, and use it to invert the matrix or show that it is singular.*

25. $\begin{bmatrix} 7 & 4 \\ 2 & 3 \end{bmatrix}$ **26.** $\begin{bmatrix} 6 & 4 \\ 3 & 2 \end{bmatrix}$

27. $\begin{bmatrix} -2 & 2 \\ 2 & 3 \end{bmatrix}$ **28.** $\begin{bmatrix} 3 & 5 \\ 1 & 4 \end{bmatrix}$

29. $\begin{bmatrix} 4 & 2 \\ 8 & 4 \end{bmatrix}$ **30.** $\begin{bmatrix} 6 & 4 \\ 7 & 5 \end{bmatrix}$

31. (i) Prove that the following matrices are inverses.

$$\begin{bmatrix} 3 & 2 \\ 1 & 1 \end{bmatrix} \quad \begin{bmatrix} 1 & -2 \\ -1 & 3 \end{bmatrix}$$

(ii) Use part (i) to solve the following systems.

$$\text{(a)}\ \begin{aligned} 3x + 2y &= 4 \\ x + y &= 1 \end{aligned} \qquad \text{(b)}\ \begin{aligned} x - 2y &= -1 \\ -x + 3y &= 2 \end{aligned}$$

32. (i) Prove that the following matrices are inverses.

$$\begin{bmatrix} 2 & -2 & 1 \\ 1 & 0 & 1 \\ 1 & -3 & 0 \end{bmatrix} \quad \begin{bmatrix} 3 & -3 & -2 \\ 1 & -1 & -1 \\ -3 & 4 & 2 \end{bmatrix}.$$

(ii) Use part (i) to solve the following systems.

$$\text{(a)}\ \begin{aligned} 2x - 2y + z &= 3 \\ x + z &= 2 \\ x - 3y &= 1 \end{aligned} \qquad \text{(b)}\ \begin{aligned} 3x - 3y - 2z &= 2 \\ x - y - z &= -1 \\ -3x + 4y + 2z &= 2 \end{aligned}$$

33. Suppose a system of equations has 0 for every right-hand side. Can it be inconsistent?

34. Prove Theorem 8.5 in the case where $a = 0$ (and $c \neq 0$).

8.5 Adjacency Matrices

Representing a relation by a matrix

Suppose $A = \{a_1, \ldots, a_m\}$ and $B = \{b_1, \ldots, b_n\}$ are two finite sets, and let α be a relation from A to B. We define the *adjacency matrix* of α to be the $m \times n$ matrix M_α with (i, j) entry

$$\alpha_{ij} = \begin{cases} 1 \text{ if } a_i \alpha b_j, \\ 0 \text{ otherwise.} \end{cases}$$

Obviously this definition depends on the order in which we take the elements of the two sets. If you change the order of the elements of A, this will reorder the rows of M_α, while changing the order of the elements of B will reorder the columns of the matrix.

Sample Problem 8.15 *Suppose $A = \{1,2\}$, $B = \{1,2,3\}$ and α is the relation from A to B defined by*

$$\alpha = \{(1,1),(1,3),(2,2),(2,3)\}.$$

What is the adjacency matrix of α?

Solution.

$$M_\alpha = \begin{bmatrix} 1 & 0 & 1 \\ 0 & 1 & 1 \end{bmatrix}.$$

Practice Exercise. For the same sets A and B, what is the adjacency matrix of the relation

$$\pi = \{(1,2),(1,3),(2,1),(2,2)\}?$$

Conjunction of adjacency matrices

Again suppose $A = \{a_1, \ldots, a_m\}$ and $B = \{b_1, \ldots, b_n\}$, and consider also a third set $C = \{c_1, c_2, \ldots, c_p\}$. Suppose α is a relation from A to B and β is a relation from B to C. Then M_β is an $n \times p$ matrix with (i, j) entry

$$\beta_{ij} = \begin{cases} 1 \text{ if } b_i \beta c_j, \\ 0 \text{ otherwise.} \end{cases}$$

We define the *conjunction* of M_α and M_β, written $M_\alpha \vee M_\beta$, to be the $m \times p$ array with (i,j) entry

$$\gamma_{ij} = \begin{cases} 1 \text{ if } \sum_{i=1}^{n} \alpha_{ik}\beta_{kj} \geq 1, \\ 0 \text{ otherwise.} \end{cases} \tag{8.2}$$

(If M_α is $m \times n$ and M_β is $n' \times p$, where $n \neq n'$, then $M_\alpha \vee M_\beta$ is not defined.)

Sample Problem 8.16 *Let A, B and α be as in Sample Problem* 8.15. *Further define $C = \{1,2,3,4\}$ and*

$$\beta = \{(1,2),(1,3),(2,1),(2,4),(3,3)\}.$$

Find $M_\alpha \vee M_\beta$.

Solution. The matrix of β is

$$M_\beta = \begin{bmatrix} 0 & 1 & 1 & 0 \\ 1 & 0 & 0 & 1 \\ 0 & 0 & 1 & 0 \end{bmatrix}.$$

To find $M_\alpha \vee M_\beta$ it suffices to find the ordinary matrix product $M_\alpha M_\beta$ and replace any non-zero element by 1.

$$M_\alpha M_\beta = \begin{bmatrix} 0 & 1 & 2 & 0 \\ 1 & 0 & 1 & 1 \end{bmatrix} \text{ and } M_\alpha \vee M_\beta = \begin{bmatrix} 0 & 1 & 1 & 0 \\ 1 & 0 & 1 & 1 \end{bmatrix}.$$

Practice Exercise. If A, B, C and α are as stated and

$$\tau = \{(1,1),(1,3),(2,2),(3,1),(3,4)\},$$

find $M_\alpha \vee M_\tau$.

Notice that $\alpha\beta = \{(1,2),(1,3),(2,1),(2,3),(2,4)\}$ and $M_\alpha \vee M_\beta = M_{\alpha\beta}$. We prove that this happens in general.

Theorem 8.6 *Let A, B, C, α and β be as before. Then:*

(i) $M_{\alpha\beta} = M_\alpha \vee M_\beta$;

(ii) $M_{\alpha^{-1}} = M_\alpha^T$, the transpose of M_α;

(iii) $M_{\alpha\beta}^T = M_\beta^T \vee M_\alpha^T$.

Proof. (i) Let $\mu = \alpha\beta$, so that $M_\mu = M_{\alpha\beta}$ is the $m \times p$ array (σ_{ij}), where

$$\sigma_{ij} = \begin{cases} 1 \text{ if } a_i \mu c_j, \\ 0 \text{ otherwise.} \end{cases}$$

We see that the conjunction $M_\alpha \vee M_\beta$ is also an $m \times p$ array. Suppose its (i,j) entry is defined by γ, as in (8.2). Now $\gamma_{i,j} = 1$ if and only if

$$\sum_{i=1}^{n} \alpha_{ik}\beta_{kj} \geq 1,$$

that is, if and only if $\alpha_{ik}\beta_{kj} = 1$ for at least one value of k. This is true if and only if $\alpha_{ik} = 1 = \beta_{kj}$ for at least one k, or equivalently if and only if $a_i\alpha b_k$ and $b_k\beta c_j$ for at least one k. But this is just the statement that $a_i\alpha\beta c_j$, so $\sigma_{i,j} = 1$.

Conversely, $\sigma_{i,j} = 1$ if and only if $a_i\alpha\beta c_j$, which is equivalent to the existence of at least one b_k such that $a_i\alpha b_k$ and $b_k\beta c_j$. But now $\alpha_{ik} = 1 = \beta_{kj}$, so $\gamma_{i,j} = 1$.

Hence $M_\alpha \vee M_\beta = M_{\alpha\beta}$.

(ii) Let $M^T = (\varepsilon_{ij})_{n\times m}$, where $\varepsilon_{ij} = \alpha_{ji}$. Then M_α is the correct size array for $\alpha^{-1} : B \to A$, as $M_{\alpha^{-1}}$ is $n \times m$. Suppose the (i,j) element of $M_{\alpha^{-1}}$ is δ_{ij}; then $\delta_{ij} = 1$ if and only if $b_i\alpha^{-1}a_j$, or equivalently $a_j\alpha b_i$, so $\delta_{ij} = 1$ if and only if $\varepsilon_{ij} = 1$, which implies $M^T = M_{\alpha^{-1}}$.

(iii)
$$\begin{aligned} M^T_{\alpha\beta} &= \text{ by (ii)} \\ &= M_{\beta^{-1}\alpha^{-1}} \text{ by Theorem 4.2} \\ &= M_{\beta^{-1}} \vee M_{\alpha^{-1}} \text{ by (i)} \\ &= M^T_\beta \vee M^T_\alpha \text{ by (ii).} \end{aligned}$$
□

We observe some of the properties of these matrices. If α is a relation on a set A with m elements, then M_α is $m \times m$. If α is reflexive, then M_α has 1's down its diagonal; if α is symmetric, then M_α is symmetric in the usual matrix sense ($a_{ij} = a_{ji}$); if α is transitive and $M_{\alpha\alpha} = M_\alpha \vee M_\alpha$ has 1 in the (i,j) position, then so does M_α.

Adjacency matrices of graphs

If the relation α is represented both by the adjacency matrix M_α and by a graph G, then M_α is also called the *adjacency matrix of the graph G*. This is such a fundamental idea that we define it separately.

Suppose G is a graph with v vertices $\{x_1, x_2, \ldots, x_v\}$. Then the *adjacency matrix* M_G of G is the $v \times v$ matrix M_G with (i,j) entry

$$m_{ij} = \begin{cases} 1 & \text{if } x_i \sim x_j, \\ 0 & \text{otherwise.} \end{cases}$$

The adjacency matrix depends on the order in which the vertices are taken. However, changing the order of the vertices results in a matrix that can be derived from the original by carrying out a reordering of the rows, and then carrying out an identical reordering of the columns. Many properties of matrices are left unchanged by operations like this.

The adjacency matrix of a graph is necessarily symmetric.

If G has e edges $\{y_1, y_2, \ldots, y_e\}$, then the *incidence matrix* N_G of G is the $v \times e$ matrix with (i,j) entry

$$n_{ij} = \begin{cases} 1 & \text{if vertex } x_i \text{ is incident with edge } a_j, \\ 0 & \text{otherwise.} \end{cases}$$

Sample Problem 8.17 *What are the adjacency and incidence matrices of $K_{2,3}$?*

Solution. We take the vertices of $K_{2,3}$ to be the sets $\{1,2\}$ and $\{3,4,5\}$. Taking the vertices in numerical order, the adjacency matrix is

$$\begin{bmatrix} 0 & 0 & 1 & 1 & 1 \\ 0 & 0 & 1 & 1 & 1 \\ 1 & 1 & 0 & 0 & 0 \\ 1 & 1 & 0 & 0 & 0 \\ 1 & 1 & 0 & 0 & 0 \end{bmatrix}.$$

If we take the edges in order $(1,3)$, $(1,4)$, $(1,5)$, $(2,3)$, $(2,4)$, $(2,5)$, the incidence matrix is

$$\begin{bmatrix} 1 & 1 & 1 & 0 & 0 & 0 \\ 0 & 0 & 0 & 1 & 1 & 1 \\ 1 & 0 & 0 & 1 & 0 & 0 \\ 0 & 1 & 0 & 0 & 1 & 0 \\ 0 & 0 & 1 & 0 & 0 & 1 \end{bmatrix}.$$

Practice Exercise. What are the adjacency and incidence matrices of K_4?

The powers of the adjacency matrix of a graph G give information about G. Suppose $M_G = (m_{ij})$. Write $x_1, x_2, \ldots$ for the vertices of G, and assume $i \neq j$. If $m_{ik} = m_{kj} = 1$, this means there are edges $x_i x_k$ and $x_k x_j$, so there is a path from x_i to x_j passing through x_k. The number of two-edge paths from x_i to x_j equals the number of vertices x_k for which this is true. The adjacency matrix has entries 0 and 1, so we have actually observed that $m_{ik}m_{kj} = 1$ when there is an $x_i x_k x_j$ path and 0 otherwise. So $\sum m_{ik}m_{kj}$ equals the number of different two-edge paths from x_i to x_j. The ith diagonal entry is $\sum m_{ik}m_{ki} = \sum m_{ik}^2$. As $m_{ik} = 1$ or 0, $m_{ik}^2 = m_{ik}$, so $\sum m_{ik}m_{ki}$ equals the number of times $m_{ik} = 1$, the degree of x_i.

Theorem 8.7 *M_G^2 has the degrees of the vertices on the diagonal, and the number of paths of length 2 joining the relevant vertices in its other positions.*

For n larger than 2, the off-diagonal entries of M_G^n give the number of walks between the relevant vertices (not necessarily paths; for larger n, repeated vertices can occur). The situation with the diagonal entries is more complicated (see, for example, Exercise 8.5.23).

Exercises 8.5

In Exercises 1 to 7, write down the adjacency matrix of the relations. These relations were all defined in Section 4.1.

1. Exercises 4.1.1 and 4.1.2, relations α, β.

2. Exercises 4.1.3 and 4.1.4, relations γ, δ.

3. Exercises 4.1.5 and 4.1.6, relations ε, ϕ.

4. Exercise 4.1.7, relations α, β.

5. Exercise 4.1.8, relations ρ, σ.

6. Exercises 4.1.1 and 4.1.2, relations $\alpha\beta, \beta\gamma$.

7. Exercise 4.1.8, relations $\rho\sigma, \sigma\rho$.

8. Suppose α is a relation between finite sets A and B. What is the relationship between M_α and M_{α^*}?

9. Let $A = \{a_1, a_2, \dots, a_m\}$, $B = \{b_1, b_2, \dots, b_n\}$, $C = \{c_1, c_2, \dots, c_p\}$ and $D = \{d_1, d_2, \dots, d_q\}$ be finite sets and α, β and γ be relations from A to B, B to C and C to D respectively. Show that $M_\alpha \vee (M_\beta \vee M_\gamma) = (M_\alpha \vee M_\beta) \vee M_\gamma$
   (i) using the fact that composition of relations is associative;
   (ii) working directly from the definition of matrix conjunction.

10. Let α and β be relations from A to B. What can you say about M_α and M_β if $\beta \subseteq \alpha$?

11. Suppose α is a relation from A to itself. What can you say about M_α if α is

   (i) irreflexive? (ii) antisymmetric? (iii) atransitive?

In Exercises 12 *to* 18, *write down the adjacency matrix and the incidence matrix of the graph.*

12. The complete graph K_5

13. The cycle C_6

14. The path P_5

15. The wheel W_5

16. The union $C_3 \cup C_3$

17. The star $K_{1,5}$

18. The Petersen graph

In Exercises 19 *and* 20, *use the adjacency matrix to answer the question.*

19. Find the number of paths of length 2 joining two vertices in the complete graph K_6.

20. Find the number of paths of length 2 joining two vertices in the wheel W_6. (You will need to consider separately the various types of pairs of vertices.)

21. Suppose G is a regular graph. What does this tell you about M_G?

22. Suppose G is any graph. What is the relationship between M_G and $M_{\overline{G}}$? Compare your answer to Exercise 8.

23. What information do the diagonal elements of M_G^2 give about the graph G?

9

Number Theory and Cryptography

When we think of secret information, the first image in our minds may be the spy in a dirty raincoat lurking around corners in the sleazy part of a central European city. But nowadays secret messages are more likely to be sent by computer, or broadcast. The emphasis is no longer on keeping the messages secret. The enemy will almost certainly be able to read your messages; your aim is to stop your adversaries from understanding them.

Encryption, or secret writing, has always been important in warfare. In antiquity, when very few people could read at all, quite elementary precautions could keep a message safe. Over the years more and more sophisticated techniques have been developed.

In general, breaking a code (or *decryption*) involves a great many trial-and-error computations. With the development of the computer there has been a spectacular increase in the speed at which such computations can be performed, and better and better techniques of encryption have been developed.

Several of the techniques of encryption and decryption involve elementary number theory, so we begin by studying primes, factors, divisors and modular arithmetic. We survey classical methods of encryption, and then look at the modern study of public-key cryptography.

9.1 Some Elementary Number Theory

Primes and divisors

We start by recalling some basic ideas that we encountered in Section 1.1. An integer a is a *multiple* of an integer b if there exists some integer c such that $a = bc$, and we say b and c are *factors* or *divisors* of a. A *prime number* is a positive integer a, other than 1, that has no divisors other than 1 and a itself. The integers are made up of four disjoint classes:

- the *zero*, 0;
- the *units*, 1 and -1;
- the *primes*, consisting of the prime numbers and their negatives;
- the *composite* numbers, all other integers.

A positive composite number can be written as a product of prime numbers. For example, the composite number 45 can be expressed as $3 \cdot 3 \cdot 5$. More importantly, this is the only way to express 45 as a product of primes: multiply two copies of the prime 3 and one copy of the prime 5. This is true for every positive integer, and is called the *Fundamental Theorem of Arithmetic*:

Every positive integer can be expressed in exactly one way as the product of prime numbers, provided the order of the factors is disregarded.

The expression is called the *prime factor decomposition* of the number concerned. For example, we say the prime factor decomposition of 45 is $3 \cdot 3 \cdot 5$, or $3^2 \cdot 5$.

The Fundamental Theorem of Arithmetic is also called the *Unique Factorization Theorem*. It is very important for understanding how the integers work. Unique factorization is the reason we do not regard 1 as a prime: if it were, then we could write

$$45 = 3 \cdot 3 \cdot 5 = 1 \cdot 3 \cdot 3 \cdot 5 = 1 \cdot 1 \cdot 3 \cdot 3 \cdot 5 = \cdots$$

and the representation would not be unique. It is also why we defined prime numbers to be *positive* integers; we do not want to worry about factorizations like

$$45 = 3 \cdot 3 \cdot 5 = (-3) \cdot (-3) \cdot 5.$$

In case you were wondering about 1: it is the product of *no* primes, and can only be expressed that way.

Sample Problem 9.1 *Express the numbers* 48, 33, 153, 59 *as products of primes.*

Solution. By repeated division we see that $48 = 2 \cdot 24$, $24 = 2 \cdot 12$, $12 = 2 \cdot 6$, $6 = 2 \cdot 3$; 3 is prime, so $48 = 2^4 \cdot 3$. Similarly $33 = 3 \cdot 11$; $153 = 3^2 \cdot 17$; 59 is prime.

Practice Exercise. Express the following numbers as products of primes: 36, 132, 47, 55.

Sample Problem 9.2 *List all positive divisors of* 228.

Solution. $228 = 2^2 \cdot 3 \cdot 19$ so the divisors are all the numbers of the form $2^x 3^y 19^z$, where x is 0, 1 or 2, y is 0 or 1 and z is 0 or 1. So the list is 1, 2, 4, 3, 6, 12, 19, 38, 76, 57, 114, 228.

Practice Exercise. Find all positive divisors of 246.

To work out the prime factor decomposition of a number a, it is necessary to divide by all the primes up to a. However, this process can be speeded up: for example, if a does not end in 5 or 0, it is not divisible by 5. Similarly, any number divisible by 2 must end in 0, 2, 4, 6 or 8. Another useful test is to add all digits of a; if this sum is not divisible by 3, then a is not divisible by 3, and if the sum is not divisible by 9, then a is not divisible by 9.

To test whether the number a is *prime*, it is not necessary to test whether a has a prime factor greater than $\sqrt{a}$. (Can you see why?)

Greatest common divisors and the Euclidean algorithm

If a and b are any two positive integers, the *greatest common divisor* or GCD of a and b is the largest integer that is a divisor of both of them. For example, the greatest common divisor of 6 and 9 is 3. Two numbers with greatest common divisor 1 are called *coprime*, or *relatively prime*. We denote the greatest common divisor of a and b by $\gcd(a,b)$, or, if there is no danger of confusion, simply (a,b).

As we saw in Section 1.1, one way to find the greatest common divisor is to write down the prime factor decomposition of each number, and compare them. However, there is a more general technique available. We shall look at it now.

Suppose a and b are any two positive integers. If a is not a multiple of b, then dividing b into a will result in a *quotient* and a *remainder*: there exist unique integers q and r, where $q > 0$ and $0 \le r < b$, such that

$$a = qb + r; \tag{9.1}$$

q is the quotient and r the remainder on division by b.

Now suppose d is any common divisor of a and b. Then d is also a divisor of r, because $r = a - qb$ and d divides each term on the right. We use this simple fact to construct a method to find the greatest common divisor of two positive integers.

Given positive integers a and b with GCD d, suppose dividing b into a as in (9.1) results in quotient q_1 and remainder r_1. Then

$$r_1 = a - q_1 b,$$

so obviously d must divide r_1. So d is a common divisor of b and r_1. If r_1 is non-zero we repeat the step, using b and r_1, to get a new remainder, r_2 say, and continue:

$$\begin{aligned} a &= q_1 b + r_1 \\ b &= q_2 r_1 + r_2 \\ &\dots \\ r_1 &= q_2 r_2 + r_3 \\ &\dots \end{aligned}$$

At each step $q_i > 0$, $0 \le r_{i+1} < r_i$ and d must divide the new remainder.

In this way we get a sequence of non-negative integers $r_1, r_2, r_3, \dots$ with the properties that d divides every element and $r_i > r_{i+1}$ for each i. Such a process cannot continue indefinitely, because there are not infinitely many positive integers less than r_1, so it must sometime happen that $r_{n+1} = 0$ (so that the division cannot be repeated). We eventually get

$$r_{n-1} = q_n r_n,$$

so r_n divides r_{n-1}. Since $r_{n-2} = q_{n-1} r_{n-1} + r_n$ it follows that r_n divides r_{n-2}. And so on. We finally find that r_n divides a and b, then r_n must divide their GCD, d. But d divides rn. So $r_n = d$.

This technique for finding greatest common divisors is called the *Euclidean algorithm.*

Sample Problem 9.3 *Use the Euclidean algorithm to find the greatest common divisor of* 864 *and* 291.

Solution.

$$\begin{aligned} 864 &= 2 \cdot 291 + 282, \\ 291 &= 1 \cdot 282 + 9, \\ 282 &= 31 \cdot 9 + 3, \\ 9 &= 3 \cdot 3, \end{aligned}$$

so $\gcd(862, 291) = 3$.

Practice Exercise. Use the Euclidean algorithm to find the greatest common divisor of 584 and 284.

The idea of a greatest common divisor can be extended to sets of three or more numbers. To find the GCD of a, b, c, first find $\gcd(a, b)$, then find the greatest common divisor of this new number with c. It will be the required answer.

Sample Problem 9.4 *Use the Euclidean algorithm to find the greatest common divisor of* 378, 336 *and* 490.

Solution.

$$\begin{aligned} 378 &= 1\cdot 336+42, \\ 336 &= 8\cdot 42, \end{aligned}$$

so $\gcd(378,336)=42$.

$$\begin{aligned} 490 &= 11\cdot 42+28, \\ 42 &= 1\cdot 28+14, \\ 28 &= 2\cdot 14, \end{aligned}$$

so $\gcd(378,336,490)=\gcd(42,490)=14$.

Practice Exercise. Use the Euclidean algorithm to find the greatest common divisor of 144, 156 and 162.

Reversing the Euclidean algorithm

The Euclidean algorithm can be reversed. Each equation of the form

$$r_{i-1} = q_i r_i + r_{i+1}$$

can be inverted to give

$$r_{i+1} = r_{i-1} - q_i r_i.$$

Starting from the equations

$$\begin{aligned} r_n &= r_{n-2} - q_{n-1} r_{n-1}, \\ r_{n-1} &= r_{n-3} - q_{n-2} r_{n-2}, \end{aligned}$$

we obtain

$$\begin{aligned} r_n &= r_{n-2} - q_{n-1}(r_{n-3} - q_{n-2} r_{n-2}) \\ &= (1+q_{n-2}) r_{n-2} - q_{n-1} r_{n-3}. \end{aligned}$$

Using

$$r_{n-2} = r_{n-4} - q_{n-3} r_{n-3}$$

we get

$$\begin{aligned} r_n &= (1+q_{n-2})(r_{n-4} - q_{n-3} r_{n-3}) - q_{n-1}(r_{n-3}) \\ &= (1+q_{n-2}) r_{n-4} - (1+q_{n-2}+q_{n-1})(r_{n-3}). \end{aligned}$$

The algebra of these equations is not important; the main point is that each time we express r_n as a function of earlier terms. We eventually obtain

$$\gcd(a,b) = r_n = z_1 a + z_2 b$$

for some integers z_1 and z_2. This is important enough to state as a theorem (you will see why in the next section).

Theorem 9.1 *The greatest common divisor of two positive integers a and b equals the sum of integer multiples of a and b.*

This sort of expression is called an *integer linear form* in a and b.

Sample Problem 9.5 *Express the greatest common divisor of* 864 *and* 291 *as an integer linear form in* 864 *and* 291.

Solution. From Sample Problem 9.3 we know that $\gcd(862, 291) = 3$. Reversing the equations in that problem,

$$\begin{aligned} 3 &= 282 - 31 \cdot 9 \\ &= 282 - 31 \cdot (291 - 282) \\ &= 32 \cdot 282 - 31 \cdot 291 \\ &= 32 \cdot (864 - 2 \cdot 291) - 31 \cdot 291 \\ &= 32 \cdot 864 - 95 \cdot 291, \end{aligned}$$

which is the required form.

Practice Exercise. Express the greatest common divisor of 584 and 284 as an integer linear form in 584 and 284.

Exercises 9.1

1. List all positive divisors of the numbers 63, 64, 288.
2. List all positive divisors of the numbers 35, 128, 162.
3. Which of the following numbers are divisible by 3: 10873, 44444, 123456, 51804?
4. Which of the following numbers are divisible by 9: 41728, 62901, 654321, 71730?
5. Which of the following numbers are prime: 195, 257, 1001, 419?
6. List all primes less than 100.

In Exercises 7 *to* 10, *find the greatest common divisors of the pair of numbers by finding the prime factor decompositions of the numbers.*

7. 70, 120
8. 168, 504
9. 180, 600
10. 260, 455

In Exercises 11 *to* 18, *find the greatest common divisors of the pair of numbers, using the Euclidean algorithm. Then express the GCD as an integer linear form in the two numbers.*

11. 60, 84

12. 234, 470

13. 480, 1800

14. 84, 180

15. 120, 144

16. 292, 244

17. 210, 861

18. 64, 160

19. Check your answers in Exercises 9.1.7 to 9.1.10 using the Euclidean algorithm.

In Exercises 20 *to* 24, *find the greatest common divisors of the set of numbers.*

20. 252, 308, 504

21. 1386, 1170, 1890

22. 450, 696, 432

23. 105, 231, 273

24. 595, 910, 1155

25. We said that, when testing to see whether a is prime, it is not necessary to check whether it has a prime factor greater than $\sqrt{a}$. Why is this true?

9.2 Modular Arithmetic

Basic definitions

Suppose you look at a clock and want to know what time it will be five hours from now. If it is currently 10 o'clock, you do not say "15 o'clock," but rather "3 o'clock." You might say "AM" or "PM," but that is extra information that is not on the clockface. In the same way, 44 hours from now it will be 6 o'clock; what day, and whether it will be morning or evening, is not given by clockface arithmetic.

What you do when interpreting a clock is to subtract multiples of 12 when necessary. In fact, all you are concerned with is the *remainder* after division by 12. For example, $10+44=54=4\times 12+6$; we ignore the 12's and say "6 o'clock."

Another way of expressing clockface arithmetic is to say calculations are carried out *modulo 12*, and we might write

$$44+10=6(\text{mod } 12).$$

Instead of an equality sign, we also write $44+10\equiv 6(\text{mod } 12)$, and say the sum is *congruent to* 6. The number 12 is called the *base* or the *modulus*. It is usual to express the answer to a modulo 12 problem as one of the numbers in the standard range $0,1,2,\ldots,11$, although we could equally well write 12 instead of 0. (In the particular case of telling the time, we say "12 o'clock," not "0 o'clock.")

Another method of telling time is the military one, or 24-hour clock. This indicates the difference between morning and evening, and is in fact a modulo 24 arithmetic. (In the military clock we say "00:30 hours," not "24:30 hours.")

Binary arithmetic, the arithmetic of even and odd, is about remainders modulo 2. The addition is as follows.

0+0=0	Even + Even = Even
0+1=1	Even + Odd = Odd
1+0=0	Odd + Even = Odd
1+1=0	Odd + Odd = Even

Binary arithmetic comes up in many electrical applications because there are often two states to be considered: " current flowing" or "no current." As an example, think about turning a light on or off. On many lamps, there is a knob that you twist clockwise. If the light is on, twisting the knob changes it to "off." Similarly, twisting from "off" gives "on." If 1 means "on" and 0 means "off," and if the process of turning the knob is interpreted as "add 1" (which makes sense if you think of the lamp being off originally), we have

$$0+1=1, \quad 1+1=0,$$

which is precisely arithmetic modulo 2.

In the same way, one could talk about modular arithmetic modulo any positive integer (although "arithmetic modulo 1" is not very interesting: the answer to any calculation must be 1). The standard range for arithmetic modulo m is $0, 1, \ldots,$ $m-1$, and these numbers are often called *residues* modulo m. If we simply write $a(\text{mod } m)$, we usually interpret this as the member of the standard range that is congruent to a modulo m (but we often write -1 for $m-1$). Congruential equations, where the role of the equals sign is taken by $\equiv$, are usually called *congruences*.

Sample Problem 9.6 *Solve the congruence* $x^2 \equiv 1(\text{mod } 5)$.

Solution. The easiest technique is to test all the possibilities in the standard range: there are only five of them. We find $0^2 = 0$, $1^2 = 1$, $2^2 = 4$, $3^2 = 9 \equiv 4$, and $4^2 = 16 \equiv 1$. So the answer is "$x \equiv 1$ or 4."

Practice Exercise. Solve the equation $x^2 \equiv 2(\text{mod } 7)$.

Sample Problem 9.7 *Carry out the calculations* $3^3 + 2(\text{mod } 5)$, $3 \times 13 - 7(\text{mod } 9)$.

Solution. Modulo 5, $3^3 = 27$ and $27 + 2 = 29$, So the first answer is 4. In the second problem, $3 \times 13 = 39 \equiv 3(\text{mod } 9)$, and $3 - 7 = -4 \equiv 5$.

Practice Exercise. Carry out the calculations $5^2 + 2(\text{mod } 7)$, $4 \times 14 - 9$ $(\text{mod } 12)$.

In the first example above, the process of getting rid of 5's and coming down to a number in the standard range is called *reducing modulo the base*.

The relation $x \equiv y(\bmod m)$ is an equivalence relation on the set of integers. The equivalence class of x is

$$[x] = \{y : x \equiv y(\bmod m)\},$$

which is usually called the *congruence class* or *residue class* of x.

Inverses

The most obvious difference between modular arithmetic and ordinary arithmetic is the existence of multiplicative inverses. In the ordinary arithmetic of integers, only the units, 1 and -1, have inverses, and each is its own inverse. In arithmetic modulo 7, for example, every non-zero residue has an inverse: $2 \times 4 = 8 \equiv 1$, so 2 has inverse 4 and 4 has inverse 2; we write $2^{-1} = 4$ and $4^{-1} = 2$. Similarly $3 \times 5 \equiv 1$ and $6 \times 6 \equiv 1$, so $3^{-1} = 5$, $5^{-1} = 3$ and $6^{-1} = 6$. And, of course, $1^{-1} = 1$.

The situation is different in other moduli, and another significant change from ordinary arithmetic is seen. Consider the arithmetic modulo 12. There are several numbers with inverses: 1, of course, and 11 (which is -1); and $5 \times 5 \equiv 7 \times 7 \equiv 1$, so both 5 and 7 are their own inverses. But 2, 3, 4, 6, 8, 9 and 10 have no inverses. In fact $2 \times 6 \equiv 0$, a phenomenon that can never happen in ordinary arithmetic. We say two numbers are *zero-divisors* $(\bmod m)$ if they are not zero modulo m but their product is zero. We shall soon show that a number cannot both be a zero-divisor and have an inverse.

Sample Problem 9.8 *Find all inverses and zero-divisors in the arithmetic modulo* 10.

Solution. Of course $1^{-1} = 1$ and $9^{-1} = 9$. $2 \times 5 = 10 \equiv 0$ and similarly 4×5, 6×5 and 8×5 are all $\equiv 0$, so 2, 4, 5, 6 and 8 are zero-divisors. Finally, $3 \times 7 \equiv 1$, so 3 and 7 are mutual inverses.

Practice Exercise. Find all inverses and zero-divisors in the arithmetic modulo 6.

The key to inverses and zero-divisors is the greatest common divisor, as the following theorem shows.

Theorem 9.2 *Suppose a is an integer not congruent to* $0\,(\bmod m)$.

(i) *If the GCD (a,m) of a and m is* 1, *then a has an inverse* $(\bmod m)$.

(ii) *If (a,m) is not* 1, *then a is a zero-divisor* $(\bmod m)$.

Proof. From Theorem 9.1, there exist integers b and c such that $(a,m) = ba + cm$. Since any multiple of m is zero $(\bmod m)$, $(a,m) = ba\,(\bmod m)$. If $(a,m) = 1$, then b is an inverse for a.

Conversely, suppose $(a,m) = d > 1$. The GCD divides both a and m, so we can find integers x and y such that $dx = a$ and $dy = m$. If x had any common factor e with m, greater than 1, then ed would be a common factor of a and m, greater than d, so this is impossible. We also know that y is not a multiple of m from the fact that $dy = m$ and $d > 1$. Now

$$ay = dxy = my \equiv 0 \pmod{m},$$

so a is a zero-divisor (mod m) (and for that matter so is y). □

This theorem shows that every integer not congruent to zero modulo m either has an inverse or is a zero-divisor (mod m), but cannot be both.

Sample Problem 9.9 *Which of the following are zero-divisors modulo 224? If the number is not a zero-divisor, find its inverse.*

$$16, \quad 53, \quad 63, \quad 84, \quad 97$$

Solution. Either by the Euclidean algorithm or by factoring, we find that $(16,224) = 16$, $(53,224) = 1$, $(63,224) = 7$, $(84,224) = 28$ and $(97,224) = 1$. So 16, 63 and 84 are zero-divisors. For 53 and 97, we look at the Euclidean algorithm calculations:

$$\begin{aligned}
224 &= 4 \times 53 + 12 \\
53 &= 4 \times 12 + 5 \\
12 &= 2 \times 5 + 2 \\
5 &= 2 \times 2 + 1 \\
1 &= 5 - 2 \times 2 &&= 5 - 2 \times (12 - 2 \times 5) \\
&= 5 \times 5 - 2 \times 12 &&= 5 \times (53 - 4 \times 12) - 2 \times 12 \\
&= 5 \times 53 - 22 \times 12 &&= \times 53 - 22 \times (224 - 4 \times 53) \\
&= 93 \times 53 - 22224
\end{aligned}$$

so $53^{-1} = 93$.

$$\begin{aligned}
224 &= 2 \times 97 + 30 \\
97 &= 3 \times 30 + 7 \\
30 &= 4 \times 7 + 2 \\
7 &= 3 \times 2 + 1 \\
1 &= 7 - 3 \times 2 &&= 7 - 3 \times (30 - 4 \times 7) \\
&= 13 \times 7 - 3 \times 30 &&= 13 \times (97 - 3 \times 30) - 3 \times 30 \\
&= 13 \times 97 - 42 \times 30 &&= 13 \times 97 - 42 \times (224 - 2 \times 97) \\
&= 97 \times 97 - 42 \times 224
\end{aligned}$$

so $97^{-1} = 97$.

Practice Exercise. Which of the following are zero-divisors mod 132?

$$49, \quad 55, \quad 66, \quad 91$$

If the number is not a zero-divisor, find its inverse.

In the preceding sample problem, we also saw a case where an element other than 1 or -1 can equal its own inverse — another departure from ordinary arithmetic.

Simultaneous congruences

We now discuss finding all integers that simultaneously solve a set of congruences. To be more precise, suppose each of the positive integers $m_1, m_2, \ldots, m_n$ is relatively prime to each other (we say they are *pairwise relatively prime*) and we wish to find all integers x such that

$$\begin{aligned} x &\equiv a_1 (\text{mod } m_1) \\ x &\equiv a_2 (\text{mod } m_2) \\ &\cdots \\ &\cdots \\ x &\equiv a_n (\text{mod } m_n). \end{aligned} \qquad (9.2)$$

We write $M = m_1 \times m_2 \times \cdots \times m_n$, and define positive integers M_i, $i = 1, 2, \ldots, n$, by $M_i = M/m_i$ (in other words, $M = M_i m_i$). Consider the n congruences

$$M_i x \equiv 1 (\text{mod } m_i). \qquad (9.3)$$

In each case $(M_i, m_i) = 1$, so there is exactly one solution to each congruence (9.3). Suppose the solution to $x \equiv a_i (\text{mod } m_i)$ is $x \equiv x_i (\text{mod } m_i)$. Now define

$$X = M_1 x_1 a_1 + M_2 x_2 a_2 + \cdots + M_n x_n a_n$$

and substitute X for x in $x \equiv a_i (\text{mod } m_i)$. Since $M_j \equiv 0 (\text{mod } m_i)$ unless $j = i$, and $M \equiv O (\text{mod } m_i)$, and $M_i x_i$ is congruent to $1 (\text{mod } m_i)$,

$$X \equiv a_i (\text{mod } m_i).$$

This is true for every i, so X satisfies all the congruences (9.2), and every integer congruent to $X (\text{mod } M)$ is also a solution.

But this class of integers is the only simultaneous solution of the set of congruences, for if Y also satisfies (9.2), then $X \equiv Y (\text{mod } m_i)$ for each i, and since the m_i are relatively prime in pairs, $X \equiv Y (\text{mod } M)$. Hence we have shown

Theorem 9.3 (Chinese Remainder Theorem) *If the n positive integers m_1, $m_2, \cdots, m_n$ are relatively prime in pairs, the n congruences $x \equiv a_i (\text{mod } m_i)$ have one and only one simultaneous solution modulo $M = m_1 \times m_2 \times \cdots \times m_n$.*

Sample Problem 9.10 *Find all integers that give the remainders* 2, 6 *and* 5 *when divided by* 5, 7 *and* 11, *respectively.*

Solution. In the notation used above,

$$M = 385, \;\; M_1 = 77, \;\; M_2 = 55. \;\; M_3 = 35.$$

$77x \equiv 1 (\text{mod } 5)$, or $2x \equiv 1 (\text{mod } 5)$, has the solution $x \equiv 3 (\text{mod } 5)$,
$55x \equiv 1 (\text{mod } 7)$, or $6x \equiv 1 (\text{mod } 7)$, has the solution $x \equiv 6 (\text{mod } 7)$,
$35x \equiv 1 (\text{mod } 11)$, or $2x \equiv 1 (\text{mod } 11)$, has the solution $x \equiv 6 (\text{mod } 11)$.
Hence,

$$X \equiv 77 \times 3 \times 2 \; + \; 55 \times 6 \times 6 \; + \; 35 \times 6 \times 5 (\text{mod } 385),$$

or $X \equiv 27 \pmod{385}$.

Practice Exercise. Find all integers that give the remainders 3, 2 and 4 when divided by 5, 7 and 11, respectively.

Exercises 9.2

In Exercises 1 *to* 10, *carry out the calculations modulo* 5.

1. $3+5$

2. 4×3

3. 3^4-1

4. $4-(4+3)$

5. $8-2$

6. $2\times 3\times 4$

7. $3\times(4-2)$

8. 2^4+3^4

9. $14+2^2$

10. $5-4^4$

In Exercises 11 *to* 20, *carry out the calculations modulo* 7.

11. $2+5+1$

12. 2×4

13. $6\times(4-2)$

14. $(2+3)^4+1$

15. $5-8$

16. $5\times 4\times 4$

17. $1+2+2^2+2^3$

18. $(2+3)\times(4+6)$

19. $3-(2+3+4)$

20. $5^4\times 4^4$

In Exercises 21 *to* 30, *carry out the calculations modulo* 6.

21. $7\times(5+2)$

22. $2\times(5+3)+4$

23. 3×4

24. $(7-3)^3$

25. $9\times 7\times 5$

26. $3-5$

27. $3\times(3^2+1)$

28. $1+2+2^2+2^3$

29. $(2+3)\times(5-2)$

30. 5×4

In Exercises 31 *to* 40, *carry out the calculations modulo* 2.

31. 2×5

32. $3\times(3-2)$

33. $(6+3)^4$

34. $2-3+1$

35. $5\times(3-4)$

36. $(1+1+1)^4+1$

37. $1-1+2$

38. $5\times3\times1$

39. $1-(4-2)$

40. $(2+3)^4+1$

41. Find all inverses and zero-divisors in the arithmetic modulo 14.

42. Find all inverses and zero-divisors in the arithmetic modulo 15.

In Exercises 43 *to* 52, *carry out the calculations modulo* 9.

43. 2×7

44. 3×5^{-1}

45. $(4+3)^{-1}$

46. $2-3+7$

47. $13+12$

48. $(4\times6)^4$

49. $1+2+2^2+2^{-1}$

50. $5-7$

51. $(2+5+4)^{-1}$

52. $(5+7)^4$

In Exercises 53 *to* 62, *carry out the calculations modulo* 12.

53. $(7\times5)^{-1}$

54. $3\times(3-6)$

55. 5×5^{-1}

56. $5-9$

57. 11^3

58. $5\times3\times7$

59. $1+5+5^2+5^{-1}$

60. $(2+7)-(4-6)$

61. $7^{-1}\times(3+5+7)$

62. $5^4\times4^4$

In Exercises 63 *to* 66, *use the Chinese remainder theorem to solve the set of congruences.*

63.
$$\begin{aligned} x &\equiv 3(\text{mod } 5)\\ x &\equiv 4(\text{mod } 7)\\ x &\equiv 1(\text{mod } 11) \end{aligned}$$

64.
$$\begin{aligned} x &\equiv 1(\text{mod } 3)\\ x &\equiv 2(\text{mod } 5)\\ x &\equiv 1(\text{mod } 7) \end{aligned}$$

65.
$$\begin{aligned} x &\equiv 1(\text{mod } 5)\\ x &\equiv 5(\text{mod } 7)\\ x &\equiv 2(\text{mod } 13) \end{aligned}$$

66.
$$\begin{aligned} x &\equiv 1(\text{mod } 3)\\ x &\equiv 4(\text{mod } 7)\\ x &\equiv 4(\text{mod } 11) \end{aligned}$$

67. Prove that the arithmetic of the Boolean algebra $\mathcal{B}_2$, defined in Section 3.1, is not the same as arithmetic modulo 2.

68. Verify that $x\equiv y(\text{mod } m)$ is an equivalence relation on the set of integers.

9.3 An Introduction to Cryptography

Secret writing

The idea of *encoding* data arises in many ways. Words are encoded as strings of binary digits whenever you use a word processor, so that a computer can work with them. Another important use of encoding is to compress data; this is often done when data is sent over telephone lines, and in fact we use data compression whenever we write an abbreviation. In both these examples, it is easy to decode the information.

In the rest of this chapter we look at a different reason for encoding — secrecy. Sometimes you encode something and tell only certain selected people how to decode it. This form of secret encoding is called *encryption*, and the study of it is *cryptography*. It has been widely used in military operations, but nowadays it is very common in commerce. For example, the use of a PIN number with your bank card is a form of encryption.

The original message, in everyday language, is often called a *plaintext*, and the encoded version is the corresponding *ciphertext*. ("Cipher" is another word for "code.") The process of moving from ciphertext to plaintext (the opposite of encryption) is called *decryption*. The special information required by your ally to decrypt your message is called a *key*.

Suppose you send a secret message. If your enemy intercepts it, he will try to figure out what it means by working out how you did the encoding — usually, he tries to find the key. This is called *cryptanalysis*, or codebreaking. The study of cryptography and cryptanalysis is collectively called *cryptology*.

There are many mystery stories in which cryptography plays an important part. If you want to read some of them, I recommend *The Adventure of the Dancing Men*, a Sherlock holmes story by Sir Arthur Conan Doyle, Edgar Allen Poe's *The Gold Bug*, or *The Nine Tailors* by Dorothy L. Sayers.

Physical secrecy

Some 2500 years ago the Spartans during their wars used a special device to transport their secret messages. They took a staff called a *scytale* (pronounced "sitterlee"), and wound a long, thin strip of parchment around it at an angle. They then wrote their message on the parchment, one letter on each width of parchment. When the strip is unwound, the letters appear to be completely random (try it yourself). But when the roll is wound again around another staff of equal diameter to the original staff, the message makes sense, but nonsense appears if you use the wrong size staff.

As a very simple example, suppose you wanted to send the message "help is coming." You would wrap a piece of parchment around your staff, and write your message on it in a few lines. Suppose you decided to write it in three lines: the first line would be "HELP," the second "ISCO," and the third "MING" (no need to include spaces — in fact, if you include spaces between words, it might help your enemy to decode your message.) We'll call the number of lines, three in this case, the *circumference* of the scytale. You end up with something like Figure 9.1. The strip will actually say "HIMESILCNPOG," but when your allies wrap it around their own staff (the same diameter as yours, so that the scytale comes out to have circumference 3) they get the welcome news "HELPISCOMING."

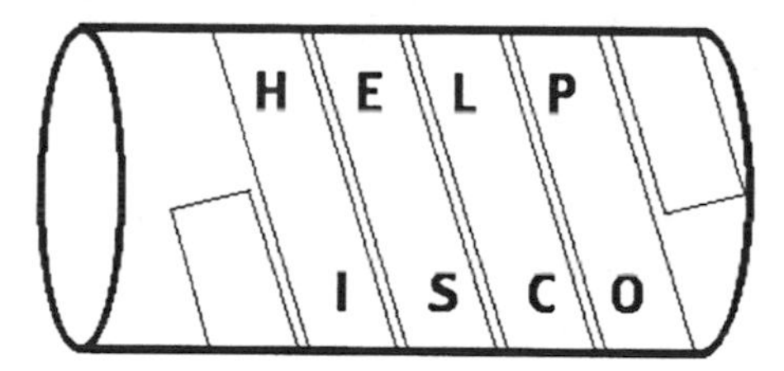

Figure 9.1: A scytale

However, suppose your enemy intercepts the message. If she uses a staff of the same diameter, you are in trouble; but if a slightly larger staff is used, so that the circumference is 4, the enemy will see the picture in Figure 9.2, and try to interpret "HSNIIPMLOECG."

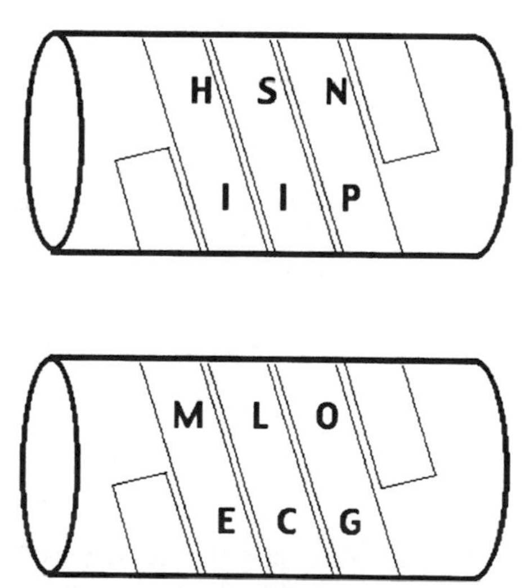

Figure 9.2: Misinterpreting the scytale

Nowadays this would not be a very secret method of writing, as it can be easily deciphered by moving around the edges of the roll; but very few people knew how to read at the time, and this ensured the secrecy of the scytale method. If your enemy could read, but did not know about the scytale method, the strip would probably be regarded as nonsense, or else as some strange language.

Sample Problem 9.11 *Interpret the scytale message*

```
MARSSEAVLESENEISANYLVEGDPOECEEES
SRSCOTBEHOPTETADLHCMVEEEAEBDHIU
```

Solution. To decode this we try various circumferences. If we try circumference 5 (and rearrange our message in five rows) we get complete nonsense, as follows.

$$\begin{array}{ccccccccccccc}
M & E & S & S & V & O & E & C & H & T & C & E & H \\
A & A & E & A & E & E & S & O & O & A & M & A & I \\
R & V & N & N & G & C & S & T & P & D & V & E & U \\
S & L & E & Y & D & E & R & B & T & L & E & B & \\
S & E & I & L & P & E & S & E & E & H & E & D &
\end{array}$$

However, if we arrange it in six rows, the message becomes clear.

$$\begin{array}{ccccccccccc}
M & A & N & Y & P & E & O & P & L & E & H \\
A & V & E & L & O & S & T & T & H & E & I \\
R & L & I & V & E & S & B & E & C & A & U \\
S & E & S & E & C & R & E & T & M & E & \\
S & S & A & G & E & S & H & A & V & B & \\
E & E & N & D & E & C & O & D & E & D &
\end{array}$$

Practice Exercise. Interpret the scytale message

```
TBSTNHAAHTERREGBIEFAAAARTRNTOE
```

Other physical cryptography techniques were used. Histiacus, a Greek ambassador in Persia, sent secret information home to his country by shaving a slave's head, branding a message on his skull and letting the hair grow again. He then sent him to Greece. Others used to drug their slaves and brand messages on their backs. The slaves knew nothing about the important messages they were carrying. In both cases, it was unlikely that the slave would live very long after the message was delivered.

The Caesar cipher

An early user of cryptographic techniques was the famous Roman commander-in-chief and statesman Julius Caesar (100–44 B.C.). In the second century A.D. Suetonius (*The Twelve Caesars* I, 56) describes:

> his private letters to friends, the more confidential passages of which he wrote in cipher: to understand their apparently incomprehensible

> meaning one must number the letters of the alphabet from 1 to 22, and then replace each of the letters that Caesar has used with the one which occurs three numbers lower — for instance, D stands for A.

As you can see, the Romans used only 22 letters in their alphabet. But, for convenience, let's assume Caesar used the English alphabet and used the same alphabetical order as we do. We obtain Caesar's cipher if we write the ciphertext alphabet beneath the plaintext alphabet, shifted three positions to the left. Notice that the alphabet wraps around at the end.

plaintext:	*a*	*b*	*c*	...	*w*	*x*	*y*	*z*
ciphertext:	D	E	F	...	Z	A	B	C

We encipher a plaintext letter by replacing it by the letter beneath it. For instance, the message *the plaintext* becomes `WKH SODLQWHAW`. (Caesar did not use the trick of omitting blanks.) Deciphering is equally easy; we replace a ciphertext letter by the plaintext letter above it. So `OHWWHU` is deciphered to the message *letter.*

Sample Problem 9.12 *What is the Caesar cipher for* the house by the stream?

Solution. `WKH KRXVH EB WKH VWUHDP.`

Practice Exercise. What is the Caesar cipher for *enter the maze?*

Caesar chose to go forward three letters for no particular reason; one could of course get the ciphertext alphabet by shifting any number of positions. Since the English alphabet consists of twenty-six letters, we obtain exactly twenty-six such ciphers; they are called *additive ciphers*. The key is the number of places you shift; Caesar's key was 3. People sometimes denote the key by the corresponding letter of the alphabet, the letter to which *A* is transformed, and in that terminology we would say Caesar's key is `D`. The additive cipher with key 0 (or `A`) is no good — the ciphertext is exactly the same as the plaintext — so there are twenty-five useful additive ciphers.

Sample Problem 9.13 *The phrase* `RFS TW BTRFS` *was produced by an additive cipher. All the words were regular English words. What was the plaintext?*

Solution. This is very easy because the spaces have not been removed. To save work, we attack the shortest word. To decode `TW` we write out two copies of the alphabet, one starting with T and the other with W, as shown.

```
TUVWXYZABCDEFGHIJKLMNOPQRS
WXYZABCDEFGHIJKLMNOPQRSTUV
```

TW must have come from some pair of letters that lie one over the other. The only English words that appear are *be*, *lo* and *or*; one might also accept *ad*, even though it is an abbreviation. If the original was *be*, then to get TW the key must have been 18, and the plaintext was *zna be jbzna*, which is nonsense. If it was *lo*, the key was 8 and the plaintext was *jxk lo tljxk*; if it was *ad*, the key was 19 and the plaintext was *ymz ad iaymz*; if it was *or*, the key was 5 and the plaintext was *man or woman*. Only the last makes sense, so it was the plaintext.

Practice Exercise. The phrase JGTG CPF PQY was produced by an additive cipher. All the words were regular English words. What was the plaintext?

In the movie *2001: A Space Odyssey*, the computer is called HAL; this can be obtained by moving the letters IBM one step to the right (or twenty-five steps to the left). The producers claimed that this was a coincidence.

The Vigenère method

Vigenère invented an encryption method that combines several additive ciphers. In modern terminology we could say that the key is a string of numbers, each one between 0 and 25. The easiest way to remember this is to convert it into a word, using $A = 0, B = 1, \ldots$. For example, the key $1, 17, 14, 22, 13$ is equivalent to the keyword BROWN. Then the plaintext is written out, with the keyword underneath it, repeated as necessary. For example, to encrypt the message *the computer has become very useful* we write

```
THECOMPUTERHASBECOMEVERYUSEFUL
BROWNBROWNBROWNBROWNBROWNBROWN
```

Now each letter is encrypted using the additive cipher whose key is below it. In the example, T would become U, H would become Y, and so on. The message becomes

UYS YBNGIPRS YOO OFTCIR WVFU HTVTQY.

In fact, the ciphertext would not usually be broken into words. You would probably receive

UYSYBNGIPRSYOOOFTCIRWVFUHTVTQY,

or perhaps it would arrive arbitrarily broken into four- or five-letter groups in order to make copying errors less likely:

UYSY BNGI PRSY OOOF TCIR WVFU HTVT QYZZ

where the last two letters have been added simply to make the last set of letters the right size. (Presumably the recipient of a message will have no trouble ignoring a couple of nonsense letters at the end.)

Exercises 9.3

1. What is the difference in meaning between the words *encode* and *encrypt*?

In Exercises 2 *to* 9, *decrypt the following scytale message.*

2. PSNRMLEDEIESMAEAEORS

3. WALOANLFKDTFESHEUMEEPEC

4. WRLHSEETRNASRCWOIEMTREYEWIKATISSNFRGIUS

5. HRSOETWMWMIEALMNEAYSRMMCOUH

6. TNIOHEHSNFOTEWVTDWSEEHORPRN
IFIAETSSTRTOMEITHRECNAESTRG

7. ISDIGBTATNOEILHPOISLAODNAYTSFWTAASOAR
CSERNUKISTTTNNSUOHOGINFUWLOEANLENMWIE
MOUIVDAFSFEGNATERE

8. CTNITOHTNHDEFTEEMOEWBORLORSMLRETOIL
AIFGDKMSETIPENONOCCDGRREA ITEIYSATN

9. LGOABIAIRFRENPKEHSEGHEAIENIIMTS
ESNCAGTMACSNEOSDRUYNRTLYBOEYOY
PTFRCHLTLTAAAAOEHLEVCGTESSEKRY.

10. In the scytale method, what is the *key*?
    (A) the staff
    (B) the diameter of the staff
    (C) the strip of parchment
    (D) the plaintext

In Exercises 11 *to* 14, *how would Caesar encode the message (if he spoke English)?*

11. *send troops*

12. *birthday greetings*

13. *the end is near*

14. *all sales are final*

In Exercises 15 *to* 18, *assume Caesar spoke English. What command was he sending?*

15. DWWDFN

16. PRYH QRUWK

17. UHWUHDW

18. EULQJ JROG

In Exercises 19 *to* 22, *encrypt the following messages using an additive cipher with key* 7.

19. *the moon has risen*

20. *sell all your gold stocks*

21. *seven knights are approaching*

22. *the rent is due*

In Exercises 23 *to* 26, *decrypt the following messages using an additive cipher with key* 5.

23. IT STY UFXX LT

24. MJWJ YTIFD LTSJ YTRTWWTB

25. GWJFI FSI HNWHZXJX

26. F HMNHPJS NS JAJWD UTY

In Exercises 27 *to* 30, *the message was encoded with an additive cipher. Assuming they consist of ordinary English words, find the key and the plaintext.*

27. NBLF MPWF OPU XBS

28. E XEPI SJ XAS GMXMIW

29. AMVL QV BPM KTWEVA

30. VLPSOH SOHDVXUHV

In Exercises 31 *to* 34, *use the Vigenère method to encrypt the plaintext* mary had a little lamb *with the given keyword.*

31. FERRY

32. PREVIEW

33. CUP

34. BANANA

9.4 Substitution Ciphers

Generating substitution ciphers

In an additive cipher, each letter is changed into another fixed letter. For example, in the Caesar cipher, every A becomes a D, every B becomes an E, and so on. To generalize this, we make a fixed substitution for each letter, but use a more complicated rule than simple addition. You could write out the complete alphabet, and the list of substitutions, such as

plaintext:	*a*	*b*	*c*	*d*	*e*	*f*	*g*	*h*	*i*	*j*	*k*	*l*	*m*
ciphertext:	D	A	F	B	Q	K	M	C	Z	Y	E	G	H
plaintext:	*n*	*o*	*p*	*q*	*r*	*s*	*t*	*u*	*v*	*w*	*x*	*y*	*z*
ciphertext:	W	X	P	R	S	I	T	L	J	N	V	U	O

In this code the message *an elephant* becomes `DWQGQPCDWT`. (Spaces and punctuation are usually ignored.) There are arithmetical ways to break codes such as this, based on the frequency of letters. For example, E is the most common letter in English, so if you think your message is written in English, try putting E for the most commonly occurring letter. This method only works for reasonably long messages; for short ones, you need the key (the whole substitution table).

Substitution ciphers with a keyword

One simplification of the substitution method involves the use of a keyword. The second row of the table is constructed by writing first the keyword and then the rest of the alphabet in order. The keyword should not be too short; it should contain no repeated letters, and preferably at least one of the letters should be rather late in the alphabet. For example, keyword *CRAZY* produces the table

plaintext:	*a*	*b*	*c*	*d*	*e*	*f*	*g*	*h*	*i*	*j*	*k*	*l*	*m*
ciphertext:	C	R	A	Z	Y	B	D	E	F	G	H	I	J
plaintext:	*n*	*o*	*p*	*q*	*r*	*s*	*t*	*u*	*v*	*w*	*x*	*y*	*z*
ciphertext:	K	L	M	N	O	P	Q	S	T	U	V	W	X

Sample Problem 9.14 *Which of the following would be a good keyword for a substitution code?* UP; COPPER; RAYON; FACED.

Solution. UP is too short. COPPER contains a repeated letter. RAYON is good. FACED is too close to the beginning of the alphabet; most letters would be unchanged.

Practice Exercise. Which of the following would be a good keyword for a substitution code? BIRTHDAY; APPLE; TEA; BEAD.

This method has the advantage that the encryption method is easy to remember, provided you choose a memorable keyword. However, long words with no repeated letters are not common. So a *reduced keyword* (or *reduced key phrase*) is used. Every time a letter is repeated, it is deleted from the key. This allows longer words, and phrases to be used. For example, Americans have no trouble remembering the name *George Washington*; it gives the reduced key GEORWASHINT and the table

plaintext:	*a*	*b*	*c*	*d*	*e*	*f*	*g*	*h*	*i*	*j*	*k*	*l*	*m*
ciphertext:	G	E	O	R	W	A	S	H	I	N	T	B	C
plaintext:	*n*	*o*	*p*	*q*	*r*	*s*	*t*	*u*	*v*	*w*	*x*	*y*	*z*
ciphertext:	D	F	J	K	L	M	P	Q	U	V	X	Y	Z

Letter frequencies

Here is a table of the frequency of letters in everyday English. Out of every thousand letters, the expected number of times a letter will appear is:

A	72	*B*	11	*C*	33	*D*	40
E	126	*F*	30	*G*	18	*H*	33
I	76	*J*	2	*K*	3	*L*	36
M	25	*N*	76	*O*	74	*P*	27
Q	3	*R*	83	*S*	58	*T*	90
U	30	*V*	13	*W*	14	*X*	5
		Y	21	*Z*	1		

so the most common letters are E, followed by T, then R, and then I, N, O and A are of approximately equal frequency.

Of course, this is experimental data from extensive counting, and another long count could give slightly different frequencies. Also, the frequencies will vary when different sorts of prose are considered. As an example, the relative frequency of letters in Lincoln's Gettysburg Address per thousand letters is

A	88	*B*	12	*C*	27	*D*	50
E	144	*F*	23	*G*	24	*H*	70
I	59	*J*	0	*K*	3	*L*	36
M	11	*N*	67	*O*	81	*P*	14
Q	1	*R*	69	*S*	38	*T*	109
U	19	*V*	21	*W*	24	*X*	0
		Y	9	*Z*	0		

An example

We give a detailed example of how frequency data is used. Consider the ciphertext

```
ZJR KRBRYMT DYQRYRQ ZJR WRPDBQ
NYIKMQR ZD MZZMPO MZ ZJYRR
ZJIYZG MU ZJR ZJIYQ NYIKMQR
DBR JDAY TMZRY DB ZJR TRSZ
ZJR SDAYZJ ZD ORRH IB YRWRYCR
```

In this example, and in the exercises for this section, we include spaces (but not punctuation), to make the decryption a little easier. But this would not usually happen in the real world.

First we calculate the letter frequencies. There are 116 letters, and the frequencies are

A	2	*B*	5	*C*	1	*D*	8
E	0	*F*	0	*G*	1	*H*	1
I	5	*J*	10	*K*	3	*L*	0
M	8	*N*	2	*O*	2	*P*	2
Q	6	*R*	22	*S*	2	*T*	3
U	1	*V*	0	*W*	2	*X*	0
		Y	13	*Z*	17		

The two most common letters are R and Z, so we first guess that R stands for e and Z stands for t. Making these substitutions, we obtain

```
tJe KeBeYMT DYQeYeQ tJe WePDBQ
NYIKMQe tD MttMPO Mt tJYee
tJIYtG MU tJe tJIYQ NYIKMQe
DBe JDAY TMteY DB tJe TeSt
tJe SDAYtJ tD OeeH IB YeWeYCe
```

(our substitutions are shown as lower case letters). The most commonly occurring two-letter combination in English is th, and RJ occurs eight times in the message. If R represents t then probably J stands for h. So

```
the KeBeYMT DYQeYeQ the WePDBQ
NYIKMQe tD MttMPO Mt thYee
thIYtG MU the thIYQ NYIKMQe
DBe hDAY TMteY DB the TeSt
the SDAYth tD OeeH IB YeWeYCe
```

Y occurs 13 times, and if we substitute r for Y we get the word three in the message; it is very hard to imagine any other translation of thYee. So:

```
the KeBerMT DrQereQ the WePDBQ
NrIKMQe tD MttMPO Mt three
thIrtG MU the thIrQ NrIKMQe
```

```
DBe hDAr TMter DB the TeSt
the SDArth tD OeeH IB reWerCe
```

The third word may well be `ordered`. So replace `D` by `o` and `Q` by `d`. This makes sense, because `D` occurred eight times, a relatively high frequency, and `o` is a common letter. The message becomes

```
the KeBerMT ordered the WePoBd
NrIKMde to MttMPO Mt three
thIrtG MU the thIrd NrIKMde
oBe hoAr TMter oB the TeSt
the SoArth to OeeH IB reWerCe
```

It looks as though `thIrtG` might be the word `thirty`. So we make those substitutions.

```
the KeBerMT ordered the WePoBd
NriKMde to MttMPO Mt three
thirty MU the third NriKMde
oBe hoAr TMter oB the TeSt
the SoArth to OeeH iB reWerCe
```

Now replace `B` by `n` and `M` by `a` to form the small words `on`, `one` and `at`. (We choose `on` rather than `of` because `one` is a word but `ofe` is not; neither `in` nor `it` were available because `i` has been used already.)

```
the KeneraT ordered the WePond
NriKade to attaPO at three
thirty aU the third NriKade
one hoAr Tater on the TeSt
the SoArth to OeeH in reWerCe
```

In order to make various recognizable words, we replace `K` by `g`, `T` by `l`, `W` by `s` and `P` by `c`. Then `N` becomes `b` and `O` becomes `k`.

```
the general ordered the second
brigade to attack at three
thirty aU the third brigade
one hoAr later on the leSt
the SoArth to keeH in reserCe
```

The final substitutions are `v` for `C`, `m` for `U`, `u` for `A`, `f` for `S`, and `p` for `H`. The plaintext is

```
the general ordered the second
brigade to attack at three
thirty am the third brigade
one hour later on the left
the fourth to keep in reserve
```

and the encryption table was

plaintext:	*a*	*b*	*c*	*d*	*e*	*f*	*g*	*h*	*i*	*j*	*k*	*l*	*m*
ciphertext:	*M*	*N*	*P*	*Q*	*R*	*S*	*K*	*J*	*I*		*O*	*T*	*U*
plaintext:	*n*	*o*	*p*	*q*	*r*	*s*	*t*	*u*	*v*	*w*	*x*	*y*	*z*
ciphertext:	*B*	*D*	*H*		*Y*	*W*	*Z*	*A*	*C*			*G*	

A few letters are not filled in because they were never used.

Exercises 9.4

In Exercises 1 *to* 4, *use a substitution cipher to encrypt the message* encipher is another word for encrypt, *with the given keyword.*

1. DROVE

2. BRAZEN

3. EXTRA

4. BOTANIC

In Exercises 5 *to* 8, *use the reduced keyphrase method to encrypt the message* encipher is another word for encrypt, *with the given keyphrase.*

5. GIANT LIZARD

6. EXTRATERRESTRIAL

7. MOTHER'S WORRY

8. DIXIELAND JAZZ

9. A substitution cipher produces the ciphertext KLKEVQOLLMPAOOFTER. Assuming the message was in simple English, what was the keyword?

(i) PROW

(ii) AZURE

(iii) BOXCAR

(iv) OXEN

In Exercises 10 *to* 17, *decide whether the given word would be a good keyword for a substitution code.*

10. POTATO

11. AORTA

12. IMAGINE

13. DAYLONG

14. AZIMUTH

15. PIE

16. DAZZLE

17. LAZY

18. Encrypt the following messages using the reduced key phrase with key *seven lazy days*:

(i) *It is common to illustrate sets and operations on sets by diagrams.*

(ii) *Find a boolean expression corresponding to the following circuit.*

(iii) *Three boys and four girls are to sit along a bench.*

(iv) *Prove that any two diagonal matrices commute.*

In Exercises 19 *to* 22, *repeat Exercise 18 using the given key phrase.*

19. *merry christmas*

20. *feet of clay*

21. *fox and geese*

22. *dazed and confused*

23. In the following sentence from a book on the history of mathematics, the most common letter is not a vowel. Decipher the quotation.

PWY QPCBN CZ ZYBDXPQ JXQP PWYCBYD AQ AVYIPBATXUJN JAVHYO FAPW PWY WAQPCBN CZ DXPWYDXPATQ

24. ONBBCSNNE AI T ZBTHQ SOQXQ ZQNZBQ UXNW ANST WAIPTLQ QTHO NPOQX UNX WNRAQ IPTXI

The above show business quotation has been enciphered using a substitution where P was enciphered as Z. To save you counting, letter frequencies are

A	4	*B*	4	*C*	1	*D*	0
E	1	*F*	0	*G*	0	*H*	2
I	4	*J*	0	*K*	0	*L*	1
M	0	*N*	9	*O*	4	*P*	3
Q	9	*R*	1	*S*	3	*T*	6
U	2	*V*	0	*W*	3	*X*	5
		Y	0	*Z*	3		

What is the plaintext?

25. Decrypt CDKLI QLT KLRDBIL ABI JFM HM HV MPL ECFMP APC QFVM GHZPM BTK KHL

26. Decrypt ABC DEFACG HAIACH JFKK ECMCL NC I OFMFKFPCG OQDEALR DEAFK JC HSCEG TQLC VQL NQQWH ABIE JC GQ QE OBCJFEX XDT

27. Decrypt EKD AZEFLFHE YDWFDMDH DMDNOEKFCP KD NDBTH AC EKD QBSXDE AI B CDG YAAX

9.5 Modern Cryptography

The RSA system

In the classic cryptographic systems, knowing the method of encryption allows you to decrypt a coded message. Modern developments have centered on cases where this is not true — even when the encryption method is known, decryption takes a prohibitively long time.

Several cryptographic systems with this property have been developed. For our purposes, it is sufficient to discuss one of these. We shall examine a system that was invented by Ronald L. Rivest, Avi Shamir and Leonard M. Adelman. It is usually called RSA.

We shall describe their system by giving a simple example. Your friend, the person to whom you wish to send a secret message, must choose two prime numbers — let us call them p and q. The *least common multiple* of $p-1$ and $q-1$ — the smallest positive integer such that both $p-1$ and $q-1$ divides it — will also be important, so we'll give it a name; call it m. Your friend then chooses an integer r, greater than 1, which is relatively prime to m (so no proper divisor of r can also be a divisor of $p-1$ or $q-1$). Your friend also has to find a number s such that $r \cdot s \equiv 1 \pmod{m}$. (This can be done using the Euclidean algorithm.)

For the purposes of our example, let us suppose your friend chose $p = 7$ and $q = 17$, so that $p-1 = 6, q-1 = 16$ and $m = 48$. Then r can have no proper divisor that also divides 6 or 16. 2, 3, 4 and 6 are impossible, but 5 or 7 are allowable, so let's say she chose 5. The number m must satisfy $5 \cdot s \equiv 1 \pmod{48}$. So $5 \cdot s$ must be one of the numbers $1, 49, 97, 145, 193, \ldots$. As $5 \cdot s$ must be divisible by 5, the product is 145, and $s = 29$.

Say you wish to send the message CAP. You first convert the message to numbers by replacing by 1, B by 2, ..., and Z by 26. The plaintext CAP becomes 3 1 16. To convert a number from plaintext to ciphertext, raise it to the power r and then reduce the answer modulo pq. Using $r = 5, pq = 7 \cdot 17 = 119$:

$3^5 = 243 \equiv 5 \pmod{119}$, so 3 is encoded 5;

$1^5 = 1 \equiv 1 \pmod{119}$, so 1 is encoded 1;

$16^5 \equiv 67 \pmod{119}$, so 16 is encoded 67. (We'll show how this was calculated in a moment.)

Therefore the ciphertext for CAP is 5 1 67.

When your friend receives the message 5 1 67, she raises each number to the power s, which is 29 in this case. 5^{29} is a very large number — bigger than 70 million million million, but it is easily shown to be 3 modulo 119 (see the next subsection). So 5 is decoded as 3. Similarly 1 decodes to 1 and 67 to 16, and CAP is recovered.

Formally, we would say that encryption is carried out by converting the message to a sequence of numbers and applying the *encryption rule* $x \mapsto x^r \pmod{pq}$; sometimes one refers to the *encoding function* $E(x) = x^r \pmod{pq}$, or calls r the *encryption exponent*. Similarly, decryption is carried out by applying the decoding function $D(x) = x^s \pmod{pq}$ and converting back from numbers to a message.

This method is not restricted to the 26 letters of the alphabet, represented by the numbers 1 to 26. The only restriction is that the number of symbols must be smaller than the modulus pq, or else there would not be enough residues to represent all the symbols. But this is no problem. In the real world, RSA encryption is used with moduli that run to hundreds of digits.

In practice, you would not transform CAP into the sequence 3 1 16. Instead you would run the numbers together, putting in a 0 so that A is written as 01 (every letter becomes two digits, or else you could not tell whether 12 represented AB or L), and encrypt the number 30116 by a single application of the encryption function. In general, one could make the whole message into one big number (either ignoring spaces between words, or putting 27 for a space), break it up into a sequence of smaller numbers — 10-digit or 50-digit numbers perhaps — and encrypt these numbers. As stated above, the only restriction is that the smaller numbers, the ones that are individually encrypted, must be smaller than the modulus m, in order to avoid ambiguity.

Sample Problem 9.15 *Suppose messages are constructed by substituting* 01 *for* A, 02 *for* B, ..., *and then running the digits together. What numbers represent* ZENITH *and* FOREST*?*

Solution. ZENITH is 260514092008; FOREST is 61518051920.

Practice Exercise. What number represents ALPHABET?

Binary expressions and powers

We chose CAP to show you that even in a very small example, big numbers arise. To calculate 16^5 on a calculator, we find $16^2 \cdot 16^2 \cdot 16$. Now $16^2 = 256$, and by subtracting 119 (twice) we find $256 \equiv 18$. Then $18 \cdot 256 \equiv 18 \cdot 18 \equiv 324 \equiv 86$. (Or you could say $18 \cdot 256 = 4608$. To reduce mod 119, first divide and find $4608/119 = 38.72\ldots$. Subtract 119 from 4608 thirty-eight times (you actually calculate $4608 - (38 \cdot 119)$) and get 86. Then $86 \cdot 16 = 1376 \equiv 67$.)

This is an example of using binary representations to calculate large powers of numbers. Suppose $r = (a_n a_{n-1} \ldots a_1 a_0)_2$, or

$$r = 2^n a_n + 2^{n-1} a_{n-1} + \cdots + 2a_1 + a_0.$$

Then any number b satisfies

$$b^r = b^{2^n a_n} \times b^{2^{n-1} a_{n-1}} \times \cdots \times b^{2a_1} \times b^{a_0}.$$

Now each a_i is either 1 or 0, so we have expressed b as a product of some of the terms

$$b^{2^n}, b^{2^{n-1}}, \ldots, b^2, b.$$

Calculating these powers of b is relatively easy: start from b, square it, square the result, and so on.

When the calculations are being carried out modulo n, the arithmetic is made even easier by the fact that you can reduce modulo n at each step.

Sample Problem 9.16 *Calculate* 4^{19} *and* $4^{19} \pmod{82}$.

Solution. $4^1 = 4, 4^2 = 16, 4^4 = 16^2 = 256, 4^8 = 256^2 = 65536$, $4^{16} = 65536^2 = 4294967296$, so

$$4^{19} = 4 \times 16 \times 4294967296 = 274877906944.$$

One could reduce this answer modulo 82, but it is easier to proceed as follows: $4^1 = 4, 4^2 = 16, 4^4 = 16^2 = 256 \equiv 10 \pmod{82}, 4^8 \equiv 10^2 = 100 \equiv 18$, $4^16 \equiv 18^2 = 324 \equiv -4$, so

$$4^{19} \equiv 4 \times 16 \times (-4) = -256 \equiv 72 \pmod{82}.$$

Practice Exercise. Calculate 3^{13} and $3^{13} \pmod{19}$.

Notice that the calculation of 4^{19} took six multiplications. Multiplying it out directly would take 18 multiplications. Of course, computer programs are available that do these calculations automatically, and for very large numbers the saving in time is significant.

We now perform the calculation that was required in the RSA example:

Sample Problem 9.17 *Find* $5^{29} \pmod{119}$.

Solution. We use the decomposition $29 = 16+8+4+1$. We calculate $5^2 = 25$, $5^4 = 625 \equiv 30 \pmod{119}$, $5^8 = (5^4)^2 \equiv 30^2 \equiv 67 \pmod{119}$, $5^{16} \equiv 67^2 \equiv 86$, $\pmod{119}$ and

$$\begin{aligned}
5^{29} &= 5^{16} \cdot 5^8 \cdot 5^4 \cdot 5 \pmod{119} \\
&\equiv 86 \cdot 67 \cdot 30 \cdot 5 \pmod{119} \\
&\equiv 86 \cdot 67 \cdot 150 \pmod{119} \\
&\equiv 86 \cdot 67 \cdot 31 \pmod{119} \\
&\equiv 86 \cdot 2077 \pmod{119} \\
&\equiv 86 \cdot 54 \pmod{119} \\
&\equiv 4644 \equiv 3 \pmod{119}.
\end{aligned}$$

Practice Exercise. Find $7^{29} \pmod{53}$.

Public key cryptosystems

RSA can be used to construct a *public key cryptosystem.* Suppose several different people wish to send secret messages to the same recipient. One example is a brokerage house, where the customers want to send their brokers orders to buy or sell; another is military intelligence, where different operatives all report to the same central office. More generally, we can think of an electronic mail service, in which anybody might wish to send a message to any member of the service. We refer to those who can send and receive messages as *members* of the system. There may also be other *users* who receive, but do not send, messages.

Nowadays we must assume that any outsider can read any computer system; hackers have broken into even highly secret government and military computers. So how can we keep outsiders from understanding our messages, even though they can read the messages? The answer must lie in encryption.

Here is one method, using the RSA scheme. The recipient could publish the key — the product pq and the power r. This is enough information for a member of the public to encrypt a message. However, when the ciphertext is received, the least common multiple of $p-1$ and $q-1$ is needed for decryption. In theory, this can be calculated if you know pq; you just break this number down into its factors p and q, and then you know $p-1$ and $q-1$. For example, it takes only a fraction of a second for a computer to find that $119 = 7 \cdot 17$. But breaking a large number down into factors is a very hard computation. It is possible to choose primes p and q so large that, if you are given only their product, it would take a computer many years to find p and q. Your enemy doesn't have years to spend, so only the proper recipient can decrypt the ciphertext.

Other methods have also been developed, and RSA is just one example. But all the systems work in this way. Every possible recipient has an encryption method. This is "published": they might actually publish the key in a directory, or users might be given access to a computer program that encrypts the message when the recipient is specified, and then delivers it. In any event, member A somehow provides a (public) function, call it e_A, that encrypts messages. Then A, and only A, has access to a (private) decryption function, d_A say, that decrypts. If the plaintext message m is delivered to the system, then A receives the message $e_A(m)$, and then calculates $d_A(e_A(m))$, which equals m.

Suppose you subscribe to an electronic news service. In this case, there is only one member of the system, and all other users only need to decrypt messages. The system will select an encryption function for you, and send details of the decryption method to your computer. These details will again be encrypted, so that all details of the system are kept secret by the news provider.

Security of RSA

The most important feature of any public key cryptosystem is that, even though you know the function e_A, it must be an extremely long computational task to calculate the function d_A.

In discussing any computational problem, it is important to know how long it takes to perform the computation. Usually larger problems take more time. For example, we saw in Section 3.3.4 that a circuit to add two binary digits needs four gates, while our circuit to add three binary digits needed nine gates. If your computer took the same amount of time to utilize any gate, increasing the size of the problem from two to three (the number of binary digits to be added) will multiply the computation time by $\frac{9}{4}$.

Suppose stating a computational problem requires inputting n binary digits to a computer. The number of steps required to carry out the computation will depend on the size of n. Say the solution requires $f(n)$ steps. The function $f(n)$ is called the *complexity* of the algorithm. Sometimes one particular instance of a problem can be solved more quickly than another, so $f(n)$ will be the largest solution-time for size n problems (the *worst case*).

If $f(n)$ is a polynomial function of n (or smaller), we say the algorithm is *polynomial.* If no polynomial algorithm is known for solving a problem, it follows that increasing the size of the problem will cause a very great increase in the time needed to solve it†. Some problems, such as finding the greatest common divisor of two integers or testing whether a graph is Eulerian, have polynomial algorithms. There are many other problems, such as the Traveling Salesman problem, for which no polynomial algorithm is known.

Suppose you are given the public information, m and r, for an RSA system. In order to decipher messages, you need the number s. In order to calculate s you need to know not only r but also $(p-1)(q-1)$, and in order to calculate $(p-1)(q-1)$ you need to know the factors p and q, that is you need to decompose m into its prime factors. But no polynomial algorithm is known for factoring integers, and many people believe that no such algorithm exists. The difficulty of factoring becomes much greater as the prime factors become larger. So we can make an RSA system safe against codebreakers by making the modulus a product of two factors, and making those factors large. Current suggestions are that p and q should be at least 100 decimal digits in length.

Of course, it is always possible that somebody will discover a polynomial algorithm to factor integers, and that would make RSA insecure. Many mathematicians believe this to be theoretically impossible, but no proof is known.

Exercises 9.5

1. Consider an RSA scheme with $p = 5$ and $q = 11$.

(i) What are the possible r values if $r < 10$?

(ii) If $r = 3$ is chosen, use the system to encrypt the plaintext "23."

---

†This informal discussion of complexity is very superficial, and is only intended to give you a rough idea of why RSA is secure. The interested reader should consult books on cryptography or computational theory.

2. Consider an RSA scheme with $p = 7$ and $q = 11$.
   (i) What are the possible r values if $r < 10$?
   (ii) If $r = 5$ is chosen, use the system to encrypt the plaintext "13."

3. An RSA encryption system has $pq = 143$ and $r = 43$. Find p, q and s.

4. An RSA encryption system has $pq = 143$ and $r = 23$. Find p, q and s.

5. Use the RSA scheme with $pq = 85$ and $s = 3$ to decrypt the message "13."

In Exercises 6 *to* 14, *assume you are using an RSA scheme with* $p = 13$ *and* $q = 23$. *Could the given number be chosen as a value for* r?

6. 6
7. 5
8. 9
9. 33
10. 31
11. 7
12. 2
13. 35
14. 55

Evaluate the expressions in Exercises 15 *to* 32.

15. $2^{45} \pmod{11}$
16. $3^{19} \pmod{23}$
17. $2^{21} \pmod{44}$
18. $3^{25} \pmod{37}$
19. $4^{12} \pmod{19}$
20. $2^{17} \pmod{81}$
21. $2^{25} \pmod{7}$
22. $5^{33} \pmod{29}$
23. $12^{50} \pmod{35}$
24. $7^{28} \pmod{17}$
25. $2^{81} \pmod{13}$
26. $3^{54} \pmod{61}$
27. $2^{53} \pmod{11}$
28. $17^{35} \pmod{31}$
29. $4^{37} \pmod{51}$
30. $3^{29} \pmod{77}$
31. $9^{23} \pmod{79}$
32. $6^{41} \pmod{33}$

33. You are using an RSA system with $r = 3$ and $pq = 55$.
   (i) What is the decryption key for this system?
   (ii) You code A as 01, B as 02, and so on, and send each symbol separately. How would you encrypt these messages?

   a. `MAN DEAD`
   b. `FIFTEEN`
   c. `EVIL`
   d. `TOMORROW`

34. Repeat the preceding problem using an RSA system with $r = 5$ and $pq = 51$.

35. Using an RSA system with $s = 13$ and $pq = 51$ you receive the message 04 01 05 01. Decrypt it (01 = A, etc).

36. Using an RSA system with $s = 5$ and $pq = 51$, decrypt the message 04 20 01 04 20 05 04.

37. An encoded message 08 15 15 25 01 25 49 20 20 49 has been intercepted. It was sent to someone whose public key encoding function is given by

$$m \mapsto m^{33} (\text{mod } 55).$$

(i) Find s so that

$$x \mapsto x^s (\text{mod } 55)$$

is the decryption function.

(ii) Use this decryption function to find numbers that correspond to the message.

(iii) It is assumed that the person who sent the encoded message used the correspondence

$$a \mapsto 01, b \mapsto 02, c \mapsto 03, \ldots, z \mapsto 26$$

to convert letters to numbers and then used the above function to encode the numbers. Find the message.

38. An RSA scheme uses modulus 91 and encryption rule $m \mapsto m^{29}$.

(i) What are the prime factors (p and q) of the modulus?

(ii) What is the decryption rule?

(iii) You receive the ciphertext

$$72\ 88\ 88\ 72 \quad 13\ 21\ 72\ 88 \quad 33\ 51\ 51\ 33.$$

If the encryptor is using the correspondence

$$1 \mapsto 1, \ldots, 9 \mapsto 9, 0 \mapsto 10, A \mapsto 11, \ldots \mapsto 36$$

to turn English letters and digits into numbers, what was the original plaintext?

9.6 Attacks on the RSA system

If the RSA system is secure, what is the point of attacking it?

Even though the system itself seems impervious to attack, it is possible that it has been implemented in such a way that it can be broken. We would then say that the RSA system has been *compromised*. In this section we shall discuss some faulty implementations of RSA.

Using the Chinese remainder theorem

Consider a public key system using the RSA scheme that has been set up as suggested in the preceding section. Member i has two primes p_i and q_i, and an exponent r_i which is relatively prime to $(p_i - 1)(q_i - 1)$. He releases to the public the information (m_i, r_i), where $m_i = p_i q_i$. He keeps secret the information (p_i, q_i, s_i) where s_i is the inverse of r_i modulo $(p_i - 1)(q_i - 1)$:

$$s_i \equiv r_i^{-1} (\text{mod } (p_i - 1)(q_i - 1)).$$

Suppose a user (such as a system operator or a military commander) needs to communicate the same message to most or all members of the system. For the convenience of this user, it has been suggested that one should use a common value r for all the r_i and that it should be a small number. For example, one might set each r_i equal to 3. This would in fact be cheaper and quicker to operate. However, we shall now show that this system would not be secure.

Let us suppose you wish to send the message A to users U_1, U_2 and U_3. (We assume that A is an integer that can be translated back to a verbal message by some prearranged system.) You would proceed to calculate the three encrypted messages C_1, C_2, and C_3, and transmit C_i to U_i. The C_i are remainders calculated as follows:

$$\begin{aligned} A^3 &\equiv C_1 (\text{mod } m_1), \\ A^3 &\equiv C_2 (\text{mod } m_2), \\ A^3 &\equiv C_3 (\text{mod } m_3). \end{aligned} \tag{9.4}$$

We can assume the moduli m_1, m_2 and m_3 are relatively prime, because otherwise the system is compromised (see Sample Problem 9.19). If a codebreaker solves the equations (9.4) using the Chinese remainder theorem, she obtains the solution $A^3 = X$ where X is the unique integer such that

$$A^3 \equiv X (\text{mod } m_1 m_2 m_3), 0 \le X < m_1 m_2 m_3.$$

In order for the recipients to decode messages unambiguously, all messages must be positive integers smaller than the modulus. So $A < \min(m_1, m_2, m_3)$. This means $A^3 < m_1 m_2 m_3$, and in fact

$$A^3 = X$$

is an *equality* between integers. So A equals the cube root of X (the ordinary cube root, in ordinary arithmetic). There is a polynomial algorithm for extracting roots in ordinary integer arithmetic, so A can be calculated easily by the codebreaker. (There is no known polynomial algorithm for extracting roots in *modular* arithmetic.)

Suppose we apply the same analysis to a system with common exponent r and n users. We obtain the equation

$$A^r \equiv X (\text{mod } m_1 m_2 \cdots m_n), 0 \le X < m_1 m_2 \cdots m_n.$$

In order to achieve the *integer* equation $A^r = X$ we need $A^r < m_1 m_2 \cdots m_n$. This will certainly be true if $n \geq r$. So the problem can be avoided by choosing r greater than n. But this will not provide the suggested savings in a large system.

Here is another way around this attack. Say there are 100 users of the system. Select a random arrangement π of the integers $\{1, 2, \ldots, 100\}$. Instead of a message of (say) 993715, send to user i the message $99371500 + \pi(i)$. When a user decrypts to A, he takes $\lfloor \frac{A}{100} \rfloor$ as the message.

The following example illustrates this method. The moduli are far too small for a practical RSA system — they involve the six smallest primes with which 3 can be used as an exponent (remember that if p is one of the primes, then $p - 1$ and the exponent must be coprime). However, observe that some very large numbers arise.

Sample Problem 9.18 *In an RSA-based public key cryptosystem, three users have moduli $m_1 = 46$, $m_2 = 77$ and $m_3 = 85$. A message is sent to all three using the common encryption exponent 3, and the coded messages received are $C_1 = 37$, $C_2 = 62$, and $C_3 = 68$ respectively. What was the original message?*

Solution. We have the following equations.

$$\begin{aligned} A^3 &\equiv 37 \pmod{46} \\ A^3 &\equiv 62 \pmod{77} \\ A^3 &\equiv 68 \pmod{85}. \end{aligned}$$

In the notation of the Chinese remainder theorem in Section 9.2, $A_1 = 6545, A_2 = 3910, A_3 = 3542, A = 301070, a_1 - 37, a_2 = 63, a_3 = 68$. The numbers x_1, x_2, x_3 are the solutions to

$$\begin{aligned} 6545x &\equiv 1 \pmod{46} \\ 3910x &\equiv 1 \pmod{77} \\ 3542x &\equiv 1 \pmod{85} \end{aligned}$$

respectively. Using the Euclidean algorithm we find $x_1 = 39, x_2 = 9, x_3 = 3$. Therefore

$$\begin{aligned} A^3 &\equiv 6545 \times 39 \times 37 + 3910 \times 9 \times 62 + 3542 \times 3 \times 68 \\ &= 12348783 \equiv 4913 \pmod{301070}. \end{aligned}$$

Therefore $A^3 = 4913$, and $A = 17$.

Practice Exercise. In the same system, the coded messages received were 21, 35 and 3 respectively. What was the original message?

Sample Problem 9.19 *Suppose a public key system, of the kind we are discussing, has two moduli m_1 and m_2 that are not coprime. Why is the system insecure?*

Solution. We assume that each modulus is a product of two primes; say $m_1 = pq_1$ and $m_2 = pq_2$, where p is the common factor. All of the moduli are common knowledge, so a codebreaker can easily test all pairs of them for common factors, and will find p. She can then calculate q_1 and q_2. This enables her to quickly find the inverses of r_1 modulo m_1 and r_1 modulo m_1, and decode the messages to users 1 and 2.

Belonging to an exponent

Suppose a and m are relatively prime integers. In the arithmetic modulo m, the sequence of powers $a^1, a^2, a^3, \ldots$ cannot all be different, because the arithmetic has only a finite number of elements, so there will exist positive integers d and e for which $a^e \equiv a^{d+e} (\text{mod } m)$. Therefore

$$m \mid a^{d+e} - a^e = a^e(a^d - 1).$$

Since m has no common factor with a, m must divide $a^d - 1$, so $a^d \equiv 1 (\text{mod } m)$ for some values of d. The smallest such positive integer d is called the *exponent* of a modulo m, and we say a belongs to the exponent $d (\text{mod } m)$. For example, modulo 7, $4^1 \equiv 4$, $4^2 \equiv 2$, $4^3 \equiv 1$, so 4 belongs to the exponent $3 (\text{mod } 7)$.

Sample Problem 9.20 *Find the exponent of* 11 *modulo* 84.

Solution. Calculating modulo 84,

$$\begin{array}{rcrcr} 11^2 & = & 121 & \equiv & 37 \\ 11^3 & \equiv & 407 & \equiv & 71 \\ 11^4 & \equiv & 781 & \equiv & 25 \\ 11^5 & \equiv & 275 & \equiv & 23 \\ 11^6 & \equiv & 253 & \equiv & 1 \end{array}$$

so 11 belongs to exponent $6 (\text{mod } 84)$.

Practice Exercise. Find the exponent of 5 modulo 48.

The result of the following example will be used shortly.

Sample Problem 9.21 *Find the exponent of* 49 *modulo* 214684.

Solution. We see

$$\begin{array}{rcrcr} 49^2 & = & 2401 & & \\ 49^3 & = & 117649 & & \\ 49^4 & = & 5764801 & \equiv & 183017 \\ 49^5 & \equiv & 8967833 & \equiv & 165789 \\ 49^6 & \equiv & 8123661 & \equiv & 180353 \\ 49^7 & \equiv & 8837297 & \equiv & 35253 \\ 49^8 & \equiv & 1727397 & \equiv & 9925 \\ 49^9 & \equiv & 486325 & \equiv & 56957 \\ 49^{10} & \equiv & 2790893 & \equiv & 1 \end{array}$$

so the exponent is 10.

Consider an RSA cryptosystem with $p = 383$, $q = 563$, $r = 49$. The public information is $m = 215629$, $r = 49$. Using the factorization of m (i.e., the values of p and q) one finds that $(p-1)(q-1) = 214684$, $g = 56957$. So encryption $A \to C$ and decryption $C \to A$ follow the rule

$$\begin{aligned} C &\equiv A^{49} \pmod{215629} \\ A &\equiv C^{56957} \pmod{215629} \end{aligned}$$

Suppose $C_1 = C^{49}$, $C_2 = C_1^{49}$, and in general $C_{i+1} = C_i^{49}$. Then $C_i = C^x$ where $x \equiv 49^i \pmod{214684}$. We just saw that 49 belongs to the exponent 10 modulo 214684, so $49^{10} \equiv 1$ and $C^10 = C, C^9 = A$.

For example, say $A = 123456$. We see

$$\begin{aligned}
C &= (123,456)^{49} &= 1603 \\
C_1 &= C^{49} &= 180661 \\
C_2 &= C_1^{49} &= 109265 \\
C_3 &= C_2^{49} &= 131172 \\
C_4 &= C_3^{49} &= 98178 \\
C_5 &= C_4^{49} &= 56372 \\
C_6 &= C_5^{49} &= 63846 \\
C_7 &= C_6^{49} &= 146799 \\
C_8 &= C_7^{49} &= 85978 \\
C_9 &= C_8^{49} &= 123456 &= A \\
C_{10} &= C_9^{49} &= 1603 &= C
\end{aligned}$$

So you could try to break an RSA encryption by forming the sequence C^1, C^2, C^3, ... If this sequence contains a repetition, say $C^{i+1} = C$, then C_i will equal the original message A.

The "solution" to this problem is to choose a value r that belongs to a large exponent modulo $(p-1)(q-1)$. However, there is no known polynomial algorithm for finding an exponent.

Fermat factorization

Finally, some choices of two primes are such that their product is easy to factorize. This happens if the two primes are close in value. Fermat proposed the following method for factoring an odd positive integer n into two approximately equal factors.

Suppose the factorization of n is $n = ab$, where $a > b$. Write $x = \frac{1}{2}(a+b)$, the number midway between a and b, and $y = \frac{1}{2}(a-b)$. Then $a = x+y$, $b = x-y$, and $n = (x+y)(x-y)$. This is the desired factorization. Moreover, $n = x^2 - y^2$, so $y = \sqrt{x^2 - n}$. If we could guess x, then we could calculate y and find the calculation.

Fermat factorization proceeds as follows. First find $\sqrt{n}$. If this is an integer, we have found the desired factorization ($n = x^2$ and $y = 0$). Otherwise, we set $x = \lceil \sqrt{n} \rceil$ and calculate $\sqrt{x^2 - n}$. If this is an integer, set $y = \sqrt{x^2 - n}$. Otherwise, increase x by 1 and test $\sqrt{x^2 - n}$ again. If we continue in this way, eventually a factorization will be found. Even if n is a prime, the method will eventually give the factorization $n = n \times 1$.

This factorization algorithm is not polynomial, and will take a very long time if the two factors are not close. However, it is often very fast. In the following example, the algorithm concludes after three steps, although one factor is 20% larger than the other.

Sample Problem 9.22 *Use Fermat factorization to factor* $n = 430681$.

Solution. The square root of n is 656.2.... This is not an integer, so we start testing from $x = 657$.

$x = 657 \quad x^2 - n = 968 \quad \sqrt{x^2 - n} = 31.1\ldots$
$x = 658 \quad x^2 - n = 2283 \quad \sqrt{x^2 - n} = 47.7\ldots$
$x = 659 \quad x^2 - n = 3600 \quad \sqrt{x^2 - n} = 60.$

So we take $y = 60$. The factorization is $n = (x-y)(x+y)$, or

$$430681 = 599 \times 719.$$

Practice Exercise. Use Fermat factorization to factor 449329.

In view of the Fermat factorization problem, it is recommended that the two primes chosen for RSA should be relatively different in size. If a 230-digit modulus is used, primes of about 100 digits and 130 digits respectively have been recommended.

Exercises 9.6

In an RSA-based public key cryptosystem, three users have moduli $m_1 = 46$, $m_2 = 77$ *and* $m_3 = 85$. *A message is sent to all three using the common encryption exponent* 3. *The coded messages received are shown in Exercises* 1 *to* 6. *In each case find the original message.*

1. $C_1 = 2, C_2 = 15, C_3 = 16.$

2. $C_1 = 5, C_2 = 6, C_3 = 59.$

3. $C_1 = 17, C_2 = 64, C_3 = 60.$

4. $C_1 = 26, C_2 = 34, C_3 = 28.$

5. $C_1 = 43, C_2 = 22, C_3 = 56.$

6. $C_1 = 6, C_2 = 50, C_3 = 2.$

In Exercises 7 *to* 10, *find the exponent of the given integer for the given modulus.*

7. $37 \pmod{66}$

8. $13 \pmod{66}$

9. $13 \pmod{84}$

10. $5 \pmod{84}$

In Exercises 11 *to* 16, *use the "exponent" method to decrypt the coded message.*

11. Exponent 3, modulus 55, coded message 48

12. Exponent 3, modulus 55, coded message 18

13. Exponent 7, modulus 143, coded message 63

14. Exponent 7, modulus 143, coded message 42

15. Exponent 5, modulus 117, coded message 92

16. Exponent 17, modulus 365, coded message 37

In Exercises 17 *to* 26, *use Fermat factorization to factor the given integer.*

17. 396611

18. 436021

19. 282943

20. 321389

21. 301033

22. 70303

23. 294677

24. 788131

25. 862091

26. 499561

9.7 Other Cryptographic Ideas

Signature systems

The public-key encryption schemes described above can be used to protect your mail so that others cannot read and understand it, but there is another problem: how can you be sure who really sent the message?

In a written message, we usually expect the writer to sign the document. This is how banks verify checks and lawyers verify wills. But the traditional "pen-on-paper" signature does not apply to electronic information transfer. We need some sort of electronic "signature," a means of authenticating a document.

Suppose you are using a system in which every member has made public an encryption rule and has a private decryption rule. For example, if the system is based on RSA, the public information consists of an exponent, a modulus, and instructions to raise that integer to the exponent and reduce modulo the modulus. (Of course there are also some instructions on how to convert a message to an integer, but we shall assume this is done automatically.) For simplicity let us denote A's encryption rule as $m \mapsto e_A(m)$ and her decryption rule by $c \mapsto d_A(c)$.

We assume that A wishes to send message a to B. We assume that B will know the message comes from A — if this is not to be public knowledge, it can be included in the message a. A first sends the message $e_B(a)$. She then sends $d_A(b)$, where b is some other message (B's name, or the first part of a, or possibly — if added security is needed — the whole of a). B first calculates $d_B(e_B(a))$, recovering the original message. Then he calculates $e_A(d_A(b))$ as a check on authenticity. (We assume he knows the message comes from A — if this is not to be public knowledge, it can be included in the message a.)

Avoiding the key exchange

Suppose two people wish to communicate using a cryptosystem, but they do not want to publish their encryption methods. This is still possible; we shall describe a method called the *Massey–Omura scheme*, also known as *Shamir's no-key algorithm*. The two participants are Alice (A) and Bob (B). Each chooses an encryption scheme. We again denote A's encryption rule as $m \mapsto e_A(m)$ and her decryption rule by $c \mapsto d_A(c)$.

Say Alice wishes to send message a to Bob. She first sends the message $e_A(a)$. Bob does not know Alice's secret key, so he cannot decipher the message. Instead, he *reencrypts* it with his own encryption method. So

> Alice sends $e_A(a)$ to Bob;
> Bob sends $e_B(e_A(a))$ to Alice.

The message received by Alice is incomprehensible to anybody. Alice proceeds to decode it with her own secret key, then sends it back to Bob. Bob then decodes the message with his own key. So

> Alice sends $d_A(e_B(e_A(a)))$ to Bob;
> Bob calculates $d_B(d_A(e_B(e_A(a))))$.

In general, $e_B(d_A(e_B(e_A(a))))$ will be a string of nonsense. But suppose the two functions d_A and e_B commute. Then $d_A(e_B(x)) = e_B(d_A(x))$ for any x, so

$$d_B(d_A(e_B(e_A(a)))) - d_B(e_B(d_A(e_A(a)))) = d_B(e_B(a)) = a$$

and Bob recovers the message.

Here is an analogous situation, which may clarify what is going on here. Suppose Alice wants to send a suitcase to Bob, She does not trust the employees of the courier service, so she locks the suitcase. Of course, Bob does not have the key. He puts a chain around the suitcase, attaches a padlock, and ships it back to Alice. She then unlocks the suitcase — she still cannot open it, because of the chain — and sends it back to Bob, who removes the chain.

The only drawback to this system is that two functions d_A and e_B will not necessarily commute. For example, the RSA scheme cannot be used. (See Exercise 9.7.1.) However, RSA is suitable if the two moduli are the same. If both Alice and Bob use modulus m, then $d_A(x) = x^s (\bmod m)$ and $e_B(x) = x^r (\bmod m)$ for some integers r and s, and for any x

$$d_A(e_B(x)) = d_A(x^r) = (x^r)^s = (x^s)^r = e_B(x^s) = e_B(d_A(x)),$$

so the functions commute. So Alice and Bob would need to have some communication, but the information they exchange in public — the modulus — is not enough to allow others to send messages.

Tossing a coin

We conclude with another example of the need to authenticate information. The problem is: how is it possible to simulate a random process with a 50% probability of success or failure. If two people meet face to face, they can flip a coin. But this is not possible if the two participants are corresponding by telephone or computer. Assuming that they do not trust each other, what is to stop A from tossing a head, and then, when B calls heads, saying that the toss was a tail? Instead of authenticating the coin toss, we shall outline a fair procedure that has a 50% chance that A wins and a 50% chance that B wins.

First A selects two large primes p and q, and sends the product pq (but *not* the factorization) to B. To "win the toss," B must factor pq. What he does is select an integer x in the range $1 \leq x \leq pq$, and send x^2 to A. Now A selects a number y such that $y^2 \equiv x^2 (\bmod pq)$ (a "square root" of x^2) and sends it to B.

If $y^2 \equiv x^2$, then

$$pq \mid y^2 - x^2,$$
$$pq \mid (y-x)(y+x),$$

and there are four possibilities, according to whether p divides $y-x$ or $y+x$ and q divides $y-x$ or $y+x$. So we get four possiblecases:

$$\begin{aligned} y &\equiv x(\bmod p) \\ y &\equiv x(\bmod q) \end{aligned} \qquad (9.5) \qquad\qquad \begin{aligned} y &\equiv -x(\bmod p) \\ y &\equiv x(\bmod q) \end{aligned} \qquad (9.7)$$

$$\begin{aligned} y &\equiv x(\bmod p) \\ y &\equiv x(\bmod q) \end{aligned} \qquad (9.6) \qquad\qquad \begin{aligned} y &\equiv -x(\bmod p) \\ y &\equiv -x(\bmod q) \end{aligned} \qquad (9.8)$$

Each of these has a unique solution modulo pq : (9.5) has solution $x \equiv y$, (9.8) has $x \equiv -y$, while (9.6) and (9.7) relate x to y in other ways. It is easy to check that they are all different. A cannot tell which of the four possibilities equals x, so she chooses one at random.

If she chooses the solution corresponding to case (9.5) or (9.8), and sends it back to B, this does not help B to factor pq. However, say she chooses the solution to (9.7). Then

$$p \mid y+x, \quad q \mid y-x.$$

Since $y \not\equiv x$, p does not divide $y-x$. So $(y-x, pq) = q$. So B can factor pq; similarly he can factor pq if (9.6) is chosen.

Notice that A must send a number N that has two prime factors. If N is itself a prime number (or a power of a prime), then B always loses. This problem can be handled by allowing B to test whether N is prime. Ordinarily, this would be as difficult as factoring N, and we have already pointed out that factoring takes a long time. But by using a probabilistic test for primality, B can very quickly (and with a very small probability of error) determine whether or not N is prime.

If A sends a number with more than two prime factors, she decreases her chance of winning — with n factors there are 2^n roots, and the number of cases in which A wins is only 2. Similarly B can simply deny that he has the ability to factor. So either player can cheat *in the opponent's favor*, but not for his/her own gain.

Exercises 9.7

1. A uses an RSA cryptosystem with modulus 33 and encryption exponent 7. B uses an RSA cryptosystem with modulus 35 and encryption exponent 5.

(i) Show that A's decryption exponent is 3.

(ii) Find $[2^5 \pmod{35}]^3 \pmod{33}$.

(iii) Find $[2^3 \pmod{33}]^5 \pmod{35}$.

(iv) Show that the functions d_A and e_B, defined by

$$d_A(x) = x^3 \pmod{33} \text{ and } e_B = x^5 \pmod{35},$$

do not commute.

2. Repeat the preceding exercise with the message 2 replaced by 5.

3. Calculate $[5^3 \pmod 7]^4 \pmod{11}$ and $[5^4 \pmod{11}]^3 \pmod 7$.

4. Calculate $[2^3 \pmod 5]^5 \pmod 7$ and $[2^5 \pmod 7]^3 \pmod 5$.

Solutions to Practice Exercises

Chapter 1

1.1 $\{x : x^3 - 5x^2 + 11x + 6 = 0\}$, the set of the first three positive whole numbers and $\{1,2,3\}$ are some answers.

1.2 $72 = 2^3 \cdot 3^2, 84 = 2^2 \cdot 3 \cdot 7, (72,84) = 2^2 \cdot 3 = 12.56 = 2^3 \cdot 7, 42 = 2 \cdot 3 \cdot 7, (56,42) = 2 \cdot 7 = 14.$

1.3 $1, 0, 1, x^3, -.5, .04$

1.4 $t^{-2}/t^{-3} = t^3/t^2 = t$; $y^{5-2} = y^3$; $(4x^{-2})(3x^4) = 12x^{4-2} = 12x^2$.

1.5 $3^2 = 9$ so $\log_3 9 = 2$; $5^3 = 125$ so $\log_5 125 = 3$; $\sqrt{4} = 2$ so $\log_4 2 = \frac{1}{2}$.

1.6 $9\sqrt{3} = 3^2 \cdot 3^{\frac{1}{2}}$ so $\log_3 9\sqrt{3} = \frac{5}{2}$ and $\log_3 (9\sqrt{3})^5 = 5 \log_3 9\sqrt{3} = \frac{25}{2}$or 12.5.

1.7 $2, -2, 5, -4, 3.1, 4.4.$

1.8 $\sum_{i=3}^{5} i(i-1) = 3 \cdot 2 + 4 \cdot 3 + 5 \cdot 4 = 6 + 12 + 20 = 38; \sum_{i=2}^{6} i = 2 + 3 + 4 + 5 + 6 = 20.$

1.9 $1 + 3 + 5 + 7 + 9 = \sum_{i=1}^{5} 2i - 1$; $8 + 27 + 64 + 125 = \sum_{i=2}^{5} i^3$.

1.11 $3 + 7 + \ldots + 43 = \sum_{i=1}^{11} 4i - 1 = 4 \cdot \sum_{i=1}^{11} i - 11 = 4 \cdot \frac{12 \cdot 11}{2} - 11 = 264 - 11 = 253.$

1.12 $(144)_5 = 5^2 + 4 \cdot 5 + 4 = 49, (203)_7 = 2 \cdot 7^2 + 3 = 101, (112)_3 = 3^2 + 3 + 2 = 14.$

1.13 $(.242)_5 = 2\cdot 5^{-1} + 4\cdot 5^{-2} + 2\cdot 5^{-3} = .4 + .16 + .016 = .576$ so (using the preceding answer) $(144.242)_5 = 49.576$.

1.14
$$\begin{array}{r|l} 5 & 54 \\ 5 & 10 + 4 \\ & 2 + 0 \end{array} \quad 54 = (204_5 \qquad \begin{array}{r|l} 6 & 103 \\ 6 & 17 + 1 \\ & 2 + 5 \end{array} \quad 103 = (251)_6$$

1.15

$$\begin{aligned} 2\cdot(.40625) &= 0.8125,\ \text{first digit } 0 \\ 2\cdot(.8125) &= 1.625,\ \text{second digit } 1 \\ 2\cdot(.625) &= 1.25,\ \text{third digit } 1 \\ 2\cdot(.25) &= 0.5,\ \text{fourth digit } 0 \\ 2\cdot(.5) &= 1.0,\ \text{fifth digit } 1 \end{aligned}$$

So $.40625 = (.01101)_2 = .01101B$.

1.16 $53 = (125)_6$.

$$\begin{aligned} 6\cdot(.12) &= 0.72,\ \text{first digit } 0 \\ 6\cdot(.72) &= 4.32,\ \text{second digit } 4 \\ 6\cdot(.32) &= 1.92,\ \text{third digit } 1 \\ 6\cdot(.92) &= 5.52,\ \text{fourth digit } 5 \\ 6\cdot(.52) &= 3.12,\ \text{fifth digit } 3 \\ 6\cdot(.12) &= 0.72,\ \text{sixth digit } 0\ldots \end{aligned}$$

The whole process recurs. So $.12 = .(\overline{04153})_6$, $53.12 = (125.\overline{04153})_6$.

1.17 $5B3.76 = 0101\ 1011\ 0011.0111\ 0110 = 101\ 1011\ 0011.0111\ 011$

1.18 $10\ 1101\ 1011.01 = 0110\ 1101\ 1011.0100 = 6DB.4$

1.19 $1.750\times 10^1, -1.112\times 10^1, 1.401\times 10^0$

1.20 (1) -8.050×10^2, (2) 1.088×10^2, (3) 1.144×10^3, (4) 3.077×10^{-1}

1.21 $1.043\times 10^2 + 3.223\times 10^3 = (.1043 + 3.223)\times 10^3 = 3.327\times 10^3$

1.22 $6.041\times 10^2 + 3.303\times 10^3 = (.6041 + 3.303)\times 10^3 = 3.907\times 10^3$; $7.007\times 10^{-4} + 4.644\times 10^{-4} = (7.007 + 4.644)\times 10^{-4} = 11.651\times 10^{-4} = 1.165\times 10^{-3}$

1.23 $(4.640\times 10^2)\times(3.020\times 10^4) = (4.640\times 3.020)\times 10^6 = 14.0128\times 10^6 = 1.401\times 10^7$ in normalized form, noting that the problem was set in a length 4 system.

1.24 $46 = 101110_2$. Since it is positive, the representation is just 00101110.

Since -51 is negative, the two's complement is used. $51 = 110011_2$, which, completed to eight places, is 00110011. The one's complement is 11001100, so the two's complement is 11001101.

1.25 As 01111001 has first digit 0, it represents a positive number.

$$\begin{aligned} 1111001_2 &= 2^6 + 2^5 + 2^4 + 2^3 + 1 \\ &= 64 + 32 + 16 + 8 + 1 \\ &= 121. \end{aligned}$$

Since its first digit is 1, 10101101 represents a negative number. To compute it, reverse the construction process: *subtract* 1 to get 10101100 and then *complement* to get 01010011. So the number represented is -1010011_2 and

$$\begin{aligned} 1010011_2 &= 2^6 + 2^4 + 2 + 1 \\ &= 64 + 16 + 2 + 1 \\ &= 83. \end{aligned}$$

So the strings represent 121 and -83.

1.26 $8 = 1000_2$ and $4 = 100_2$. So the representation of 8 is 00001000 and that of 4 is 00000100. Thus 4 has complement 11111011, and two's complement $11111011 + 1 = 11111100$; and -4 is represented by 11111100. Now add:

$$\begin{array}{r} 00001000 \\ 11111100 \\ \hline 100000100 \end{array} \qquad\qquad \begin{array}{r} 8 \\ -4 \\ \hline 4 \end{array}$$

The first digit (*carry digit*) is ignored, as it takes us outside the 8-digit block available for integers. So the answer is represented by 00000100 which equals 100_2 or 4, as expected.

1.27 12.25: First, $12 = 1100_2$. Next convert .25 to binary.

$$\begin{aligned} 2 \times .25 &= 0.5 \\ 2 \times .5 &= 1. \end{aligned}$$

So the binary form of 12.25 is 1100.01_2, or 1.10001×2^3. The exponent is 3, which has excess 127 form $130 = 10000010_2$. So the exponent is stored as 10000010. The sign bit is 0. So the $IEEE754$ form is

0100 0001 0100 0100 0000 0000 0000 0000.

-0.3: Converting 0.3 to binary, we get

$$\begin{aligned} 2 \times .3 &= 0.6 \\ 2 \times .6 &= 1.2 \\ 2 \times .2 &= 0.4 \\ 2 \times .4 &= 0.8 \\ 2 \times .8 &= 1.6 \\ 2 \times .6 &= 1.2 \end{aligned}$$

and the expression now repeats:

$$0.3 = .0\overline{1001} = 1.\overline{0011} \times 2^{-2}.$$

The sign bit is 1. The exponent is -2, which has excess 127 form $125 = 1111101_2$, or 01111101. So the *IEEE*754 expression is

1011 1110 1001 1001 1001 1001 1001 1001.

112.5: $112 = 1110000_2$ and $.5 = .1_2$, so in base 2, $112.5 = 1110000.1 = 1.1100001 \times 2^6$. The sign bit is 0. The exponent is 6, which has excess 127 form $133 = 10000101_2$. So the *IEEE*754 expression is

0100 0010 1110 0001 0000 0000 0000 0000.

1.28 Converting 0.195625 to binary, we get

$$\begin{aligned} 2 \times .195625 &= 0.39125 \\ 2 \times .39125 &= 0.7825 \\ 2 \times .7825 &= 1.565 \\ 2 \times .565 &= 1.125 \\ 2 \times .125 &= 0.25 \\ 2 \times .25 &= 0.5 \\ 2 \times .5 &= 1. \end{aligned}$$

so $0.195625 = .0011001_2 = 1.1001_2 \times 2^{-3}$. The exponent -3 has excess 127 form 01111100, so the *IEEE*754 expression is

0011 1110 0100 1000 0000 0000 0000 0000,

or, in hexadecimal form, 3*E*480000.

Chapter 2

2.1 $\sim q$ means "the wind is not blowing," $p \vee q$ means "the sun is shining or the wind is blowing (maybe both)," and $\sim p \wedge q$ is "The sun is not shining but the wind is blowing."

2.2 $p \wedge r \wedge \sim q$ means "Joseph and Donna are here but Nancy is not';' $q \vee r$ means "either Nancy or Donna is here (maybe both)" (or "at least one of Nancy and Donna is here").

2.3 For $p \wedge \sim q$ to be true, the card must be a heart but not an honor — that is, the draw is the heart 2, 3, 4, 5, 6, 7, 8 or 9. $p \vee \sim q$ is true when the card is either a heart or is not an honor, so the draw is one of the minor cards — any 2, 3, 4, 5, 6, 7, 8 or 9 — or the heart ace, king, queen, jack or ten, a total of 37 draws.

2.4 It is convenient to write $s = (\sim p \wedge \sim r)$.

p	q	r	$\sim p$	$\sim r$	s	$(q \vee s)$	$(p \vee (q \vee s)$
T	T	T	F	F	F	T	T
T	T	F	F	T	F	T	T
T	F	T	F	F	F	F	T
T	F	F	F	T	F	F	T
F	T	T	T	F	F	T	T
F	T	F	T	T	T	T	T
F	F	T	T	F	F	F	F
F	F	F	T	T	T	T	T

2.5

p	q	$\sim q$	$(p \vee \sim q)$	$(q \rightarrow p)$	$(p \vee \sim q) \equiv (q \rightarrow p)$
T	T	F	T	T	T
T	F	T	T	T	T
F	T	F	F	F	T
F	F	T	T	T	T

2.6

p	q	$p \vee q$	$(p \wedge q)$	$\sim(p \vee q)$	$(p \wedge q) \wedge \sim(p \vee q)$
T	T	T	T	F	F
T	F	T	F	F	F
F	T	T	F	F	F
F	F	F	F	T	F

2.7 $\{x,y,z\}$, $\{x,y\}$, $\{x,z\}$, $\{y,z\}$, $\{x\}$, $\{y\}$, $\{z\}$, and $\emptyset$

2.8 $\mathbb{Z} \backslash \mathbb{Z}^+ = \{0,-1,-2,\ldots\}$, $\mathbb{Z} \cap \mathbb{Z}^+ = \mathbb{Z}^+$, $(\mathbb{Z}^+ \backslash \mathbb{E}) \cup \Pi = \{\text{odd positive integers}\} \cup \{2\}$

2.10 Suppose $x \in R \cap (S \cup T)$. Then $x \in R$ and $x \in (S \cup T)$, so either $x \in S$, and $x \in R$, so $x \in R \cap S$); or $x \in T$, and $x \in R$, so $x \in R \cap T$. Either have $x \in (R \cap S)$ or $x \in (R \cap T)$, so $x \in (R \cap S) \cup (R \cap T)$. So

$$x \in R \cap (S \cup T) \Rightarrow x \in (R \cap S) \cup (R \cap T).$$

Conversely, suppose $x \in (R \cap S) \cup (R \cap T)$. If $x \notin R$, then certainly $x \notin R \cap S$,and $x \notin R \cap T$). So we can be sure $x \in R$. If $x \notin (S \cup T)$, then $x \notin S$, so $x \notin (R \cap S)$, and $x \notin T$, so $x \notin (R \cap T)$; but x must belong to one or the other of these sets, so $x \in (S \cup T)$. So we've shown $x \in R \cap (S \cup T)$. So

$$x \in (R \cap S) \cup (R \cap T) \Rightarrow x \in R \cap (S \cup T).$$

Together these prove $R \cap (S \cup T) = (R \cap S) \cup (R \cap T)$.

2.12

A	B	C	$A \cup C$	$A \cap B \cap C$	$B \cap (A \cup C)$
T	T	T	T	T	T
T	T	F	T	F	T
T	F	T	T	F	F
T	F	F	T	F	F
F	T	T	T	F	T
F	T	F	F	F	F
F	F	T	T	F	F
F	F	F	F	F	F

There is no line with T in the second-last column and F in the last.

2.14

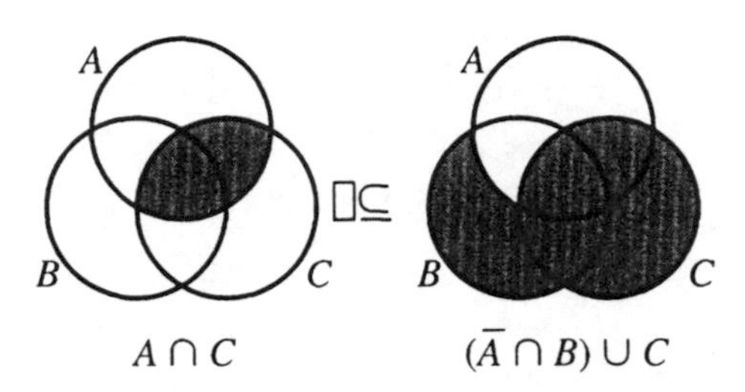

2.15 The argument is not valid: the sets might be as follows:

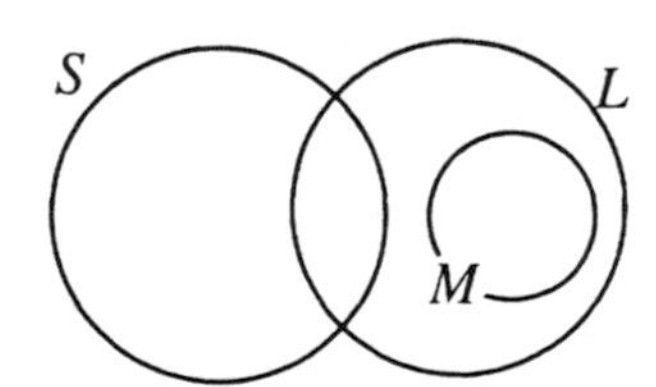

2.16

2.17 $\{(1,1),(1,4),(1,5),(2,1),(2,4),(2,5)\}$

2.18 The case $n = 1$ is $1 = 1^2$, which is obviously true, so the formula gives the correct answer when $n = 1$. Suppose it is true when $n = k$; therefore

$$1 + 3 + \cdots + (2k-1) = k^2.$$

Then

$$\begin{aligned} 1 + 3 + \cdots + (2k-1) + (2k+1) &= k^2 + (2k+1) \\ &= (k+1)^2, \end{aligned}$$

and the formula is proven correct when $n = k+1$. So, by induction, we have the required result.

2.20 Suppose the proposition $P(n)$ means $n^2 \geq 2n+1$. Then $P(3)$ means $9 \geq 7$, which is true. Now suppose k is an integer greater than or equal to 3, and $P(k)$ is true: $k^2 \geq 2k+1$. Then $(k+1)^2 = k^2+2k+1 \geq (2k+1)+(2k+1)$ (assuming $P(k)$) $= 4k+2$, and when $k \geq 3$, obviously $4k+2 \geq 2(k+1)+1$ (in fact this is true for every positive integer k). So $P(k+1)$ is true. Therefore $P(k)$ implies $P(k+1)$, and the result follows by induction.

2.21 Suppose $P(n)$ means 7 divides $3^{2n}-2^n$. Then $P(1)$ is true because $3^2-2^1 = 7$. Now suppose k is any positive integer, and $P(k)$ is true: say $3^{2k}-2^k = 7x$, where x is some integer. Then $3^{2k+2}-2^{k+1} = 9\cdot 3^{2k}-2\cdot 2^k = 7\cdot 3^{2k}$ $+2\cdot 3^{2k}-2\cdot 2^k = 7\cdot 3^{2k}+2\cdot 7x$, which is divisible by 7. So the result follows by induction.

2.22 $f_{n+4} = f_{n+3}+f_{n+2} = (f_{n+2}+f_{n+1})+f_{n+2} = 2f_{n+2}+f_{n+1} = 2(f_{n+1}+f_n)+f_{n+1} = 3f_{n+1}+f_n$. So f_{n+4} is a multiple of 3 if and only if f_n is. If $P(n)$ means f_{4n} is divisible by 3, then $P(1)$ is true ($P(1)$ means f_4 is divisible by 3; in fact $f_4 = 3$), and $P(k)$ true implies $P(k+1)$ true, so $P(n)$ is true for all positive integers, by induction. If $Q(n)$ means f_{4n-1} is not divisible by 3, then $P(1)$ is true ($Q(1)$ means f_3 is not divisible by 3; in fact $f_3 = 2$), and $Q(k)$ true implies $Q(k+1)$ true, so $Q(n)$ is true for all positive integers, by induction. Similarly we may prove that f_{4n-2} and f_{4n-1} are not divisible by 3. The four statements together give the required answer.

Chapter 3

3.1 See the laws of logic in Section 2.1. Rules B1, B2, B3 and B4 for Boolean algebras are the commutative and associative laws for $\vee$ and $\wedge$, the distributive laws, and the identity laws, respectively. We noted in that section that $p\wedge\sim p$ is a contradiction, and it is also easy to see that $p\vee\sim p$ is a tautology; these give B5.

3.2 $xy'z'x$.

3.3
$$\begin{aligned} x &= x1 && \text{by B4(b)} \\ &= x(x+x') && \text{by B5(a)} \\ &= xx+xx' && \text{by B3(b)} \\ &= xx+0 && \text{by B5(b)} \\ &= xx && \text{by B4(a).} \end{aligned}$$

3.4 We prove the first law. The second follows by duality.

$$\begin{aligned} x+xy &= x1+xy && \text{by B4(b)} \\ &= x(y+y')+xy && \text{by B5(a)} \\ &= xy+xy'+xy && \text{by B3(b)} \\ &= xy+xy+xy' && \text{by B1(a)} \\ &= x(y+y)+xy' && \text{by B5(a)} \\ &= xy+xy' && \text{by Theorem 3.2(a)} \\ &= x(y+y') && \text{by B3(b)} \\ &= x1 && \text{by B5(a)} \\ &= x && \text{by B4(b).} \end{aligned}$$

3.5 $yz+yz = y(z+z)$ (by the Distributive law) $= yz$ (by the Absorption law applied to $z+z$).

3.6 $\beta(1,0,1) = 10+011' = 10+010 = 0+0 = 0$; $\beta(1,1,0) = 11+101' = 1+0 = 1$; $\beta(0,1,1) = 01+110' = 01+111 = 0+1 = 1$

3.7 $xy'x'z' = xx'y'z'$ (commutative) $= 0y'z' = 0$; $x'yz'y = x'yyz'$ (commutative) $= x'yz'$ which is prime.

3.8 Step 1: $(x'y)'(x'+xyz') = (x+y')(x'+xyz')$ by De Morgan's laws.
Step 2: $(x+y')(x'+xyz') = x(x'+xyz')+y'(x'+xyz') = xx'+xxyz'+y'x'+y'xyz'$.
Step 3: $xx' = 0$; $xxyz' = xyz'$; $y'x'$ is fundamental; $y'xyz' = y'yxz' = 0$. So the expression equals $xyz'+x'y'$.
Step 4: Neither term includes the other, so we are finished.

3.9 $xy = xy1 = xy(z+z') = xyz+xyz'$, so

$$\begin{aligned} xy+\beta &= xy+xyz+xy'z+xyz'+x'y'z' \\ &= xy+xyz+xyz'+xy'z+x'y'z' \\ &= xyz+xyz'+xy'z+x'y'z' \\ &= \beta \end{aligned}$$

The only fundamental products included in xy are x and y. To show that $x+\beta \neq \beta$, is is sufficient to observe that if $x = 1$, $y = z = 0$, we have $x+\beta = 1$, $\beta = 0$. For the form y, the substitution $x = 0$, $y = z = 1$ gives $y+\beta = 1, \beta = 0$.

3.10 $xy'+xz'+yz = xy'(z+z')+x(y+y')z'+(x+x')yz = xy'z+xy'z'+xyz'+xy'z'+xyz+x'yz = xy'z+xy'z'+xyz'+xyz+x'yz$

3.11

	y	y'
x	1	
x'		

	y	y'
x		1
x'	1	

3.12 The maps for the two forms are

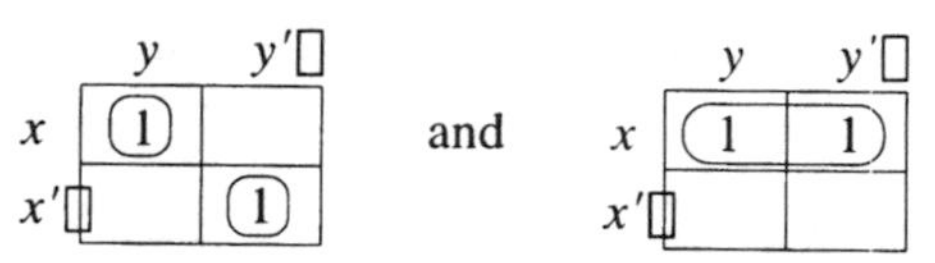

respectively. The basic rectangles are marked. There is no change in the first expression: $xy + x'y'$ is already minimal. The second map gives the minimal expression $xy + xy' = x$, as expected.

3.16 An appropriate circuit is

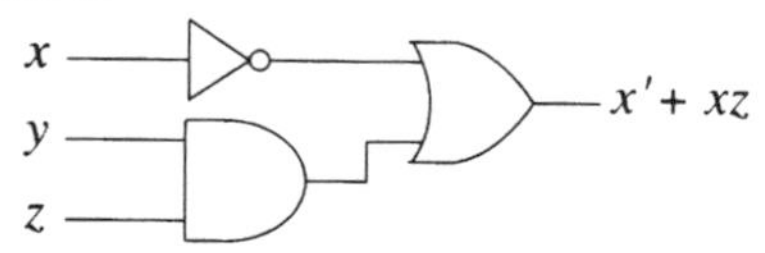

3.17 In Sample Problem 3.14 we saw that the given expression has minimal form $yz' + y'z$. A suitable circuit is

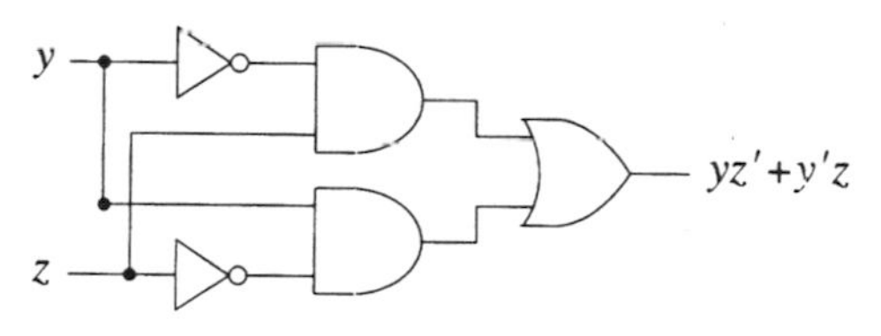

Chapter 4

4.1 $s \leq t$ corresponds to $\{(1,1),(1,2),(1,3),(1,4),(2,2),(2,3),(2,4),(3,3),(3,4)\}$. $s \geq t$ corresponds to $\{(1,1),(2,1),(2,2),(3,1),(3,2),(3,3)\}$.

4.2 $\rho\tau$ is not defined

$$\tau\sigma = \{(1,6),(2,6),(2,7)\}$$
$$\rho \cup \sigma = \{(3,6),(3,7),(4,6),(4,7),(5,7)\}.$$

4.3 $\{6,3),(7,3),(7,5)\}$

4.4 Both $=$ and $\geq$ are reflexive. $=$ is symmetric while $\geq$ is antisymmetric. Both are transitive.

4.5 Every sum of two members of A is again a member of A, except that $a+b$ is not in A whenever $a = \sqrt{2}$ or $b = \sqrt{2}$. So $a\,\delta\,b$ is true if and only if both a and b are in $\mathbb{Z}$. That is, $\delta = \mathbb{Z} \times \mathbb{Z}$. The relation δ is neither reflexive nor irreflexive, because $1\,\delta\,1$ is true and $\sqrt{2}\,\delta\,\sqrt{2}$ is false. The relation is symmetric and transitive.

4.6 The relation is obviously reflexive. To say "A has the same volume as B" and "B has the same volume as A" mean the same thing so the relation is symmetric. If A has the same volume as B and B has the same volume as C, then all three solids have the same volume, so $A\,\sigma B$ and $B\,\sigma C$ together imply $A\,\sigma C$, and the relation is transitive. So σ is an equivalence relation.

4.7 The classes are $[a] = \{a, -a\}$. (Notice that $[0]$ has only one element.) There are infinitely many of them, one for each non-negative integer.

4.8 If $a \wr b$ and $b \wr a$ are both true, then $a \mid b$ and $b \mid a$ are both true, so (as pointed out in Sample Problem 4.8) $a = b$, which contradicts $a \wr b$. So $\wr$ is asymmetric. Therefore it is antisymmetric. Transitivity follows from the transitivity of $|$. So $\wr$ is a partial order relation. Since $a \wr a$ never holds, it is strong. As $2 \wr 3$ and $3 \wr 2$ are both false, 2 and 3 are not comparable, so it is not total.

4.9 $f_1(f_1(x)) = f_1(x)$ so $f_1(f_1) = f_1$.
$f_2(f_4(x)) = f_2(1/(1-x)) = 1/[1/(1-x)] = 1-x = f_3(x)$ so $f_2(f_4) = f_3$.
$f_5(f_6(x)) = f_5((x-1)/x) = ((x-1)/x)/[(x-1)/x) - 1] =$
$((x-1)/x)/(-1/x) = (x-1)/(-1) = 1-x = f_3(x)$ so $f_5(f_6) = f_3$ also.

4.10 $f_3(x) = 1-x$. If $f_3(x) = y$, then $y = 1-x$ so $x = 1-y$ and $f_3^{-1}(y) = 1-y$. So $f_3^{-1} = f_3$.

Chapter 5

5.1 $\{3,4,5,6\}$.

5.2 The outcomes are$\{H,T\}$. This set has 2 elements, so it has $2^2 = 4$ subsets. So there are four events.

5.3 In the obvious notation, $\{H0, H1, H2, T0, T1, T2\}$.

5.4

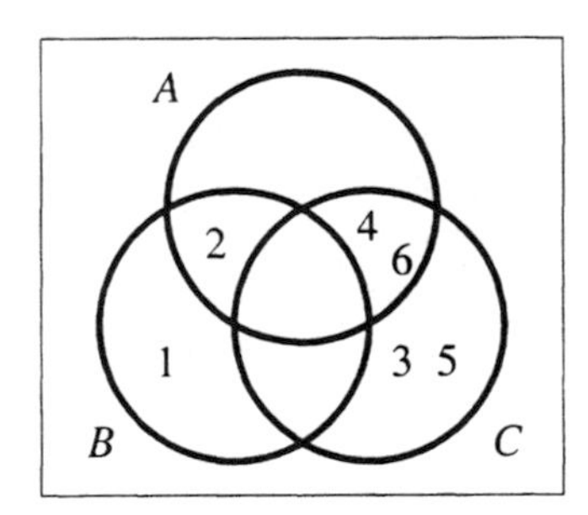

5.6

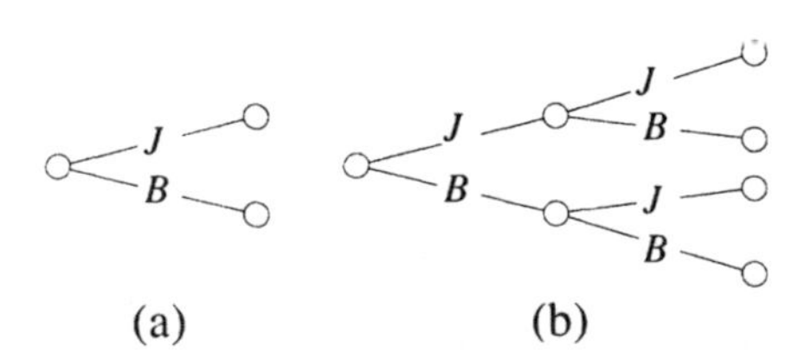

5.7 $|M \cup E| = |M| + |E| - |M \cap E| = 12 + 7 - 3 = 16$, so 16 readers were surveyed.

5.8 The figures are represented by the following diagram. As only readers were surveyed, there is no need for any "outside area" outside $M \cup E$.

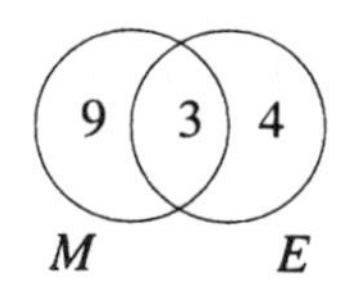

5.10

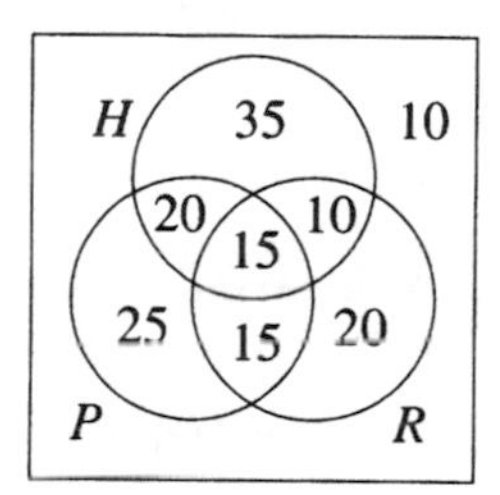

20 like horror and police procedural movies, but do not like romances. 20 like romances only. 10 like none of these three types.

5.11 Use the function $f(1) = a$, $f(2) = b$, and so on: $f(n)$ equals the n-th letter.

5.12 g is one-to-one: if x_1 and x_2 are integers and $g(x_1) = g(x_2)$, then $2x_1 = 2x_2$ so $x_1 = x_2$. g is onto: if y is any even integer, then $\frac{1}{2}y \in \mathbb{Z}$ and $g(\frac{1}{2}y) = y$.

5.13 The function f, defined by $f(x) = 2x + 1$, is a one-to-one correspondence between the countably infinite set $\mathbb{Z}$ and $\mathbb{O}$.

5.14 $6! = 6{\cdot}5{\cdot}4{\cdot}3{\cdot}2{\cdot}1 = 720$.

$(6)_4 = 6{\cdot}5{\cdot}4{\cdot}3 = 360$. Or using the formula: $(6)_4 = 6!/(6-4)! = 6!/2! = 720/2 = 360$.

5.15 Simply $n!$.

5.16 Treating the committee as an ordered set of three elements chosen from the 12-element set of members, the answer is $(12)_3$, or 1320.

5.17 The boys can be ordered in $3! = 6$ ways, and the girls can be ordered in $4! = 24$ ways. As the table is circular, it doesn't matter whether the boys are to the left or to the right of the girls. So there are $3! \times 4! = 6 \times 24 = 144$ arrangements.

5.19 There are $(10)_3$ boys' committees and $(13)_3$ girls' committees, for a total of $(10)_3 \times (13)_3 = 720 \times 1716 = 1235520$ choices.

5.21 There are three A's, two N's and one A, for a total of six letters. So the number of orderings is $6!/(3!{\cdot}2!) = 60$.

5.22 If unlimited numbers of each color were available, there would be $3^4 = 81$ solutions. It is necessary to exclude the solution with four blue marbles, *BBBB*. So the answer is $81 - 1 = 80$.

5.23 $C(9,5) = \dfrac{9\cdot 8\cdot 7\cdot 6}{4\cdot 3\cdot 2\cdot 1} = 126$, $\binom{6}{0} = \dfrac{6!}{6!\cdot 0!} = 1$.

5.24 She must choose 4 of the last 7 questions, so $\binom{7}{4} = 35$ ways.

5.25 There are $C(8,5)$ choices of where to place the 1's, so the answer is $C(8,4)$, or 70.

5.26 You can choose the mysteries in $\binom{5}{2}$ ways and the westerns in $\binom{7}{3}$ ways. So you can choose in $\binom{5}{2} \times \binom{7}{3} = 10 \times 35 = 350$ ways.

5.27 The three consonants can be chosen in $\binom{5}{3} = 10$ ways, and the vowels in $\binom{3}{2} = 3$ ways. After the choice is made, the letters can be arranged in $5! = 120$ ways. So there are $10\cdot 3\cdot 120 = 3600$ "words."

5.28 The committee contains one man or no men. With one man, the number of choices is $3 \times \binom{7}{2} = 63$ (3 ways to choose the man, $\binom{7}{2}$ to choose the women). With no men, there are $\binom{7}{3} = 35$ possibilities. So there are $63 + 35 = 98$ possibilities.

5.30 $\binom{81}{79} = \binom{81}{2} = \frac{81\cdot 80}{1\cdot 2} = 3240$.

5.31 $(x-y)^6 = \binom{6}{0}y^6 - \binom{6}{1}xy^5 + \binom{6}{2}x^2y^4 - \binom{6}{3}x^3y^3 + \binom{6}{4}x^4y^2 - \binom{6}{5}x^5y + \binom{6}{6}x^6$ $= y^6 - 6xy^5 + 15x^2y^4 - 21x^3y^3 + 15x^4y^2 - 6x^5y + x^6$.

5.32 We write $0.99^8 = (1 - 10^{-2})^8$. Then it equals

$$1 - 8\cdot 10^{-2} + 28\cdot 10^{-4} - 56\cdot 10^{-6} + \cdots$$

Every subsequent term is at most one-tenth of the one before it. So the approximate value is

$$1 - .08 + .0028 - .000056 + \cdots = 0.923 \text{ approx.}$$

5.33 There are 2^{10} possible ways to choose a subset of the ten books. However, the subsets with 8, 9 or 10 elements are not allowed. So the number is

$$2^{10} - \binom{10}{10} - \binom{10}{9} - \binom{10}{8} = 1024 - 1 - 10 - 45 = 964.$$

5.35 In the notation of the Sample Problem, we want to find $|S_6 \cup S_7 \cup S_8 \cup S_9|$. For any x, $|S_x| = 9^9$, if $x \neq y$, $|S_x \cap S_y = 8^9|$. Similarly $|S_x \cap S_y \cap S_z| = 7^9$, and so on. So, from (5.12) the sum is

$$4\cdot 9^9 - \binom{4}{2}\cdot 8^9 + \binom{4}{3}\cdot 7^9 - \binom{4}{4}\cdot 6^9$$

and the answer we require is

$$4\cdot 9^9 - 6\cdot 8^9 + 4\cdot 7^9 - 6^9.$$

5.37 Again let P_i be the set of permutations with i fixed points. No permutation can have exactly three fixed points (if three numbers are in their natural position, then the fourth must also be). So the answer is $|P_4|+|P_2| = f(4,4)+f(4,2) = \binom{4}{0}D_0\binom{4}{2}D_2 = 1+6\cdot 1 = 7$.

Chapter 6

6.1 The outcomes are equally likely, with probability $\frac{1}{6}$. Events E and G contain two outcomes each ($\{4,5\}$ and $\{3,5\}$ respectively) and F contains three ($\{2,4,6\}$) so $P(E) = \frac{1}{3}, P(F) = \frac{1}{2}, P(E) = \frac{1}{3}$.

6.2 Write XY to mean there is an X on the quarter and a Y on the nickel (where X and Y might stand for H or for T). There are four equally likely outcomes, namely HH, HT, TH, TT, and two of them (HT, TH) are in the event. So the probability is $P(E) = \frac{2}{4} = \frac{1}{2}$.

6.3 There are 52 cards in a deck, of which 12 are picture cards. So

$$P(E) = \frac{|E|}{|S|} = \frac{12}{52} = \frac{3}{13}.$$

6.4 Define the outcome xy to mean "there is an x on the first die and a y on the second." There are nine equally likely outcomes: each of x and y can be 1, 2, or 3. If the event E_j means "the sum is j," then there are five events $(E_2, E_3, E_4, E_5, E_6)$ and

$$\begin{aligned}
E_2 &= \{11\}, & P(E_2) &= \tfrac{1}{9} \\
E_3 &= \{12,21\}, & P(E_3) &= \tfrac{2}{9} \\
E_4 &= \{13,22,31\}, & P(E_4) &= \tfrac{3}{9} \\
E_5 &= \{23,32\}, & P(E_5) &= \tfrac{2}{9} \\
E_6 &= \{33\}, & P(E_6) &= \tfrac{1}{9}.
\end{aligned}$$

6.5 There are five non-blue marbles out of nine, so

$$P(E) = \frac{|E|}{|S|} = \frac{5}{9}.$$

6.6

$$\begin{aligned}
P(E \cup F) &= P(E) + P(F) - P(E \cap F) \\
&= .7 + .3 + .1 = .9 \\
P(\overline{E}) &= 1 - P(E) \\
&= 1 - .7 = .3.
\end{aligned}$$

6.7 The data can be represented as

	M	F
S	.19	.22
U	.28	.31

The required probabilities are (i) $P(S)$; (ii) $P(F)$; (iii) $P(S \cup F)$. From formulas, $P(S) = P(S \cap M) + P(S \cap F) = .19 + .22 = .41$; $P(F) = P(F \cap S) + P(F \cap U) = .22 + .31 = .53$; and $P(S \cup F) = P(S) + P(F) - P(S \cap F) = .41 + .53 - .22 = .72$. This could also be answered by adding the entries in the appropriate cells.

6.8 The possible outcomes of the first subexperiment — the selection of die — are A and B, with probabilities each $\frac{1}{2}$. If A is selected, the probability of any of the rolls 1, 2, 3, 4, 5, 6 is $\frac{1}{6}$. If B is selected, rolls 1 and 6 each have probability $\frac{1}{2}$. So there are 8 outcomes — $A1, A2, A3, A4, A5, A6, B1, B6$ — and the probability distribution is

$$\begin{aligned}
P(A1) &= \tfrac{1}{2} \times \tfrac{1}{6} = \tfrac{1}{12}\\
P(A2) &= \tfrac{1}{2} \times \tfrac{1}{6} = \tfrac{1}{12}\\
P(A3) &= \tfrac{1}{2} \times \tfrac{1}{6} = \tfrac{1}{12}\\
P(A4) &= \tfrac{1}{2} \times \tfrac{1}{6} = \tfrac{1}{12}\\
P(A5) &= \tfrac{1}{2} \times \tfrac{1}{6} = \tfrac{1}{12}\\
P(A6) &= \tfrac{1}{2} \times \tfrac{1}{6} = \tfrac{1}{12}\\
P(B1) &= \tfrac{1}{2} \times \tfrac{1}{2} = \tfrac{1}{4}\\
P(B6) &= \tfrac{1}{2} \times \tfrac{1}{2} = \tfrac{1}{4}.
\end{aligned}$$

So

$$\begin{aligned}
P(6) = P(A6) + P(B6) &= \tfrac{1}{12} + \tfrac{1}{4} = \tfrac{1}{3}\\
P(3) = P(A3) &= \tfrac{1}{12}
\end{aligned}$$

6.9 There are three ways in which exactly one success can occur: the sequences SFF, FSF and FFS. As

$$\begin{aligned}
P(SFF) &= \tfrac{1}{3} \times \tfrac{2}{3} \times \tfrac{2}{3} = \tfrac{4}{27}\\
P(FSF) &= \tfrac{2}{3} \times \tfrac{1}{3} \times \tfrac{2}{3} = \tfrac{4}{27}\\
P(FFS) &= \tfrac{2}{3} \times \tfrac{2}{3} \times \tfrac{1}{3} = \tfrac{4}{27}
\end{aligned}$$

the probability of exactly one success is

$$\frac{4}{27} + \frac{4}{27} + \frac{4}{27} = \frac{12}{27}.$$

6.10 In this case each call is a Bernoulli trial with $p = \frac{1}{2}$, so the five calls in a day can be thought of as a binomial experiment with $p = \frac{1}{2}, n = 5$. So

$$\begin{aligned}
P(3\text{ successes}) &= \tbinom{5}{3}(\tfrac{1}{2})^3(\tfrac{1}{2})^2 = 10 \times \tfrac{1}{32} = \tfrac{10}{32}\\
P(4\text{ successes}) &= \tbinom{5}{4}(\tfrac{1}{2})^3(\tfrac{1}{2})^2 = 5 \times \tfrac{1}{32} = \tfrac{5}{32}\\
P(5\text{ successes}) &= \tbinom{5}{5}(\tfrac{1}{2})^3(\tfrac{1}{2})^2 = 1 \times \tfrac{1}{32} = \tfrac{1}{32}
\end{aligned}$$

and the probability of at least two successes is the sum of these:

$$\frac{10+5+1}{32} = \frac{16}{32} = \frac{1}{2}.$$

6.11 See the Sample Problem. Continuing with our experimental arithmetic:

$$n = 9: \quad \frac{3^{n+1}(3+n)}{4^n} = \frac{6561 \times 12}{262144} = .30\ldots$$

which is just too big. Clearly the next term will be smaller than .3 (if you check, you will find that the $n = 10$ term is .24...). So he needs to make 10 calls.

6.12 There are 11 marbles, so there are $\binom{11}{3}$ ways of selecting three. The number of selections with one of each color is $5 \times 4 \times 2 = 40$. So the probability is $40/\binom{11}{3} = 8/33$.

6.13 There are 10 balls, so two can be chosen in $C(10,2) = 45$ ways. Two blue balls can be selected in $C(4,2) = 6$ ways. So the probability of no white ball — two blue — is $6/45 = 2/15$, and the probability of at least one white is $1 - \frac{2}{15} = \frac{13}{15}$.

6.14 The number of spade flushes is $\binom{13}{5}$, and similarly for the other suits. So the total number of flushes is $4 \times \binom{13}{5}$. As noted in the Sample Problem, the total number of possible hands is $\binom{52}{5}$. So the probability of a flush is

$$\frac{4 \times \binom{13}{5}}{\binom{52}{5}} = \frac{4 \cdot 13 \cdot 12 \cdot 11 \cdot 10 \cdot 9}{52 \cdot 51 \cdot 50 \cdot 49 \cdot 48} = \frac{33}{16660}$$

or about one chance in 500.

6.15 There are $\binom{18}{3}$ possible committees. The number of committees with all Math majors is $\binom{6}{3}$; for Economics majors it is $\binom{5}{3}$ and for Computer Science majors, $\binom{7}{3}$ So the probability is

$$\frac{\binom{6}{3} + \binom{5}{3} + \binom{7}{3}}{\binom{18}{3}} = \frac{65}{816}.$$

6.16 There are $4! = 24$ ways the students can choose backpacks, so the probability that they all choose correctly is $\frac{1}{24}$. As we saw, the probability that none choose correctly is $\frac{3}{8}$. So the required probability is

$$1 - \frac{1}{24} - \frac{3}{8} = \frac{24 - 9 - 1}{24} = \frac{14}{24} = \frac{7}{12}.$$

6.17 After the first experiment there are 51 cards left: 12 spades and 13 hearts. So the probability of a spade as the second card is $\frac{12}{51}$; of a heart, $\frac{13}{51}$.

6.18 We write S for "a Spade" and and N "a card other than a Spade." The probabilities are

$$\begin{aligned} P(S \mid S) &= \tfrac{12}{51} \\ P(S \mid N) &= \tfrac{13}{51} \end{aligned}$$

and, by subtraction,

$$\begin{aligned} P(N \mid S) &= \tfrac{39}{51} \\ P(N \mid N) &= \tfrac{38}{51}. \end{aligned}$$

So the diagram is

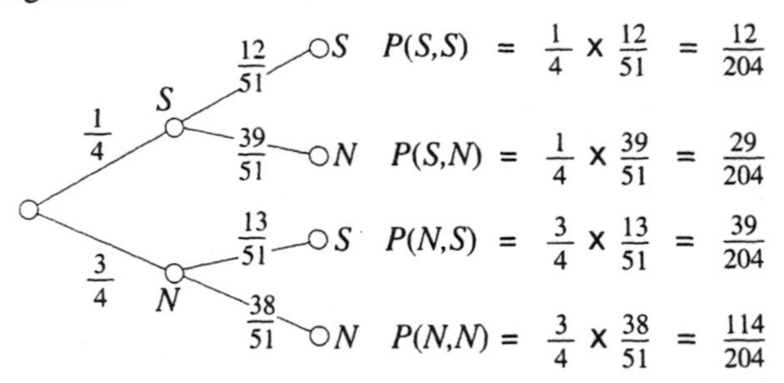

6.19 Write D for the event that the committee has both Democrats, R for both Republicans, and M for a mixed committee. There are $\binom{8}{2} = 28$ possible committees, so $|S| = 28$, while $|D| = 3$, $|R| = 10$, and $|M| = 15$. So $P(D) = \frac{3}{28}$. There are 15 committees that contain at least one Democrat, so the probability that both are Democrats, given that one is, is $\frac{3}{15}$ or $\frac{1}{5}$.

6.20 $P(D \cap R) = P(R \mid D)P(D) = \frac{1}{3}\frac{2}{5} = \frac{2}{15}$.

6.21 From the data,

$$\begin{aligned} P(A) &= \tfrac{3}{8} \\ P(B) &= \tfrac{1}{8} \\ P(C) &= \tfrac{4}{8} \\ P(A \cap B) &= 0 \\ P(A \cap C) &= \tfrac{2}{8} \\ P(B \cap C) &= 0. \end{aligned}$$

No two are independent.

6.22 In Sample Problem 6.20 we saw that $P(Y) = .6$ and $P(F \mid Y) = .02$. We also knew that $P(X) = .4$, and we have just been told $P(F \mid X) = .015$. So

$$\begin{aligned} P(F) &= P(X \cap F) + P(Y \cap F) \\ &= P(X)P(F \mid X) + P(Y)P(F \mid Y) \\ &= .4 \times .015 + .6 \times .02 \;=\; .006 + .012 \;=\; .018 \end{aligned}$$

6.23 Use B, F, H, T for *biased, fair, heads, tails.* Then

$$\begin{aligned} P(H) &= P(H \mid F)P(F) + P(H \mid B)P(B) \\ &= (\tfrac{1}{2})(\tfrac{2}{3}) + (\tfrac{2}{3})(\tfrac{1}{3}) \\ &= \tfrac{1}{3} + \tfrac{2}{9} \;=\; \tfrac{5}{9} \\ P(B \cap H) &= P(H \mid B)P(B) \;=\; \tfrac{2}{9}. \end{aligned}$$

So $P(B \mid H) = \frac{2}{9}/\frac{5}{9} = \frac{2}{5}$.

6.24 We know that $P(X) = P(Y) = .5$. From Bayes' formula,

$$\begin{aligned} P(B)P(X \mid B) &= P(X)P(B \mid X), \\ .5 \times P(X \mid B) &= .5 \times .4 \\ P(X \mid B) &= .4 \end{aligned}$$

6.25

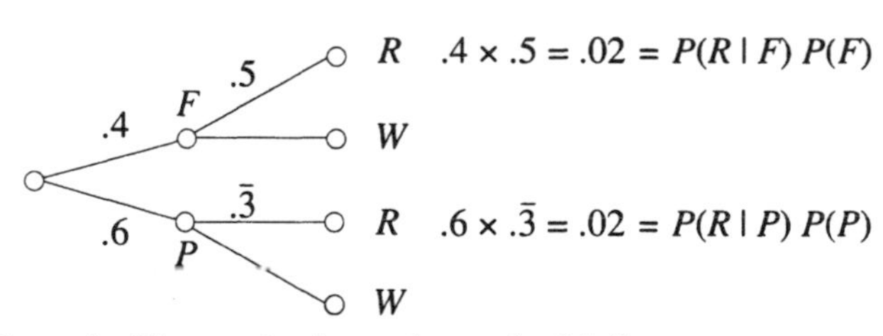

From the Figure, the denominator is .04. So

$$P(F \mid R) = \frac{P(R \mid F)P(F)}{.04} = \frac{.02}{.04} = .5.$$

6.26 We want $P(D \mid V)$. From Figure **??**,

$$P(D \mid V) = \frac{.112}{.653} = .17.$$

6.27

	H	T	
B	$\frac{2}{9}$	$\frac{1}{9}$	$\frac{1}{3}$
U	$\frac{1}{3}$	$\frac{1}{3}$	$\frac{2}{3}$
	$\frac{5}{9}$	$\frac{4}{9}$	

To find $P(B \mid H)$, look at column H, which has a total of $\frac{5}{9}$. Then look at the (B,H) entry $\frac{2}{9}$, and take the ratio

$$P(B \mid H) = \frac{P(B \cap H)}{P(H)} = \frac{\frac{2}{9}}{\frac{5}{9}} = \frac{2}{5}.$$

6.28 We use the abbreviations T (tests positive), N (tests negative), D (has the disease) and H (is healthy). We want to find $P(D \mid T)$. From the data, we know the following.

$$\begin{aligned} P(T \mid D) &= .95 & \quad P(N \mid D) &= .05 \\ P(T \mid H) &= .05 & \quad P(N \mid H) &= .95 \\ P(D) &= .05 & \quad P(H) &= .95. \end{aligned}$$

So

$$\begin{aligned} P(D \mid T) &= \frac{P(T \mid D)P(D)}{P(T \mid H)P(H) + P(T \mid D)P(D)} \\ &= \frac{(.95)(.05)}{(.05)(.95) + (.95)(.05)} \\ &= \frac{.0475}{.0475 + .0475} = \frac{.0475}{.095} = .5. \end{aligned}$$

This example shows that, in some cases, the test gives no useful information.

Chapter 7

7.1 The diagrams are

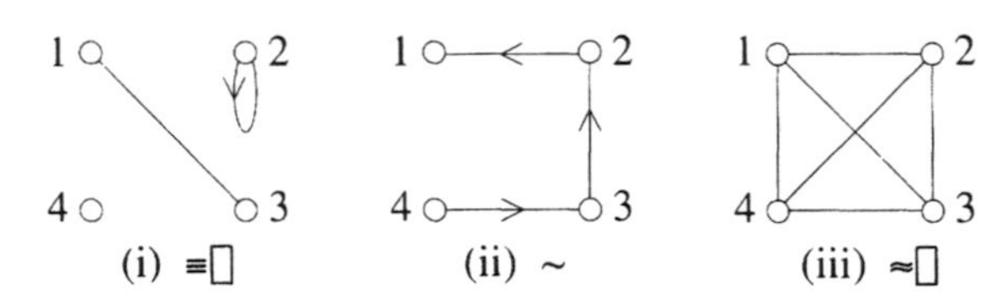

Relation $\approx$ yields a graph, $\equiv$ gives a looped graph, and $\sim$ a simple digraph.

7.2

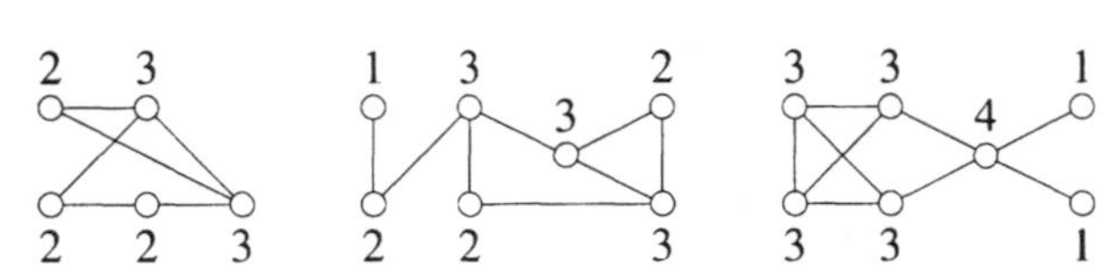

7.3 $5,3,3,1,1,1$ is impossible. For suppose there were such a graph. Write x for the vertex of degree 5 and y for the first vertex of degree 3. If $x \sim y$, there are 7 edges with one endpoint in $\{x,y\}$ and the other outside. So the sum of the degrees of the other vertices must be at least 7. But it is only 6 — too small. If x and y are not adjacent, the deficiency is even greater. The other two are graphical. Here are realizations

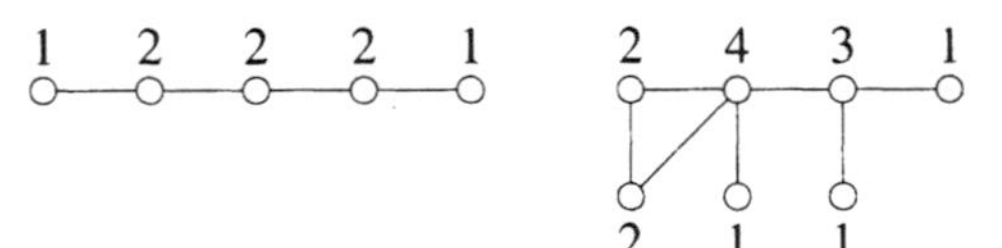

7.4

D A B C

7.5 The walk must start (or finish) at T, because there is only one bridge to that island. With a little experimentation we find the solution $TXYZXYX$.

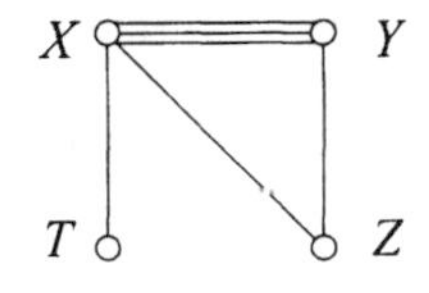

7.6 The original network is shown in Figure (i). We start from A, and randomly choose the walk $AGJKHDA$. After these edges are deleted, Figure (ii) remains.

We now start at D (alphabetically, the first vertex remaining that was in the first walk and is not yet isolated). One walk is $DBCFEHGD$, and its deletion leaves Figure (iii).

Finally, walk $FILKIHF$ uses up the remaining edges.

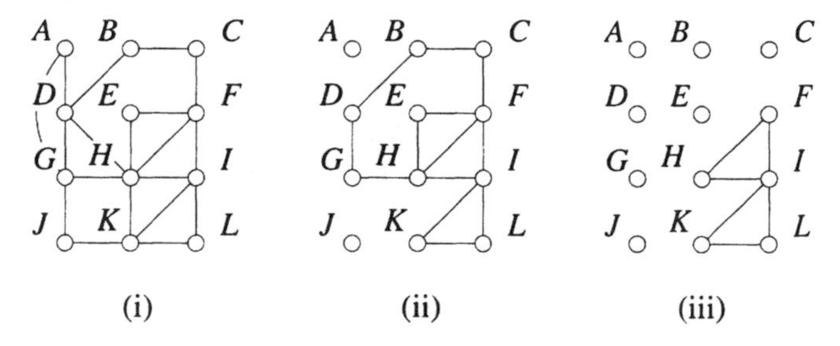

Putting these together, we get $AGJKHDBCFILKIHFEHGDA$.

7.7 The original is shown on the left. The two black vertices need another edge. They are not joined, so one edge will not suffice. So $eu(H) > 1$. The right-hand shows an Eulerization that requires just two edges. So $eu(H) = 2$, and the right-hand picture is a good Eulerization.

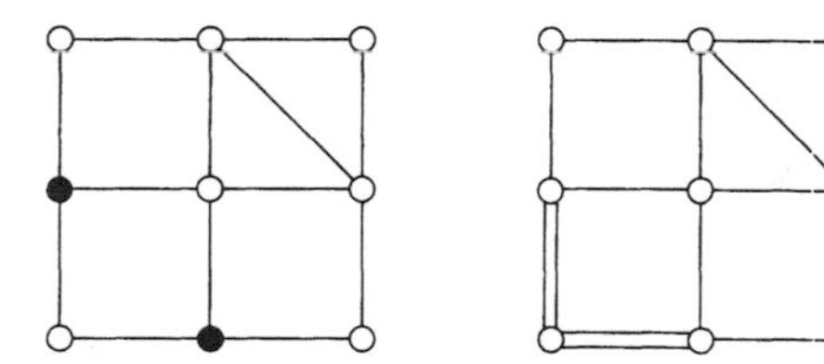

7.8 In a city's road network, a cutpoint represents an intersection such that all traffic between two areas has to pass through that intersection. As such, it might get a traffic officer assigned in peak hours, or special priority might be given to servicing its traffic lights.

7.9 C has no cutpoint or bridge, so $\kappa(C) \geq 2$ and $\kappa'(C) \geq 2$. However, it may be disconnected by removing either the two edges incident with the top left vertex, or the other endpoints of those two edges. So $\kappa(C) = \kappa'(C) = 2$.

D has a cutpoint, the central point, so $\kappa(D) = 1$. There is no bridge, but there are several ways to disconnect D by deleting two edges, so $\kappa'(C) = 2$.

7.10 C has minimum degree 2, so $\kappa(C), \kappa'(C), \delta(C) = 2, 2, 2$. D also has minimum degree 2, so $\kappa(D), \kappa'(D), \delta(D) = 1, 2, 2$.

7.11 There are four different-looking vertices, shown in black. The distance of every other vertex from the black one is shown in each case. (The top left and second from left vertices are actually equivalent, but this is not easy to see.)

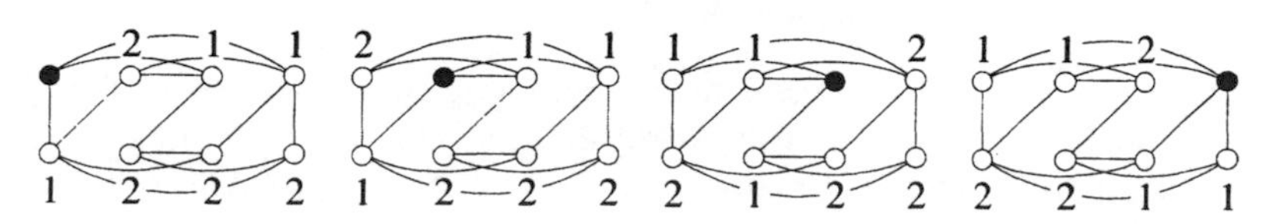

7.12 The graph (b) of Sample Problem 7.11 has $D = R = 2$. The center of W_n has eccentricity 1, while every other vertex has eccentricity 2, so $D = 2, R = 1$. In C_{2n+1}, every vertex is distance n from the two opposite vertices, and the distance is smaller for all other vertices, so $D = R = n$.

7.15 There are four types of vertex in the first tree and two in the second. The trees are shown as they were in the Sample Exercise.

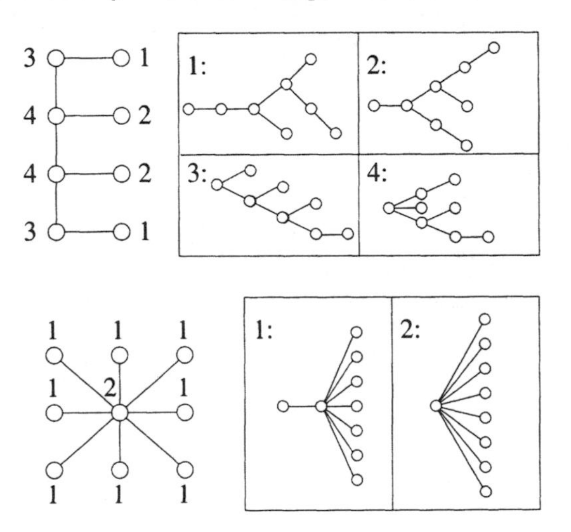

7.16 If a tree has four vertices, then the largest possible degree is 3. Moreover there are three edges (by Theorem 7.6), so from Theorem 7.1 the sum of the four degrees is 6. As there are no vertices of degree 0 and at least two vertices of degree 2, the list of degrees must be one of $3,1,1,1$ or $2,2,1,1$. In the first case, the only solution is the star $K_{1,3}$. The only case with the second degree list is the path P_4. So there are two trees.

7.17 Select $x_0 = a$. Then $S_1 = \{a\}$. Now $y_0 = d$ as $w(a,d) < w(a,b)$, so $x_1 = d$ and $S_2 = \{a,d\}$. The tree will contain edge ad.

Working from S_2, we get $y_0 = b$ and $y_1 = e$. Since ab $(= x_0y_0)$ has smaller weight than de $(= x_1y_1)$, we select $x_2 = b$. Then $S_3 = \{a,d,b\}$ and edge ab goes into the tree. Similarly, from S_3, we get $x_3 = y_2 = e$ and $S_4 = \{a,d,b,e\}$, and the new edge is be. Working from S_4, $x_4 = y_2 = c$ and we add edge bc; $S_4 - \{a,d,b,e,c\}$. The final vertex is f, and the edge is cf. So the tree has edges ad, ab, be, bc, cf and weight 14.

7.18 Candidates (i) and (iv) are Hamiltonian. Candidate (ii) is not Hamiltonian because it contains a repeated vertex, b. Candidate (iii) is not Hamiltonian because the graph contains no edge cd.

7.19 To traverse vertex a, a Hamiltonian cycle in G must contain one of the paths bad, bae or ead. If bad is included, the other edge through d might be dc or de; in the former case, neither bc nor de can be edges, and the only cycle is *badcfe*, while the latter case bars *be*, and the only cycle is *badefc*. If *bae* is included, *ad* is not an edge, so *cd* and *de* are edges, so we have the path *baedc*, and the cycle is *baedcf*. If *ead* is included, *de* is not an edge, so *dc* is an edge, and there are two possibilities, *adcbfe* and *adcfbe*. So there are five Hamiltonian cycles.

Any Hamiltonian cycle in H must contain edges *ab* and *ad*, because a has degree 2. This means *bd* is not an edge (it would form a triangle), so *de* must be in the cycle. There are two ways to finish a cycle: *bcfe* or *bfce*. So there are two cycles: *dabcfe* and *dabfce*.

7.20 The nearest neighbor algorithm, starting from Evansville, begins with EM, because it has the least cost of the three edges incident with E. The next edge must have M as an endpoint, and ME is not allowed (one cannot return to E, it has already been used), so the cheaper of the remaining edges is chosen, namely MN. The only available edge from N is NS, as E and M have already been visited, and the route is EMNSE, with cost $430.

Starting at Nashville, the first edge selected is NE, with cost $90. The next choice is EM, then MS, then SN, and the resulting cycle NEMSN costs $410. If you start at St. Louis, the first stop will be Memphis ($110 is the cheapest flight from St. Louis), then Evansville, then Nashville, costing $410. From Memphis, the cheapest leg is to Evansville, then Nashville, and finally St. Louis, for $410. So both St. Louis and Memphis yield the same cycle as the Nashville case (with different starting points, and in reverse).

To apply the sorted edges algorithm, first sort the edges in order of increasing cost: EM($80), EN($90), NM($100), MS($110), ES ($120), NS($130). Edge EM is included, and so is EN. The next choice would be MN, but this is not allowed because its inclusion would complete a cycle of length 3 (too short), so the only other choices are MS and NS, forming route EMSNE (or ENSME) at a cost of $410.

In this example the route ENMSE, with cost $420, does not arise from either algorithm.

Chapter 8

8.1 $4(2,0,-1) = (8,0,-4)$, so $4(2,0,-1)+(1,4,-3) = (8,0,-4)+(1,4,-3)$ $= (9,4,-7)$.

8.2 $\boldsymbol{u}\cdot\boldsymbol{t} = (-1,3,0)\cdot(1,2,3) = (-1\cdot 1)+(3\cdot 2)+(0\cdot 3) = (-1)+6+0 = 5$; $(2\boldsymbol{u}-3\boldsymbol{v})\cdot\boldsymbol{t} = [(-2,6,0)+(-6,6,-6)]\cdot(1,2,3) = (-8,12,-6)\cdot(1,2,3) =$ $(-8\cdot 1)+(24\cdot 2)+(-6\cdot 3) = -8+12-18 = -2$.

8.3 From $(2x,2) = (y,x)$ we get the two equations $2x = y$ and $2 = x$. The second gives $x = 2$, so the first gives $y = 2\cdot 2 = 4$.

8.4 $-A = \begin{bmatrix} -1 & -3 \\ 1 & -2 \end{bmatrix}$, $3A - B = \begin{bmatrix} 5 & 9 \\ -4 & 2 \end{bmatrix}$, $B + C$ is not defined, as B and C are of different sizes.

8.6 $B^T = \begin{bmatrix} -1 & 2 \\ 1 & 0 \end{bmatrix}$.

8.7 $CD = DC = \begin{bmatrix} 5 & 3 \\ -3 & -1 \end{bmatrix}$.

8.8 Suppose C has inverse

$$E = \begin{bmatrix} x & y \\ z & t \end{bmatrix}.$$

Then $CE = I$, so

$$\begin{bmatrix} 1 & 0 \\ 0 & 0 \end{bmatrix}\begin{bmatrix} x & y \\ z & t \end{bmatrix} = \begin{bmatrix} x & y \\ 0 & 0 \end{bmatrix} = \begin{bmatrix} 1 & 0 \\ 0 & 1 \end{bmatrix}.$$

The $(2,2)$ entries of the two matrices cannot be equal $(0 \neq 1)$, so no such E exists. Similarly, if D has inverse

$$\begin{bmatrix} x & y \\ z & t \end{bmatrix},$$

then

$$\begin{bmatrix} 0 & 1 \\ 0 & 0 \end{bmatrix}\begin{bmatrix} x & y \\ z & t \end{bmatrix} = \begin{bmatrix} z & t \\ 0 & 0 \end{bmatrix} = \begin{bmatrix} 1 & 0 \\ 0 & 1 \end{bmatrix}$$

and again we have the impossible equation $0 = 1$.

8.9 Suppose the inverse is

$$C = \begin{bmatrix} x & z \\ y & t \end{bmatrix}.$$

Then $BC = I$ means

$$\begin{bmatrix} 2 & 1 \\ 1 & 1 \end{bmatrix}\begin{bmatrix} x & z \\ y & t \end{bmatrix} = \begin{bmatrix} 1 & 0 \\ 0 & 1 \end{bmatrix}$$

which is equivalent to the four equations

$$\begin{aligned} 2x + y &= 1, \qquad & 2z + t &= 0, \\ x + y &= 0, \qquad & z + t &= 1. \end{aligned}$$

The left-hand pair of equations are easily solved to give $x = 1$ and $y = -1$, while the right-hand pair give $z = -1$ and $t = 2$. So the inverse exists, and is

$$A = \begin{bmatrix} 1 & -1 \\ -1 & 2 \end{bmatrix}$$

8.12 For A: $R3 \leftarrow R3 - R1$; $R2 \leftrightarrow R3$; $R1 \leftarrow R1 - 2R2$, $R3 \leftarrow R3 + 2R2$; $R1 \leftarrow R1 + R3$. For B: $R1 \leftrightarrow R3$; $R2 \leftarrow R2 - 2R1$, $R3 \leftarrow R3 - 3R1$; $R3 \leftarrow R3 - 2R2$.

8.13 $\det(C) = 1\cdot 4 - 1\cdot 3 = 1; C^{-1} = \begin{matrix} 4 & -1 \\ -3 & 1 \end{matrix}$.

$\det(D) = 2\cdot 2 - 4\cdot 1 = 0$; no inverse.

8.14 As observed in the Sample Problem, the matrix of coefficients has inverse

$$\begin{bmatrix} 1 & 1 & 0 \\ 1 & 0 & -1 \\ 2 & 1 & -2 \end{bmatrix}$$

Now

$$\begin{bmatrix} 1 & 1 & 0 \\ 1 & 0 & -1 \\ 2 & 1 & -2 \end{bmatrix} \begin{bmatrix} 2 \\ 2 \\ 1 \end{bmatrix} = \begin{bmatrix} 4 \\ 1 \\ 4 \end{bmatrix}$$

so the first system has solution $x = 4, y = 1, z = 4$.

$$\begin{bmatrix} 1 & 1 & 0 \\ 1 & 0 & -1 \\ 2 & 1 & -2 \end{bmatrix} \begin{bmatrix} 4 \\ -1 \\ 3 \end{bmatrix} = \begin{bmatrix} 3 \\ 1 \\ 1 \end{bmatrix}$$

and the second system has solution $x = 3, y = 1, z = 1$.

8.15 $M_\alpha = \begin{bmatrix} 0 & 1 & 1 \\ 1 & 1 & 0 \end{bmatrix}$.

8.16 From the definition of τ, the matrix of β is

$$M_\tau = \begin{bmatrix} 1 & 0 & 1 & 0 \\ 0 & 1 & 0 & 0 \\ 1 & 0 & 0 & 1 \end{bmatrix}$$

so

$$M_\alpha M_\tau = \begin{bmatrix} 2 & 0 & 1 & 1 \\ 1 & 1 & 0 & 1 \end{bmatrix} \text{ and } M_\alpha \vee M_\beta = \begin{bmatrix} 1 & 0 & 1 & 1 \\ 1 & 1 & 0 & 1 \end{bmatrix}.$$

8.17 We take the vertices of K_4 to be $\{1,2,3,4\}$. Taking the vertices in numerical order, the adjacency matrix is

$$\begin{bmatrix} 0 & 1 & 1 & 1 \\ 1 & 0 & 1 & 1 \\ 1 & 1 & 0 & 1 \\ 1 & 1 & 1 & 0 \end{bmatrix}.$$

If we take the edges in order $(1,2)$, $(1,3)$, $(1,4)$, $(2,3)$, $(2,4)$, $(3,4)$, the incidence matrix is

$$\begin{bmatrix} 1 & 1 & 1 & 0 & 0 & 0 \\ 1 & 0 & 0 & 1 & 1 & 0 \\ 0 & 1 & 0 & 1 & 0 & 1 \\ 0 & 0 & 1 & 0 & 1 & 1 \end{bmatrix}.$$

Chapter 9

9.1 $36 = 2^2 \cdot 3^2; 132 = 2^2 \cdot 3 \cdot 11; 47$ is itself prime; $55 = 5 \cdot 11$.

9.2 $246 = 2 \cdot 3 \cdot 41$, so the factors are 1, 2, 3, 6, 41, 82, 123 and 246.

9.3

$$\begin{aligned} 584 &= 2 \cdot 284 + 16 \\ 284 &= 17 \cdot 16 + 12 \\ 16 &= 1 \cdot 12 + 4 \\ 12 &= 3 \cdot 4 \end{aligned}$$

So $(584, 284) = 4$.

9.4

$$\begin{aligned} 156 &= 1 \cdot 144 + 12 & \qquad 162 &= 13 \cdot 12 + 6 \\ 144 &= 12 \cdot 12 & \qquad 12 &= 2 \cdot 6 \end{aligned}$$

So the GCD is 6.

9.5 Reverse the solution to **9.3**.

$$\begin{aligned} 4 &= 1 \cdot 16 - 12 \\ &= 1 \cdot 16 - (284 - 17 \cdot 16) &= 18 \cdot 16 - 284 \\ &= 18 \cdot (584 - 2 \cdot 284) - 284 &= 18 \cdot 584 - 37 \cdot 284. \end{aligned}$$

9.6 Modulo 7, $0^2 = 0$, $1^2 = 1$, $2^2 = 4$, $3^2 = 9 \equiv 2$, $4^2 = 16 \equiv 2$, $5^2 = 25 \equiv 4$, and $6^2 = 36 \equiv 1$. So $x \equiv 3$ or 4.

9.7 Modulo 7, $5^2 \equiv 4, so 5^2 + 2 \equiv 6$. Modulo 12, $5^2 \equiv 4$, so $5^2 + 2 \equiv 6$. Modulo 12, $14 \equiv 2$, so $4 \cdot 14 \equiv 8$, and $4 \cdot 14 - 9 \equiv -1 \equiv 11$.

9.8 $1^{-1} = 1$ and $5^{-1} = 5$. $2 \cdot 3 = 6 \equiv 0$, and $4 \cdot 3 = 12 \equiv 0$, so 2, 3 and 4 are zero-divisors.

9.9 Clearly 11 divides 132, 55 and 66, so 55 and 66 are zero-divisors.

$$\begin{aligned} 132 &= 2 \cdot 49 + 34 & \qquad 132 &= 1 \cdot 91 + 41 \\ 49 &= 1 \cdot 34 + 15 & \qquad 91 &= 2 \cdot 41 + 9 \\ 34 &= 2 \cdot 15 + 4 & \qquad 41 &= 4 \cdot 9 + 5 \\ 15 &= 3 \cdot 4 + 3 & \qquad 9 &= 5 + 4 \\ 4 &= 1 \cdot 3 + 1 & \qquad 5 &= 4 + 1 \end{aligned}$$

$1 = 4 - 3 = 4 - (15 - 3 \cdot 4) = 4 \cdot (34 - 2 \cdot 15) - 15 = 4 \cdot 34 - 9 \cdot 15 = 4 \cdot 34 - 9 \cdot (49 - 34) = 13 \cdot 34 - 9 \cdot 49 = 13 \cdot (132 - 2 \cdot 49) - 9 \cdot 49 = 13 \cdot 132 - 35 \cdot 49)$ so $49^{-1} \equiv -35 \equiv 97$.
$1 = 5 - 4 = 5 - (9 - 5) = 2 \cdot (41 - 4 \cdot 9) - 9 = 2 \cdot 41 - 9 \cdot 9 = 2 \cdot 41 - 9 \cdot (91 - 2 \cdot 41) = 20 \cdot 41 - 9 \cdot 91 = 20 \cdot (132 - 1 \cdot 91) - 9 \cdot 91 = 20 \cdot 132 - 29 \cdot 91$ so $91^{-1} \equiv -29 \equiv 103$.

9.10 $R = 385$ $R_1 = 77$ $R_2 = 55$ $R_3 = 35$. So
$77x \equiv 1 \pmod 5$, or $2x \equiv 1 \pmod 5$, has the solution $x \equiv 3 \pmod 5$,
$55x \equiv 1 \pmod 7$, or $6x \equiv 1 \pmod 7$, has the solution $x \equiv 6 \pmod 7$,
$35x \equiv 1 \pmod{11}$, or $2x \equiv 1 \pmod{11}$, has the solution $x \equiv 6 \pmod{11}$.
So $X = 77 \cdot 3 \cdot 3 + 55 \cdot 6 \cdot 2 + 35 \cdot 6 \cdot 4 = 2193 \equiv 268 \pmod{385}$.

9.11 The first word starts with T. A key of 3 makes it start TTA..., which is not promising. Key 4 makes it TNH..., not good. Key 5 makes it THE..., so we try, getting

```
THEBAR
BARIAN
SAREAT
THEFRO
NTGATE
```

the barbarians are at the front gate

9.12 HQWHU WKH PDCH

9.13 Looking at the second word:

```
CDEFGHIJKLMNOPQRSTUVWXYZAB
PQRSTUVWXYZABCDEFGHIJKLMNO
FGHIJKLMNOPQRSTUVWXYZABCDE
```

The only possibility is AND so the cipher has key 2. The message was *here and now.*

9.14 BIRTHDAY is good. APPLE has repetitions, TEA is too short, BEAD is too close to the beginning of the alphabet.

9.15 112160801020520. (Notice that the first A is coded as 1, while the second is 01; there is no need for a leading zero in the first letter.)

9.16 $3^2 = 9, 3^4 = 81, 3^8 = 6561$, so $3^{13} = 3 \times 81 \times 6561 = 1594323$.
$3^4 \equiv 5 \pmod{19}, 3^8 \equiv 25 \equiv 6$, so $3^{13} \equiv 3 \times 5 \times 6 \equiv 90 \equiv 14$.

9.17 Modulo 53, $7^2 = 49 \equiv -4$, $7^4 \equiv (-4)^2 = 16$, $7^8 \equiv 16^2 = 256 \equiv 44 \equiv -9$, $7^{16} \equiv (-9)^2 \equiv 81 \equiv 28$, so
$7^{29} = 7^{16} \cdot 7^8 \cdot 7^4 \cdot 7 \equiv 28 \cdot 44 \cdot 16 \cdot 7 =\equiv 25$.

9.18 This time we have

$$\begin{aligned} A^3 &\equiv 6545 \times 39 \times 21 + 3910 \times 9 \times 35 + 3542 \times 3 \times 3 \\ &= 6623883 \equiv 343 \pmod{301070}. \end{aligned}$$

Therefore $A^3 = 343$, and $A = 7$.

9.20 Calculating modulo 48,

$$\begin{aligned} 5^2 &= 25 \\ 5^3 &= 125 \equiv 29 \\ 5^4 &\equiv 145 \equiv 1 \end{aligned}$$

so the exponent is 4.

9.22 The square root of 449329 is 670.3.... This is not an integer, so we start testing from $x = 671$.

$x = 671$ $x^2 - n = 912$ $\sqrt{x^2 - n} = 30.1\ldots$
$x = 672$ $x^2 - n = 2255$ $\sqrt{x^2 - n} = 47.4\ldots$
$x = 673$ $x^2 - n = 3600$ $\sqrt{x^2 - n} = 60.$

So we take $y = 60$. The factorization is $449329 = 613 \times 733$.

Answers to Selected Exercises

Section 1.1

1. T **3.** F. **5.** F. **7.** F **9.** T **11.** January, June, July **13.** m, i, s, p **15.** red, white blue **17.** (i) $\mathbb{Q}, \mathbb{R}$ (ii) $\mathbb{R}$ (iii) $\mathbb{R}$ (iv) $\mathbb{R}$ (v) $\mathbb{Q}, \mathbb{R}$ (vi) $\mathbb{R}$ (vii) $\mathbb{Q}, \mathbb{R}$ (viii) $\mathbb{Q}, \mathbb{R}$ (ix) $\mathbb{Z}, \mathbb{Q}, \mathbb{R}$ (x) $\mathbb{R}$ (xi) $\mathbb{Z}^+, \mathbb{Z}, \mathbb{Q}, \mathbb{R}$ (xii) $\mathbb{R}$ **19.** $231 = 3 \cdot 7 \cdot 11, 275 = 5^2 \cdot 11, GCD = 11$ **21.** $95 = 5 \cdot 19, 125 = 5^3$, GCD = 5 **23.** $88 = 2^3 \cdot 11, 132 = 2^2 \cdot 3 \cdot 11$, GCD $= 44$ **25.** $1080 = 2^3 \cdot 3^3 \cdot 5, 855 = 5 \cdot 3^2 \cdot 19$, GCD $= 45$ **27.** 1 **29.** x^2y **31.** $\frac{1}{4x^2y^2}$ **33.** $\frac{5y^2}{z^3}$ **35.** 2 **37.** $\frac{1}{2}$ **39.** 28 **41.** 12 **43.** 78 **45.** 1.73

Section 1.2

1. $2+5+10+17+26+37 = 97$ **3.** $\sqrt{2}+\sqrt{3}+\sqrt{4} = 2+\sqrt{2}+\sqrt{3}$ **5.** $1+8+27 = 36$ **7.** $3+8+15+24 = 50$ **9.** $(10+\frac{1}{10})+(11+\frac{1}{11})+(12+\frac{1}{12}) = 33\frac{181}{660}$ **11.** $(2+1)+(3-1)+(4+1)+(5-1) = 14$ **15.** $2A+n$ **17.** $3A-B$ **19.** 39 **21.** 124 **23.** 416 **25.** $\frac{1}{6}n(2n^2+3n+7)$ **27.** $2-2^{-9}$ **29.** $\frac{1}{6}(n+1)(n+2)(2n+3)-1 = \frac{1}{6}n(2n^2+9n+13)$

Section 1.3

1. 55 **3.** 1531 **5.** 3363 **7.** 255 **9.** 95 **11.** 151 **13.** 136 **15.** 6.75 **17.** 1.01 **19.** $16.\overline{857142}$ **21.** 136.0625 **23.** $20.08\overline{3}$ **25.** 46.48 **27.** 0.25 **29.** 21.625 **31.** $14.\overline{6}$ **33.** $37.52\overline{7}$ **35.** 131.2 **37.** $5.\overline{6}$ **39.** $7.\overline{456790123}$ **41.** 101011 **43.** 110111.11 **45.** $1110.1\overline{10}$ **47.** 1111110 **49.** 1101111 **51.** 1111110.11 **53.** 11100011011 **55.** 4.04 **57.** $15.42\overline{243205}$ **59.** 0.2 **61.** 315 **63.** 5040 **65.** 22.24 **67.** $42.1\overline{4}$ **69.** $44.\overline{2}$ **71.** $6C$ **73.** $B.\overline{83E}$ **75.** $E.0\overline{7AE14}$ **77.** FF **79.** $1B$ **81.** $1.\overline{1E07C}$ **83.** $40.\overline{9}$ **85.** 924

87. $5AD.C$ **89.** $5.2E\overline{6DB}$ **91.** $76B.4$ **93.** 1101000000001 **95.** 1000100000001
97. 10101110.1111111 **99.** $111100100.\overline{10}$

Section 1.4

1. (i) 1.05×10^4 (ii) 1.10×10^{-2} (iii) 1.11×10^{-4} (iv) 1.04×10^9 **3.** 1.1740×10^2
5. 4.2857×10^{-1} **7.** 3.8350×10^3 **9.** 2.2411×10^2 **11.** -2.1200×10^1
13. 2.0830×10^2 **15.** 1.0799×10^2 **17.** -7.3480×10^1 **19.** 1.1230×10^3
21. 6.0025×10^3 **23.** 1.6264×10^4 **25.** 6.1022×10^3 **27.** -1.2960×10^{-3}
29. 9.1494×10^2 **31.** 1.3733×10^3 **33.** 3.5258×10^4 **35.** 1.4363×10^{-2}
37. 1.69×10^4 **39.** -3.33×10^1 **41.** 7.78×10^0 **43.** -4.82×10^0

Section 1.5

1. 00101110, $2E$ **3.** 01111111, $7F$ **5.** 11111111, FF **7.** 10000010, 82
9. 11110010, $F2$ **11.** 11000111, $C7$ **13.** 10000000, 80 **15.** 11111100, FC
17. 00010001, 11 **19.** 0 **21.** −118 **23.** 113 **25.** 15 **27.** −58 **29.** −125
31. 13 = 00001101 **33.** −16 = 11110000 **35.** 8 = 00001000 **37.** 40 = 00101000
39. 0100 0001 1001 1101 0000 0000 0000 0000 = $419D0000$
41. 0100 0010 0100 0000 0000 0000 0000 0000 = 42400000 **43.** 0100 0001 1101 1000 0000 0000 0000 0000 = $41D80000$ **45.** 1100 0100 1000 0001 0011 1000 0000 0000 = $C4813800$ **47.** 21×2^{-118} **49.** -47×2^{-109} **51.** 13×2^{10},
53. ± 32767, $\pm 2.15 \times 10^9$, $\pm 9.22 \times 10^{18}$ **55.** No

Section 2.1

1.

p	q	$\sim p$	$\sim p \wedge q$
T	T	F	F
T	F	F	F
F	T	T	T
F	F	T	T

3.

p	q	$p \to q$	$p \to (p \to q)$
T	T	T	T
T	F	F	F
F	T	T	T
F	F	T	T

5.

p	q	$\sim p$	$\sim p \to q$	$p \wedge (\sim p \to q)$
T	T	F	T	T
T	F	F	T	T
F	T	T	T	T
F	F	T	F	F

7.

p	q	$\sim p$	$\sim q$	$\sim p \vee \sim q$
T	T	F	F	F
T	F	F	T	T
F	T	T	F	T
F	F	T	T	T

9.

p	q	r	$p \wedge q$	$(p \wedge q) \vee r$
T	T	T	T	T
T	T	F	T	T
T	F	T	F	T
T	F	F	F	F
F	T	T	F	T
F	T	F	F	F
F	F	T	F	T
F	F	F	F	F

11.

p	q	r	$q \vee r$	$p \vee (q \vee r)$
T	T	T	T	T
T	T	F	T	T
T	F	T	T	T
T	F	F	F	T
F	T	T	T	T
F	T	F	T	T
F	F	T	T	T
F	F	F	F	F

19.

p	q	r	$q \to r$	$p \to (q \to r)$
T	T	T	T	T
T	T	F	F	F
T	F	T	T	T
T	F	F	T	T
F	T	T	T	T
F	T	F	F	T
F	F	T	T	T
F	F	F	T	T

21.

p	q	$p \vee q$	$p \wedge q$	$\sim(p \wedge q)$	$(p \vee q) \wedge \sim(p \wedge q)$
T	T	T	T	F	F
T	F	T	F	T	T
F	T	T	F	T	T
F	F	F	F	T	F

29.

p	q	$p \underline{\vee} q$
T	T	F
T	F	T
F	T	T
F	F	F

31.

p	$\sim p$	$p \to \sim p$
T	F	F
F	T	T

33.

p	q	$p \vee q$	$p \wedge q$	$(p \wedge q) \to (p \vee q)$
T	T	T	T	T
T	F	T	F	T
F	T	T	F	T
F	F	F	F	T

35.

p	q	$p \to q$	$p \wedge q$	$(p \to q) \to (p \wedge q)$	$\sim p$	$((p \to q) \to (p \wedge q)) \vee (\sim p)$
T	T	T	T	T	F	T
T	F	F	F	T	F	T
F	T	T	F	F	T	T
F	F	T	F	F	T	T

37.

p	q	$p \to q$	$q \to (p \to q)$
T	T	T	T
T	F	F	T
F	T	T	T
F	F	T	T

39.

p	q	$\sim p$	$q \to p$	$\sim p \to (q \to p)$
T	T	F	T	T
T	F	F	T	T
F	T	T	F	F
F	F	T	T	T

Section 2.2

1. $S_1 \subseteq S_3$, $S_1 \subseteq S_4$, $S_2 \subseteq S_3$, $S_2 1 \subseteq S_4$, $S_2 \subseteq S_5$, $S_3 \subseteq S_4$, $S_5 \subseteq S_4$, $S_2 = S_5$ **27.** T **29.** F **31.** F **33.** T **35.** T **37.** F

Section 2.3

1.

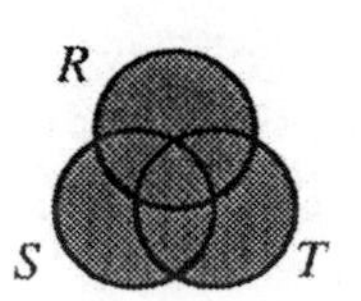

3.

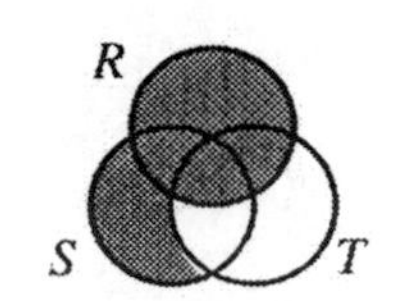

5.

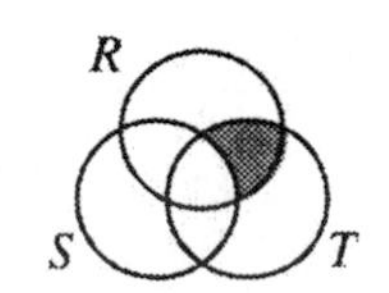

7. **21.** (i) 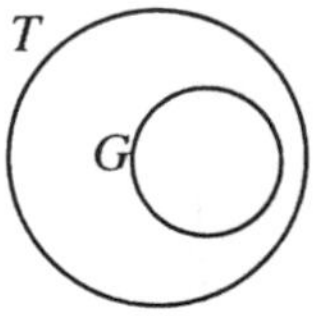 (ii)

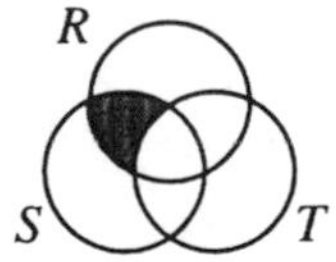

23. valid: 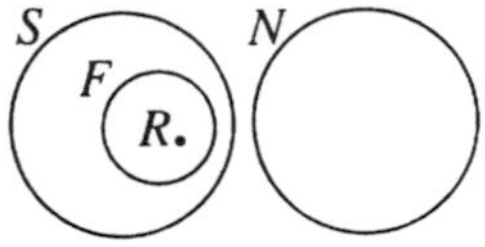 **25.** valid:

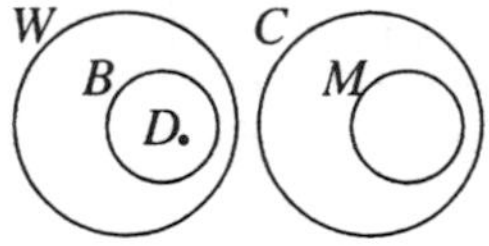

29. (i). (iv), (v) can be concluded.

Section 2.4

5. F **7.** T **9.** F **11.** T **13.** T **15.** (i) $(1,1),(1,4)$, $(1,5)$, $(2,1)$, $(2,4)$, $(2,5)$, $(3,1)$, $(3,4)$, $(3,5)$ (ii) $(1,2)$, $(1,-2)$, $(-1,2)$, $(-1,-2)$ (iii) $(1,1)$, $(1,2)$, $(1,3)$, $(3,1)$, $(3,2)$, $(3,3)$, $(5,1)$, $(5,2)$, $(5,3)$, $(7,1)$, $(7,2)$, $(7,3)$ **19.** (i) $S \times T = \emptyset$ (ii) $S = \emptyset$ or $T = \emptyset$ or $S = T$

Section 2.5

7. 1, 1, 2, 3, 5, 8, 13, 21, 34, 55, 89, 144

Section 3.1

5. $(x+y)z'$ **7.** $x'+y+z$ **9.** $x+yz'+y$ **11.** $(x+y)(y+z)(z+x)$ **19.** (15) $x(x'+y) = xy$ (16) $x(x+y)y = xy + y0$ (17) $xy'+z' = (x'z+yz)'$ (18) $x(y+xz) = xy+xz$ **25.** $(0,0)$, $(1,1)$

Section 3.2

1. (i) 0 (ii) 0 (iii) 0 (iv) 1 **3.** (i) 1 (ii) 1 (iii) 1 (iv) 0 **5.** e **7.** f **9.** d **11.** e **13.** f **15.** 1 **17.** 1 **19.** f **21.** 0 **23.** $xy'z$ **25.** xy' **27.** $xz+x'y+yz$ **29.** $xy+xz+yz$ **31.** $xy'+x'y$ **33.** $x+y$ **35.** $y+z$ **37.** xz' **39.** Yes **41.** Yes **43.** $xyz'+xy'z+xy'z'+x'yz'$ **45.** $xyz+xyz'+xy'z+x'yz+x'y'z+x'y'z'$

Section 3.3

$x'y't + x'y't + x'zt + y'zt'$

1. x **3.** $x+y'$ **5.** y' **7.** (i) xz' (ii) xz **9.** $xy+xz$ **11.** $xy'+yz'+zx'$ **13.** $xy+x'z$ **15.** $xy+yz'$ **17.** $x+y'z$ **19.** $xz'+x'z+y'z$ **21.** (i) $xy'z'$ (ii) t **23.** $xyt+x'zt+yzt'$ **25.** $y'z'+x'y'+y't'$ **27.** $xyz+xy'z'+x'yz'+x'y'zt'+yzt$ **29.** x'
31. (i) $xyz't+xy'zt'+xy'z't'+x'y'zt+x'y'zt'+x'yzt$, $xyz't+xy't'+x'zt+y'zt'$ (ii) $xyz't'+xyz't+xy'zt+xy'zt'+xy'z't'+xy'z't+x'y'zt'+x'y'z't'+x'y'z't+x'yz't'$, $xy'+xz'+y'z'+y't'+z't'$
33. (i) $xyz't'+xyz't+xy'zt+xy'zt'+x'y'zt+x'y'zt'+x'yz't'$, $xyz'+yz't'+y'z$ (ii) $xyzt+xyzt'+xy'zt+xy'zt'+xy'z't'+xy'z't+x'y'z't'+x'y'z't+x'yzt+x'yzt'$, $xy'+yz+y'z'$ **35.** $xy+x'y'+yz'$ **37.** $x'y'zt$ **39.** $x'yz+x'zt+x'yt$

Section 3.4

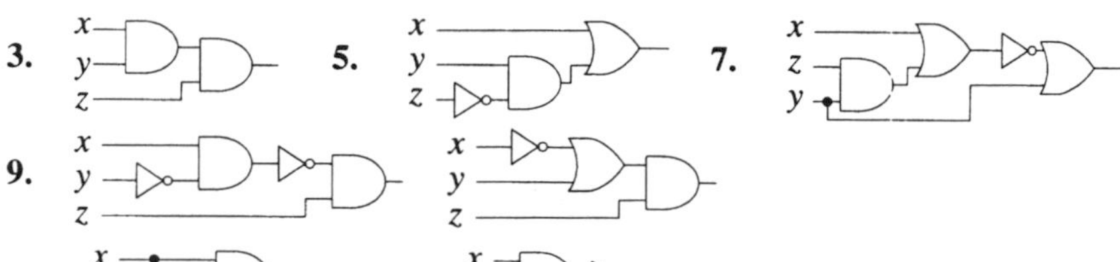

11. **13.** $xy+y'$ **15.** $[(x+y')z]'$

17. $xy+[(x+y)'+(xy')']$; $x'+y$

Section 4.1

1. $\{(1,1), (4,2), (9,3)\}$ $x\alpha^{-1}y$ means $x^2=y$ **3.** $\{(1,1), (1,2), (1,3), (1,4), (1,5), (1,6), (1,7), (1,8), (1,9), (2,2), (2,4), (2,6), (2,8), (3,3), (3,6), (3,9), (4,4), (4,8), (5,5), (6,6), (7,7), (8,8), (9,9)\}$ $x\gamma^{-1}y$ means y divides x **5.** $\{(2,8), (3,3)\}$ $x\varepsilon^{-1}y$ means $x+y^2=12$ **7.** $\alpha: \{(1,1)\}$ $\beta: \{(1,2), (2,3), (3,4)\}$ **9.** $\{(1,2), (2,2), (2,4), (3,2), (4,1), (4,3)\}$ **11.** $\{(2,2), (3,2), (3,4), (4,1), (4,2), (4,3)\}$ **25.** (i) α (ii) γ (iii) α, β (iv) α (v) – (vi) α **27.** S **29.** S **31.** ST **33.** RAT

Section 4.2

1. No (try $a=1, b=0, c=-1$) **3.** Yes **5.** Yes; weak; not total **7.** Yes; strong; total **17.** 7; 4

Section 4.3

1. No **3.** Yes **5.** Yes **7.** Yes **9.** $\{((x,y),x+y) : x,y \in \mathbb{Z}\}$ (a set of ordered pairs in which the first element, (x,y), is itself an ordered pair) **13.** not one-to-one, but onto **15.** f_4 **17.** $f^{-1}(x)=2-x$ if $x \geq 1$; $f^{-1}(x)=1/x$ if $0<x<1$

Section 5.1

1. (i) $\overline{E}$ (ii) $E\cap F$ (iii) $E+F$ (iv) $E\cup F$ (v) $\overline{E}\cap\overline{F}$ (vi) $E\backslash F$ **3.** (i) $\{BBB, BBG, BGB, BGG, GBB, GBG, GGB, GGG\}$ (ii)(a) $E=\{BBB, BBG, BGB, BGG\}$, $F=\{BBG, BGB, GBB\}$ **5.** (i) $\{22,21,12,20,11,02,10,01,00\}$ (ii)(a) $E=\{21,20, 10\}$, $F=\{20,11,02\}$, $G=\{22,20,11,02,00\}$, $E\cup F=\{21,20,10,11,02\}$, $E\cap F=\{20\}$, $E\cap G=\{20\}$, $\overline{F}\cap G=\{22,00\}$, (b) $E\cap F$: there are exactly two heads on quarters and none on nickels, $\overline{F}\cap G$: the tosses are either all heads or all tails (c) No **7.** (ii) 9 (iii) $\{BB, YY, RR\}$ **9.** Yes **11.** No **13.** Yes **15.** Yes **17.** (i) No (ii) Yes (iii) No **19.** Yes

Section 5.2

1. (i) 367 (ii) 189 (iii) 871 **3.** 25 **5.** (i) 3 (iii) 30 **7.** 2 **9.** (i) 5 (ii) 20 (iii) 18 (iv) 17 **11.** (i) 17 (ii) 57

Section 5.3

1. m **3.** 4; 2; 2 **9.**(i) $\max(m,n) \le |A \cup B| \le m+n$ (ii) $0 \le |A \cap B| \le \min(m,n)$ (iii) $\max(m-n,0) \le |A \backslash B| \le m$ (iv) $|m-n| \le |A+B| \le m+n$ **13.** No

Section 5.4

1. 120 **3.** 24 **5.** 120 **7.** 40320 **9.** 4 **11.** 20 **13.** $(10)_3 = 720$ **15.** $(13)_3 = 1716$ **17.** 48 **19.** (i) $3! \times 5! = 720$ (ii) $4! \times 4! = 576$ **21.** (i) 15! (ii) $3! \times 4! \times 5! \times 6! = 12441600$ **23.** $9!/2!2!2! = 45360$ **25.** $9!/2!2! = 90720$ **27.** $6!/2! = 360$ **29.** $10!/3!2!2! = 151200$

Section 5.5

1. 56 **3.** 126 **5.** 20 **7.** 35 **9.** 36 **11.** 28 **13.** $C(16,4) = 1820$ **15.** $C(6,2) \times C(12,3) = 3300$ **17.** $C(12,3) = 792$; $C(9,3) = 84$ **19.** $C(10,3) = 120$; $C(9,2) = 36$ **21.** (i) $C(49,5)$ (ii) 6 **23.** $2 \times \binom{6}{3} = 40$ **25.** (i) $mC(n,2) + nC(m,2)$, or $mn(m+n-2)/2$ (ii) add mn triangles that include A for a total of $mn(m+n)$

Section 5.6

1. $x^4 - 4x^3 + 6x^2 - 4x + 1$ **3.** $1 - 10z + 40z^2 - 80z^3 + 80z^4 - 32z^5$ **5.** $16x^4 + 32x^3y + 24x^2y^2 + 8xy^3 + y^4$ **7.** $x^4 + 4x^2 + 6 + 4x^{-2} + x^{-4}$ **9.** $x^3 + y^3 + z^3 + 3x^2y + 3x^2z + 3xy^2 + 3xz^2 + 3y^2z + 3yz^2 + 6xyz$ **11.** 2^8

Section 5.7

1. 44; 265 **3.** (i) 21 (ii) 31 **5.** 16 **7.** (i) 7! (ii) $7!D_7$

Section 6.1

1. (i) 5/16 (ii) 1/2 (iii) 1/2 **3.** 5/36 **5.** 1/2 **7.** 1/2 **9.** 2/5 **11.** (i) 9/19 (ii) 9/19 (iii) 1/19 (iv) 1/38 (v) 6/19 **13.** (i) 1/5 (ii) 4/5 (iii) 13/20 **15.** (i) 12, 13, 14, 15, 22, 23, 24, 25, 32, 33, 34, 35, 42, 43, 44, 45 (ii) 3/16 (iii) 1/2 **17.** (i) 2/9 (ii) 7/9 **19.** .3; .8; .2 **21.** .4; .9; .9 **23.** (ii) (a) .25 (b) .19 (c) .36 **25.** (i) .4 (ii) .1 **27.** (i) .625 (ii) .35 (iii) .125 **29.** (i) .4 (ii) .3 (iii) .25 **31.** (i) .5 (ii) .25 (iii) .2

Section 6.2

1. (i) RR: $6/15$; RW: $4/15$; WR: $4/15$; WW: $1/15$ (ii) $8/15$ **3.** (ii) $[(16 \cdot 15 \cdot 14) + 3 \cdot (16 \cdot 15 \cdot 36)]/(52 \cdot 51 \cdot 50)$ (approx. 22%) **5.** (ii) 20/56 (iii) 32/56 (iv) 1/8 **7.** (i) 8/81 (ii) 48/81 **9.** A 4-2 division is more likely **11.** $\binom{20}{16}/2^{20}$ (about 1 chance in 200) **13.** $1 - [(19/20)^{10} + 10 \cdot (19/20)^9 \cdot (1/20)]$ (approx. 8.6%) **15.** (i) $8(\frac{7}{8})^7\frac{1}{8}$ (about 40%) (ii) $8(\frac{7}{8})^7\frac{1}{8} + (\frac{7}{8})^8$ (about 74%) **17.** $1 - (.9)^4$ (just over one chance in three)

Section 6.3

1. (i) $\frac{1}{15}$ (ii) $\frac{11}{15}$ **3.** (i) $C(10,3)/C(17,3) = \frac{3}{17}$ (ii) $(10 \cdot 3 \cdot 4)/C(17,3) = \frac{3}{17}$ **5.** $\frac{3}{10}$ **7.** (i) $C(10,4)/2^{10}$ (ii) $C(10,2)/2^{10}$ (iii) $(1 + C(10,1) + C(10,2))/2^{10}$ (approx. 20.5%, 4.4% and 5.5%) **9.** (i) $1/6^4$ (ii) $25/6^4$ (iii) $50/6^4$ (approx. .1%, 1.9% and 3.9%) **11.** $10 \times 4^5/C(52,5)$ **13.** $C(5,2) \cdot C(4,2)/[C(9,4) - 6] = \frac{1}{2}$ **15.** $[\binom{6}{3} + \binom{4}{3}]/\binom{10}{3} = \frac{1}{5}$ **17.** (i) $\frac{1}{84}$ (i) $\frac{3}{14}$ **19.** $(\frac{11}{12})^{12}$ (approx. 35%)

Section 6.4

1. (i) 17/35 (ii) 11/35 (iii) 1/5 **3.** (ii) 13/24 **5.** 2/9; 2/11; 2/5; 5/11; 2/5; 5/9 **7.** 25/102; 15/34; 25/51; 25/77; 32/51; 45/77 **9.** 1/5; 2/5; 1/4; 5/8 **11.** (i) 2/5 (ii) 1/5 **13.** (i) 2/3 (ii) 17/30 (iii) 17/36 **15.** 3/4; 3/7 **17.** 1; 3/7 **19.** 2/3; 2/3 **21.** 1/3; 1/3 **23.** .9; .4 **25.** .85; .35 **27.** .97; .63 **29.** (ii) 1/4; 1/4; 1/4; 1/4; 1/16 (iii) Yes **31.** (ii) .6; .6; .6; .6; .36 (iii) Yes **33.** (i) .24 (ii) .54 (iii) .6 (iv) .25 **35.** (i) 4/9 (ii) Yes **37.** (i) Yes (ii) Yes **41.** (i) No (ii) No **43.** True **45.** False **47.** True **49.** Impossible

Section 6.5

1. .5, .5 **3.** (i) $\frac{1}{6}; \frac{1}{2}; \frac{1}{3}; \frac{3}{4}; \frac{1}{4}; 0$ (ii) $\frac{1}{5}; \frac{1}{5}; \frac{3}{5}; \frac{3}{10}; \frac{3}{10}; \frac{2}{5}$ (iii) $\frac{2}{5}; 0; \frac{3}{5}; \frac{4}{5}; \frac{1}{5}; 0$ **5.** (i) 1/97 (ii) 1/17 **7.** (i) 1.1% (ii) $\frac{27}{44}$ **9.** 2/3 **11.** 3/8; 5/8 **13.** 9/16 **15.** 8/15 **17.** 7/31 **19.** 3/143 **21.** 3/8; 1/4

Section 7.1

1. No (loops); no **3.** Yes; yes **5.** No(loops); yes **7.** Yes; no **9.** Yes; no **11.** No (loop on 1); no **13.** No (loops); no **15.** No (loops); no **17.** (i) 323303242 (ii) 1124222 (iii) 2422233 (iv) 2332332442 **19.** n; n once, 1 n times **23.** No, no **25.** Yes, yes **27.** No, no **29.** Graphical, not valid

Section 7.2

1. Yes; Yes **3.** Yes; No **5.** Yes; Yes **7.** Yes; No **9.** No **11.** Yes; Yes **13.** Yes; Yes **15.** Yes; No **19.** 2 **21.** 2 **23.** 4 **25.** 5 **27.** 3

Section 7.3

1. *sabct*; *sabt*; *sadefcbt*; *sadefct*; *safcbt*; *safct* **3.** No **5.** (i) 1; 1 (ii) 1; 2 (iii) 1; 1 **7.** 1; 2 **9.** 1; 1 **11.** 1; 1 **13.** 2; 3 **15.** 2; 2

Section 7.4

1. 4; 3; 7; 6; 5; 4 **3.** $D = 1, R = 1$ **5.** $D = 2, R = 2$ **7.** $D = 3, R = 2$ **9.** $D = 2, R = 2$ **11.** $D = 4, R = 3$ **13.** $D = 4, R = 3$ **15.** K_{n+1}; K_5 missing edges $14, 15, 25$; K_5 **17.** There is a path of weight 12 **19.** There is a path of weight 20 **21.** There is a path of weight 12

Section 7.5

1. There are six trees **5.** 9 **7.** 15 **9.** 4 **17.** MST weights: (i) 27 (ii)(a) 38 (b) 41 (c) 43 **19.** weight 28 **21.** weight 64 **23.** weight 45 **25.** weight 31

Section 7.7

1. *abcdef*, *abdcef* **3.** *abcehgfd*, *abcfdgeh*, *abdfcegh*, *abdgfceh*, *abecfdgh*, *adbcfgeh*, *adbecfgh*, *adfcbegh*, *adgfcbeh* **5.** *abcefhgd*, *abcfehgd*, *abcfhegd*, *abcfhgde*, *abcfhged*, *abdghfce*, *abecfhgd*, *abfcehgd*, *adbcfhge*, *adghfbce*, *adghfcbe* **7.** G: *abefcd* costs 115; H: *abcfed* costs 115 **9.** NN: *a* 117, *b* 117, *c* 122, *d* 117, *e* 117, SE: 117 **11.** NN: *a* 112, *b* 113, *c* 112, *d* 113, *e* 116, SE: 113

13. NN: *a* 74, *b* 76, *c* 76, *d* 74, *e* 76, SE: 76 **15.** NN: *a* 105, *b* 105, *c* 105, *d* 100, *e* 100, SE: 105 **17.** NN: *a* 158, *b* 152, *c* 158, *d* 158, *e* 152, SE: 158

Section 8.1

3. $(-2,2)$ **5.** $(9,18,3)$ **7.** $(4,1)$ **9.** $(-6,0,6)$ **11.** $(7,0,4)$ **13.** $(5,-20,10,15)$ **15.** -1 **17.** -5 **19.** 3 **21.** 6 **23.** 15 **25.** 2×4 **27.** 2×2 **29.** No **31.** 1×4 **33.** 4×4 **35.** 2×4 **37.** No **39.** 1×3 **41.** 2×3 **43.** $\begin{bmatrix} 7 & -2 \\ -3 & 2 \end{bmatrix}$ **45.** $\begin{bmatrix} 28 & -1 \\ 8 & 19 \end{bmatrix}$ **47.** $\begin{bmatrix} 3 & -1 \\ 4 & -1 \end{bmatrix}$ **49.** $\begin{bmatrix} 2 & -2 \\ -4 & 6 \end{bmatrix}$ **51.** $\begin{bmatrix} 4 \\ -6 \end{bmatrix}$ **53.** $\begin{bmatrix} 4 \\ 2 \end{bmatrix}$ **55.** $\begin{bmatrix} 7 & -2 \end{bmatrix}$ **57.** No **59.** No **61.** $\begin{bmatrix} 1 & 3 & 1 \end{bmatrix}$ **63.** No **65.** $x=2,\ y=1$

Section 8.2

1. Yes if A is square, otherwise no (O's not the same size) **3.** $AB=\begin{bmatrix} -1 & 1 \\ -1 & 7 \end{bmatrix}$, $BA=\begin{bmatrix} 4 & 7 \\ 2 & 2 \end{bmatrix}$ No **5.** $AB=\begin{bmatrix} -4 & 8 \\ -5 & 2 \end{bmatrix}$, $BA=\begin{bmatrix} 0 & -4 \\ 8 & -2 \end{bmatrix}$ No **7.** $AB=\begin{bmatrix} 0 & 10 \\ -10 & 0 \end{bmatrix}$ Yes **9.** $\begin{bmatrix} 0 & 2 \\ -2 & 0 \end{bmatrix}$, $\begin{bmatrix} -2 & 2 \\ -2 & -2 \end{bmatrix}$ **11.** $\begin{bmatrix} 2 & 2 \\ 2 & 2 \end{bmatrix}$, $\begin{bmatrix} -4 & -4 \\ -4 & -4 \end{bmatrix}$ **13.** $\begin{bmatrix} 1 & -3 & 9 \\ 1 & 10 & -4 \\ -3 & -4 & -1 \end{bmatrix}$, $\begin{bmatrix} -7 & -18 & 15 \\ 6 & 34 & -11 \\ -5 & -11 & -6 \end{bmatrix}$ **15.** (i) $\begin{bmatrix} 16 & 12 \\ 20 & 24 \end{bmatrix}$, $\begin{bmatrix} 76 & 84 \\ 140 & 132 \end{bmatrix}$ (ii) $\begin{bmatrix} 73 & 78 \\ 130 & 125 \end{bmatrix}$ **27.** First and third

Section 8.3

1. $x=3,\ y=1$, any real z **3.** $x=2-z,\ y=1+z$, any real z **5.** No solutions **7.** No solutions **9.** $x=4,\ y=3,\ z=-3$ **11.** $x=2-z,\ y=1+z$, any real z **13.** $x=2,\ y=4,\ z=-2$ **15.** $x=-3,\ y=2$ **17.** $x=\frac{1}{2}(5-3y)$, any real y **19.** No solutions **21.** $x=2,\ y=-1$ **23.** $x=1+5z,\ y=1-3z$, any real z **25.** $x=3,\ y=2,\ z=-2$ **27.** $x=\frac{2}{3},\ y=\frac{1}{6},\ z=\frac{1}{3}$ **29.** $x=3,\ y=-1,\ z=-2$ **31.** $x=\frac{5}{2},\ y=-\frac{1}{2},\ z=0$ **33.** $x=1-2z+t,\ y=2-z+2t$, any real z, any real t **35.** No solutions

Section 8.4

1. Empty **3.** Infinite **5.** Singleton **7.** $\begin{bmatrix} 0 & \frac{1}{4} \\ -\frac{1}{2} & \frac{1}{4} \end{bmatrix}$ **9.** $\begin{bmatrix} 5 & -7 \\ -2 & 3 \end{bmatrix}$ **11.** No inverse **13.** $\begin{bmatrix} -1 & 0 & 1 \\ 0 & 1 & 0 \\ 1 & 0 & 0 \end{bmatrix}$ **15.** $\begin{bmatrix} 4/7 & -1 & 2/7 \\ 3/7 & 0 & -2/7 \\ -5/7 & 1 & 1/7 \end{bmatrix}$ **17.** $\begin{bmatrix} 3/2 & 0 & \frac{1}{2} \\ -11/6 & \frac{1}{3} & -5/6 \\ 17/6 & -\frac{1}{3} & 5/6 \end{bmatrix}$ **19.** $\begin{bmatrix} 7 & -3 & -3 \\ -1 & 1 & 0 \\ -1 & 0 & 1 \end{bmatrix}$ **21.** No inverse **23.** No inverse **25.** 13; $\begin{bmatrix} 3/13 & -4/13 \\ -2/13 & 7/13 \end{bmatrix}$ **27.** -10; $\begin{bmatrix} -3/10 & 1/5 \\ 1/5 & 1/5 \end{bmatrix}$ **29.** 0; No inverse **31.** (ii)(a) $x=2,\ y=-1$ (b) $x=1,\ y=1$ **33.** No

Section 8.5

1. $\begin{bmatrix}100000000\\000000000\\000000000\\010000000\\000000000\\000000000\\000000000\\000000000\\001000000\end{bmatrix}$ $\begin{bmatrix}000000010\\000000100\\000001000\\000010000\\000100000\\001000000\\010000000\\100000000\\000000000\end{bmatrix}$ **3.** $\begin{bmatrix}000000000\\000000010\\001000000\\000000000\\000000000\\000000000\\000000000\\000000000\\000000000\end{bmatrix}$ $\begin{bmatrix}100000000\\010000000\\001000000\\000100000\\000010000\\000001000\\000000100\\000000010\\000000001\end{bmatrix}$ **5.** $\begin{bmatrix}0000\\0100\\0101\\1010\end{bmatrix}$ $\begin{bmatrix}0100\\0010\\1100\\0001\end{bmatrix}$

7. $\begin{bmatrix}0000\\0010\\0011\\1100\end{bmatrix}$ $\begin{bmatrix}0100\\0101\\0100\\1010\end{bmatrix}$ **13.** $\begin{bmatrix}010001\\101000\\010100\\001010\\000101\\100010\end{bmatrix}$ $\begin{bmatrix}100001\\110000\\011000\\001100\\000110\\000011\end{bmatrix}$ **15.** $\begin{bmatrix}011111\\101001\\110100\\101010\\100101\\110010\end{bmatrix}$

$\begin{bmatrix}1111100000\\1000010001\\0100011000\\0010001100\\0001000110\\0000100011\end{bmatrix}$ **17.** $\begin{bmatrix}011111\\100000\\100000\\100000\\100000\\100000\end{bmatrix}$ $\begin{bmatrix}11111\\10000\\01000\\00100\\00010\\00001\end{bmatrix}$ **19.** 4

Section 9.1

1. $63: 1,3,7,9,21,63$; $64: 1,2,4,8,16,32,64$; $288: 1,2,3,4,6,8,9,12,16,18,32,36,96,288$ **3.** 123456, 51804 **5.** 257, 419 **7.** 10 **9.** 60 **11.** $12 = 3 \cdot 84 - 4 \cdot 60$ **13.** $120 = 4 \cdot 480 - 1 \cdot 1800$ **15.** $24 = 1 \cdot 144 - 1 \cdot 120$ **17.** $21 = 1 \cdot 861 - 4 \cdot 210$ **21.** 18 **23.** 21 **25.** Say p divides a, $p > \sqrt{a}$. Then $a/p < \sqrt{a}$, and it is a factor of a

Section 9.2

1. 3 **3.** 0 **5.** 1 **7.** 1 **9.** 3 **11.** 1 **13.** 5 **15.** 4 **17.** 1 **19.** 1 **21.** 1 **23.** 0 **25.** 3 **27.** 0 **29.** 3 **31.** 0 **33.** 1 **35.** 1 **37.** 0 **39.** 1 **41.** 1 and 131 are their own inverses; $3^{-1} = 5$, $5^{-1} = 3$, $9^{-1} = 11$, $11^{-1} = 9$, 2, 4, 6, 7, 8, 10 and 12 are zero-divisors **43.** 5 **45.** 4 **47.** 7 **49.** 3 **51.** 5 **53.** 11 **55.** 1 **57.** 11 **59.** 0 **61.** 9 **63.** 298(mod 385) **65.** 236(mod 455) **67.** $1 + 1 = 0$ mod 2; $1 + 1 = 1$ in $\mathcal{B}_2$

Section 9.3

1. *Encrypt* always implies secrecy; *encode* need not **3.** wake up and smell the coffee **5.** how many more miles must we march **7.** it is a truth universally acknowledged that a single man in possession of a good fortune must be in want of a wife (*Pride and Prejudice,* Jane Austen) **9.** like many of the great generals of history Caesar seems to have been lacking in cryptographic subtlety (*Cryptography,* Arnold Beutelspacher) **11.** `VHQG WURRSV` **13.** `WKH HQG LV QHDU` **15.** attack **17.** retreat **19.** `AOL TVVU OHZ YPZLU` **21.** `ZLCLU RUPNOAZ HYL HWWYVHJOPUN` **23.** do not pass go **25.** bread and circuses **27.** make love not war **29.** send in the clowns **31.** `REIPF FHRCG YXCVJ FQSZZ` **33.** `OUGAB PFUAK NINYA CGQZZ`

Section 9.4

1. `EKOFMCEP FQ DKLSCEP WLPV ALP EKOPYMS` **3.** `AKTFMDAO FP EKLQDAO VLOR BLO AKTOYMQ` **5.** `THADKRTM DO GHJPRTM UJMN`

LJM THAMXKP **7.** EFTYIWEK YL MFGNWEK UGKH RGK EFTKXIN **9.** AZURE **11.** AORTA contains repetitions **13.** DAYLONG is good **15.** PIE is too short **17.** LAZY is good but short **19.** (i) TP TO RJFFJG PJ TDDQOPNMPC OCPO MGY JKCNMPTJGO JG OCPO EX YTMINMFO (ii) HTGY M EJJDCMG CWKNCOOTJG RJNNCOKJGYTGI PJ PSC HJDDJVTGI RTNRQTP (iii) PSNCC EJXO MGY HJQN ITNDO MNC PJ OTP MDJGI M ECGRS (iv) KNJUC PSMP MGX PVJ YTMIJGMD FMPNTRCO RJFFQPC **21.** (i) SR SQ XKIIKJ RK SHHTQRPFRN QNRQ FJA KLNPFRSKJQ KJ QNRQ OY ASFGPFIQ (ii) DSJA F OKKHNFJ NWLPNQQSKJ XKPPNQLKJASJG RK REN DKHHKVSJG XSPXTSR (iii) REPNN OKYQ FJA DKTP GSPHQ FPN RK QSR FHKJG F ONJXE (iv) LPKUN REFR FJY RVK ASFGKJFH IFRPSXNQ XKIITRN **23.** The story of Fermat's Last Theorem is inextricably linked with the history of mathematics (*Fermat's Enigma,* Simon Singh) **25.** Older men declare war but it is the youth who must fight and die (Herbert Hoover) **27.** The optimist believes everything he reads on the jacket of a new book

Section 9.5

1. (i) 3, 7, 9 (ii) 17 **3.** $\{p,q\} = \{11,13\}$, $s = 67$ **5.** 72 **7.** Yes **9.** No **11.** Yes **13.** Yes **15.** 10 **17.** 24 **19.** 7 **21.** 2 **23.** 4 **25.** 5 **27.** 8 **29.** 4 **31.** 25 **33.** (i) 13 (ii) a. 52 01 49 09 15 01 09 b. 51 14 51 25 15 15 49 c. 15 33 14 23 d. 25 20 52 20 02 02 20 12 **35.** DATA **37.** (i) 17 (ii) 13 05 05 20 01 20 14 15 15 14 (iii) MEET AT NOON

Section 9.6

1. 16 **3.** 15 **5.** 11 **7.** 5 **9.** 2 **11.** 27 **13.** 28 **15.** 14 **17.** 613×647 **19.** 523×541 **21.** 461×653 **23.** 467×631 **25.** 877×983

Section 9.7

1. (ii) 32 (iii) 8 **3.** 9, 1

Index